BASIC ELECTRONICS

FOR NON ELECTRICAL

ENGINEERS

(with MATLAB and Simulink Exercises)

Theodore Deliyannis

Konstantinos Giannakopoulos

Theodore Deliyannis, Emeritus Professor

Konstantinos Giannakopoulos, Postdoctoral Researcher

Electronics Laboratory
Department of Physics
University of Patras
26500, Patras, Greece

deliyan@upatras.gr
kgian1@upatras.gr

© 2012 Lulu Author. All rights reserved.

ISBN 978-1-105-70888-6

Preface

Electronics is a discipline that has progressed very fast ever since the discovery of the transistor in 1947. However, in spite of its continuous fast progress Basic Electronics has been established as an interesting and worth studying Classical Subject by most Engineers and Applied Scientists. On the other hand, Electronics has become part of our daily life; most of the devices that we use everyday have part, or the whole of their operation based on Electronics. It is considered therefore useful, for as many people as possible to gain knowledge on the fundamentals of Electronics. This would help them in handling their devices at home and/or at work with more efficiency and safety. Additionally, while science nowadays needs a multidisciplinary cooperation and since electronic devices are normally required in any experimental work, it could be most helpful for all non-electronic members of the cooperating working group to be aware of the limitations of electronics technology when they set the specifications of their requirements. Basic knowledge of Electronics could also help the non-expert to understand the language of the electronics man in such a group. This book aims to help in these directions and particularly those working in Bioscience.

The content of the book has been developed based on the material which one of the authors (T. D.) used in his teaching undergraduate students at the Department of Physics and in Postgraduate Courses in Biomedical Engineering and Medical Physics at the University of Patras, Greece. The book is divided in two parts; Part *I* deals with the fundamentals of applied electronics in a rather concise form, while Part *II* deals with basic applications of electronics to measurements of physical quantities, interesting mainly to Bio-scientists. However, the first part is independent of the second part, is presented in a general form and considered to be useful to any non-Electrical and particularly non-Electronic Engineer. It is thought that Mechanical Engineers, Computer Scientists, Physicists, Chemical Engineers and Bio-scientists, students and graduates, will benefit from studying the book, as they will be helped to better understand the operation of the electronic equipment they use in their work. No previous knowledge of Electronics is assumed and the reader of the first part will be helped to comprehend the material by following the numerical examples and solving the problems. Use of MATLAB and Simulink programs is encouraged and such examples and problems are included. An introduction to MATLAB programming has been included in Appendix *D*.

The content of the book is divided into twelve Chapters and four Appendices. The first eight chapters deal with Applied Electronics and constitute Part *I* of the book, while the last four chapters constitute Part *II* of the book. Since the applications of Electronics are tremendously numerous, we have chosen to focus on some in particular, which are related to Biomedicine. The choice has been made mainly because of the importance, which is given nowadays to the attendance of the human health for diagnostic and recuperation purposes. Thus, applications of Electronics to measurements and to biomedical instrumentation constitute Part *II* of the book.

Analytically the content of each chapter is as follows: Chapter 1 deals with circuit fundamentals defining electric signals and their representation as well as introducing signal sources, the passive elements resistance, capacitance and inductance and their use for the differentiation and integration of a signal. The concept of filtering using these simple circuits is mentioned for a first time before this is more formally dealt with in Chapter 4. The circuit analysis is explained here briefly, and further is extended in Appendix *A* for the interested reader. Also mentioned is the use of the computer for the analysis of a circuit.

In Chapter 2 various simple and most useful electronic components are introduced. These include the diode, the bipolar and *MOS* transistors as well as the optoelectronic devices. The diode is first defined as ideal and then compared to the real semiconductor diode. Its applications for rectification and waveform clipping are explained always considering the diode being ideal.

Chapter 3 deals with amplifiers. The ideal amplifier is introduced as a black box and the equivalent circuits of the four types of ideal amplifiers are presented. Then the concepts of input and output impedances are presented as characteristics of an ideal and a non ideal amplifier. Next the ideal operational amplifier is introduced again as a black box and compared to the practical one. Without looking at the internal structure of the operational amplifier this device, in its ideal version, is used in various circuits for signal conditioning and for building the important instrumentation amplifier as far as its use in biomedical instrumentation is concerned. The ideal operational transconductance amplifier is also introduced. These amplifiers in practice are considered to be integrated circuits. The concept of feedback and its effects on the amplifier characteristics is discussed. Two effects are additionally included; one to obtain an amplifier with negative input capacitance and the second concerning the amplifier stability. Based on the latter an *RC* oscillator is presented.

Chapter 4 deals with analog filters. First the characteristics of ideal and practical filters are presented. Then, low-order active filters are introduced using the operational amplifier as the active element. Resistors and capacitors are the other components. Hints of how high-order filters can be built are given.

Chapters 5 and 6 are devoted to digital electronics. Chapter 5 deals with combinational circuits and starts with a comparison of digital systems against analog. Then binary arithmetic and the representation of binary numbers by voltages are introduced. Next Boolean Algebra is presented and logic gates, treated as black boxes, are used to implement logic functions and build more advanced devices like comparators, adders, encoders, decoders, multiplexers and demultiplexers as well as error detectors.

Chapter 6 deals with digital sequential circuits. Latches and various types of flip-flops are introduced while some useful circuits like clocks, monostables and the Schmitt Trigger are also given. Flip-flops (and gates) are used to build counters and shift registers. All these are out of-the-self commercial devices available in integrated circuit form. Included is discussion on memories (*RAM* and *ROM*) and the general structure of a digital computer. There is also a presentation on the characteristics of the digital integrated circuits. Finally, an introduction to *HDL*, the Hardware Description Language, for integrated circuit design is given.

Chapter 7 deals with Data Conversion i.e. Analog-to-Digital Conversion (*A/D*) and Digital-to-Analog Conversion (*D/A*). Sampling and quantization are presented and devices for the above conversions are introduced, followed by a brief discussion on Digital Filters.

In Chapter 8 various aspects of electronic noise, its sources and ways for reducing it are discussed. Noise measures are introduced and hints for low noise measurements are given.

Chapter 9 deals with measurements of physical quantities, which are of interest to medicine. The chapter starts with a simple but general block diagram of an electronic measuring system and a review of main definitions concerning measurements. Then resistance and impedance bridges are introduced, which are useful in measurements of temperature, displacement, pressure and flow of liquids. Transducers for such measurements are given including the Hall Effect Sensors.

Chapter 10 is an introduction to biopotentials and biosignals for the sake of those readers who are not familiar with this topic. Such signals, when measured, are given the names electromyograms, electrocardiograms, electroencephalograms, electroretinograms and electrooculograms. Magnetocardiograms and magnetoencephalograms are also mentioned.

Finally the characteristics of the biosignals are summarized at the end of the chapter.

Then, in Chapter 11 the detection and measurement of biosignals is presented. Here the sensors are called electrodes, and convert the ionic currents that run inside the body to electronic currents in the metal wires, which carry them to the electronic conditioning and processing devices. The instrumentation amplifier, which was presented in Chapter 3, is adopted and properly adapted to become the biopotential amplifier. The latter is used in measuring the biosignals mentioned in Chapter 10.

Finally, Chapter 12 deals with the safety of those persons handling electronic devices and presents the principles of operation of some bio-instruments. It starts with the physiological effects of electricity, the electrical shock and the non-intentional and intentional application of electricity to human body. Some simple bio-instruments are mentioned which can be built using the knowledge the reader has acquired from studying the previous chapters.

Except for the first two chapters, integrated circuits, analog and digital, are used in all other electronic circuits throughout the book. Almost all chapters include numerical examples and unsolved problems. Use of MATLAB and Simulink programs is encouraged and for this reason, also relevant examples and exercises are included. For those not familiar with these programs a short tutorial introduction is given in Appendix D.

The authors would like to express their sincere thanks to Professor George Kostopoulos of the Department of Medicine at the University of Patras for his advice and criticism concerning the material in Chapters 10 and 11. Also they thank Jenny Carroll for correcting the errors in the language concerning the same chapters. Finally, they thank their wives for their patience, support and encouragement during the long days and nights of writing. Their help has been substantial.

Patras	T. Deliyannis
April 2012	K. Giannakopoulos

Abbreviations

ac	Alternating Current
ADC	Analog-to-Digital Converter
AL-TENS	Acupuncture-Like Transcutaneous Electrical Nerve Stimulation
ALU	Arithmetic-Logic Unit
AM	Amplitude Modulation
AP	All-Pass
ASCII	American Standard Code for Information Interchange
AV	AtrioVentricular
BAEP	Brainstem Auditory Evoked Potential
BCD	Binary Coded Decimal
BiCMOS	Bipolar Complementary Metal-Oxide Semiconductor
BJT	Bipolar Junction Transistor
BP	Band-Pass
BR	Band-Reject
CAD	Computer Aided Design
CD	Compact Disc
CM	Common Mode
CMOS	Complementary Metal-Oxide Semiconductor
CMRR	Common-Mode Rejection Ratio
CNS	Central Nervous System
CPLD	Complex Programmable Logic Device
CPU	Central Processing Unit
CRT	Cathode-Ray Tube
CS	Current Source
DAC	Digital-to-Analog Converter
dB	decibel
dc	Direct Current
DIP	Dual-In-line Package
DVD	Digital Video Disk
EAROM	Electrically Alterable Read Only Memory
ECAP	Electronic Circuit Analysis Program
ECG	Electrocardiogram
ECL	Emitter-Coupled Logic
EEG	Electroencephalogram
EM	Electromagnetic
EMG	Electromyogram

EOG	Electrooculogram
EP	Evoked Potential
EPROM	Erasable Programmable Read Only Memory
ERG	Electroretinogram
FA	Full Adder
FET	Field-Effect Transistor
FIR	Finite Impulse Response
FM	Frequency Modulation
FPGA	Field-Programmable Gate Array
FWR	Full Wave Rectifier
GB	GigaByte
GSI	Giga-Scale Integration
HA	Half Adder
HDL	Hardware Description Language
HP	High-Pass
IA	Instrumentation Amplifier
IC	Integrated Circuit
ICS	Ideal Current Source
ICU	Intensive Care Unit
IIR	Infinite Impulse Response
I/O	Input/Output
IR	Infrared
ISP	Instruction Set Processor
IVS	Ideal Voltage Source
KB	KiloByte
KCL	Kirchhoff's Current Law
KVL	Kirchhoff's Voltage Law
LA	Left Arm
LCD	Liquid Crystal Display
LDR	Light Dependent Resistor
LED	Light-Emitting Diode
LL	Left Leg
LP	Low-Pass
LSB	Least Significant Bit
LSI	Large-Scale Integration
LVDT	Linear Variable Differential Transformer
MB	MegaByte
MCG	Magnetocardiogram
MEG	Magnetoencephalogram
MESFET	Metal-Semiconductor Field-Effect Transistor
MOS	Metal-Oxide Semiconductor
MOSFET, MOST	Metal-Oxide Semiconductor Field-Effect Transistor
MPU	MicroProcessing Unit

MRI	Magnetic Resonance Imaging
MSB	Most Significant Bit
MSI	Medium-Scale Integration
NF	Noise Figure
NRDF	Non-Recursive Digital Filter
OPAMP	Operational Amplifier
OS	One Shot
OTA	Operational Transconductance Amplifier
PC	Personal Computer
PCB	Printed Circuit Board
PCM	Pulse-Code Modulation
PLA	Programmable Logic Array
PM	Phase Modulation
PNS	Peripheral Nervous System
POS	Product Of Sums
PROM	Programmable Read Only Memory
p.t.p.	peak-to-peak
RA	Right Arm
RAM	Random Access Memory
RMS	Root-Mean-Square
ROM	Read Only Memory
RTL	Register Transfer Level
RTL	Resistor-Transistor Logic
SA	SinoAtrial
SAB	Single Amplifier Biquad
SAR	Successive Approximation Register
SEP	Somatosensory Evoked Potential
S/N	Signal to Noise
SOP	Small Outline Package
SOP	Sum Of Products
SQUID	Superconductor Quantum Interference Device
SR	Shift Register
SR	Slew Rate
SSI	Small-Scale Integration
TENS	Transcutaneous Electrical Nerve Stimulation
TOP	Transistor Outline Package
TT	Twin-Tee
TTL	Transistor-Transistor Logic
ULSI	Ultra Large-Scale Integration
USB	Universal Serial Bus
VCO	Voltage-Controlled Oscillator
VCVS	Voltage-Controlled Voltage Source
VEP	Visual Evoked Potential
VHDL	Very high speed integrated circuits Hardware

	Description Language
VHSIC	Very High Speed Integrated Circuit
VLSI	Very Large-Scale Integration
VS	Voltage Source
w.r.t.	with respect to
WSI	Wafer-Scale Integration

Contents

Preface..iii
Abbreviations...vii

PART I: DEVICES AND CIRCUITS......................................1

Chapter 1: Circuit Fundamentals..3
1.1 Introduction..3
1.2 Current, Voltage, Circuit..3
1.3 Electric Signals...5
1.3.1 Signal Representation..5
      Example 1.M.1..7
      Example 1.M.2..7
1.4 Resistance, Resistor...8
      Example 1.M.3..9
1.5 Ideal Voltage and Current Sources................................10
1.6 Kirchhoff's Laws..11
      Example 1.1...12
      Example 1.M.4..14
1.7 Non-ideal Voltage and Current Sources........................15
      Example 1.M.5..17
1.7.1 Potentiometers...17
1.8 Capacitance, Capacitor...19
1.9 Differentiation and Integration of Signals.....................20
      Example 1.2...25
      Example 1.M.6..26
      Example 1.M.7..26
1.10 Inductance, Inductor..28
1.11 Circuit Analysis..29
References...30
MATLAB Problems..30
Problems...31

Chapter 2: Electronic Components – Devices....................35
2.1 Introduction..35
2.2 Ideal Diode..36
2.2.1 Rectification of *ac* Voltages....................................37
      Example 2.M.1..38
      Example 2.M.2..39
2.2.2 Full-Wave Rectifier..40

	Example 2.1..41
2.2.3	Waveform Clipping...43
	Example 2.M.3..44
	Example 2.2..46
	Example 2.M.4..47
2.3	Semiconductor Diodes..48
2.4	*dc* Power Supplies..50
2.5	Transistor..51
2.5.1	The Bipolar Junction Transistor....................................52
2.5.2	The MOSFET..53
2.6	Integrated Circuits...56
2.7	Optoelectronics...57
2.7.1	Photoconductors...57
2.7.2	Photodiodes...58
2.7.3	Phototransistor..58
2.7.4	Photovoltaics...59
2.7.5	Photoemission...59
2.7.6	Light-Emitting Diodes (*LED*).......................................59
2.7.7	Optoisolators...60
	References..60
	MATLAB Problems..61
	Problems..61

Chapter 3: Amplifiers...65

3.1	Introduction..65
3.2	The General Amplifier..66
	Example 3.1..68
	Example 3.M.1..68
3.2.1	Ideal Amplifiers..70
	Example 3.2..70
3.3	The Operational Amplifier, Opamp..............................72
3.4	Applications of the Opamp...73
3.4.1	Inverting and Non Inverting Finite Voltage Gain.............74
	Example 3.3..75
	Example 3.M.2..76
3.4.2	Summation of Voltages..77
3.4.3	Integration of a Voltage...77
3.4.4	Constant Current Source...78
3.5	Additional Characteristics of Practical Opamps............79
	Example 3.M.3..84
3.6	The Differential (Difference) Amplifier.........................84
3.7	Common-Mode Rejection Ratio, *CMRR*......................87
3.8	Instrumentation Amplifier..87
3.8.1	Input Resistance of the *IA*...90

Contents   xiii

    3.8.2   Integrated Circuit (*IC*) Instrumentation Amplifiers..................90
  3.9   Operational Transconductance Amplifier, *OTA*...................90
  3.10   Power Supply for an OPAMP, an *IA* or an *OTA*..................92
  3.11   Application of Feedback to Amplifiers..............................92
  3.12   Effects of Feedback on the Amplifier Characteristics...............94
            Example 3.M.4...95
            Example 3.M.5...96
  3.13   Amplifiers with Negative Input Capacitance.......................98
  3.14   Stability..99
  3.15   An *RC* Oscillator..100
References..102
MATLAB Problems..102
Problems..103

Chapter 4: Analog Filters..107
  4.1   Introduction..107
  4.2   Filter Characteristics...107
  4.3   Passive and Active *RC* Filters...111
  4.4   First-Order Active *RC* Filters..111
            Example 4.M.1...113
            Example 4.M.2...114
  4.5   Second-Order Active *RC* Filters......................................115
  4.5.1   A Low-Pass *SAB*..115
            Example 4.1...117
            Example 4.M.3...118
  4.5.2   A High-Pass *SAB*...119
            Example 4.M.4...121
  4.5.3   A Band-Pass *SAB*..122
            Example 4.M.5...124
  4.5.4   A Notch *SAB*..125
            Example 4.M.6...126
  4.5.5   An All-Pass *SAB*..127
            Example 4.M.7...129
  4.5.6   A Three-Opamp Biquad..130
            Example 4.2...132
  4.6   Higher-Order Active *RC* Filters......................................132
            Example 4.3...133
            Example 4.M.8...134
            Example 4.4...136
References..136
MATLAB Problems..137
Problems..137

Chapter 5: Digital Combinational Electronics 139
- 5.1 Introduction 139
- 5.2 Digital Systems Against Analog Systems 140
- 5.3 Binary System of Numbers 140
  - Example 5.M.1 144
- 5.3.1 Octal and Hexadecimal Number Systems 145
  - Example 5.M.2 146
- 5.4 Voltage Representation of Binary Numbers 147
- 5.5 Boolean Algebra 148
- 5.5.1 Theorems of Boolean Algebra 149
- 5.6 Logic Gates 151
- 5.7 Implementation of Logic Functions 153
- 5.8 Exclusive-OR and Exclusive-NOR Gates 154
  - Example 5.M.3 155
- 5.9 A Digital Comparator 157
  - Example 5.M.4 157
- 5.10 Binary Adder 158
  - Example 5.M.5 159
- 5.11 Binary Codes 160
- 5.11.1 Encoders 162
  - Example 5.M.6 163
- 5.12 Decoders 163
- 5.13 Multiplexers and Demultiplexers 164
  - Example 5.M.7 166
- 5.14 Error Detectors – Parity Checkers 167
- References 167
- MATLAB Problems 168
- Problems 169

Chapter 6: Digital Sequential Electronics 171
- 6.1 Introduction 171
- 6.2 Latches 171
  - Example 6.M.1 172
- 6.3 Flip-Flops 173
- 6.3.1 The *R-S* Flip-Flop 173
- 6.3.2 The *T* Flip-Flop 174
- 6.3.3 The *D* Flip-Flop 175
  - Example 6.M.2 176
- 6.3.4 The *JK* Flip-Flop 176
  - Example 6.M.3 177
- 6.3.5 The Race Problem in Flip-Flops. Master-Slave Method 177
- 6.3.6 Flip-Flop Characteristics 178
- 6.4 Clocks 179
- 6.5 Monostables or One-Shots 180

6.6	Schmitt Trigger	181
6.7	Counters	183
6.7.1	Ripple Counters	184
	Example 6.M.4	185
6.7.2	Synchronous Counters	187
6.8	Registers	188
6.8.1	Shift-Registers	188
6.8.2	Additional Applications of Shift-Registers	189
6.8.3	Integrated Circuit Shift-Registers	190
6.9	Memories	190
6.9.1	Random Access Memory, *RAM*	191
6.9.2	Read Only Memory, *ROM*	193
6.10	Digital Computers	194
6.11	Digital Integrated Circuits	197
6.12	Design with *HDL*	202
References		203
MATLAB Problems		204
Problems		204

Chapter 7: Data Conversion and Processing of Digital Signals 207

7.1	Introduction	207
7.2	Conversion of Analog Signals to Digital and Vice Versa	207
7.3	Sampling and Quantization	208
	Example 7.M.1	210
	Example 7.M.2	211
7.4	Digital-to-Analog Converter	213
7.5	Analog-to-Digital Conversion	215
7.5.1	Successive Approximation *ADC*	216
7.5.2	Flash *ADC*	217
7.5.3	Oversampling Converters	218
7.6	Sample and Hold Circuit	219
	Example 7.M.3	221
7.7	Digital Filters	222
References		223
MATLAB Problems		223
Problems		224

Chapter 8: Electrical Noise 225

8.1	Introduction	225
8.2	Sources of Noise in Analog Signals	225
8.2.1	External or Interference Noise Sources	226
8.2.2	Inherent Noise	226
	Example 8.1	227
	Example 8.2	228

	Example 8.M.1	229
	Example 8.M.2	229
8.3	Noise Characteristics and Measures	230
8.3.1	Total Noise Due to White and Flicker Noise	232
8.4	Reduction of Noise	233
8.5	Noise Model of an Opamp	233
8.5.1	Effect of Negative Feedback on Noise	235
8.6	Isolation Amplifiers	236
8.7	Noise in Digital Signals	237
	Example 8.M.3	237
8.8	Basic Points to Consider for Low Noise Measurements	238
References		239
MATLAB Problems		239
Problems		240

PART II: MEASUREMENTS 241

Chapter 9: Measurement of Physical Quantities 243

9.1	Introduction	243
9.2	Electronic Measuring System	243
9.2.1	Review of Some Definitions Concerning Measurements	244
9.3	Bridges	245
9.4	Temperature Measurement Transducers	247
9.4.1	Thermoresistive Sensors	248
9.4.2	Thermocouples	248
9.4.3	Thermistor	249
	Example 9.M.1	251
9.4.4	Other Temperature Sensors	252
9.5	Displacement Measurement Transducers	254
9.5.1	Capacitive Sensor	254
9.5.2	Inductive Sensor. Linear Variable Differential Transformer	256
9.6	Pressure Measurement Transducers	257
9.7	Flow Measurements	258
9.7.1	Gas Flow Measurements	259
9.7.2	Liquid Flow Measurements	259
9.8	Hall Effect Sensors	260
9.9	Optical Measurements	262
References		263
MATLAB Problems		263

Chapter 10: Biopotentials – Biosignals 265

10.1	Introduction	265
10.2	The Nervous System	265
10.2.1	The Neuron	266

10.3	The Resting Potential	267
10.4	Electrical Properties of the Axon	269
	Example 10.M.1	271
10.5	The Action Potential	274
10.6	The Electromyogram (EMG)	277
10.7	The Electrocardiogram (ECG)	278
10.8	The Electroencephalogram (EEG)	282
10.8.1	Evoked Potentials	283
10.9	The Electroretinogram (ERG) and	
	the Electrooculogram (EOG)	283
10.10	The Magnetocardiogram (MCG) and	
	the Magnitoencephalogram (MEG)	284
10.11	Characteristics of the Biosignals	284
References		285
MATLAB Problems		286
Problems		286

Chapter 11: Detection and Measurement of Biosignals ... 289
11.1	Introduction	289
11.2	A General Electronic Measuring System	290
11.3	Electrodes	291
11.3.1	Electrode-Electrolyte Interface	291
11.3.2	Polarization of Electrodes	292
11.3.3	Equivalent Circuit of an Electrode	293
11.3.4	Types of Commercial Electrodes	294
11.3.5	Skin Impedance and the Electrode-Tissue Model	295
11.4	Biopotential Amplifiers	296
11.5	ECG Detection	300
11.5.1	Right-Leg-Drive for ECG Measurement	300
11.6	Problems in Detecting EEG Signals	302
11.7	EMG Detection	303
References		304
Problems		305

Chapter 12: Bio-instruments and Safety ... 307
12.1	Introduction	307
12.2	Physiological Effects of Electricity	307
12.3	Electrical Shock	308
12.3.1	Macroshock	309
12.3.2	Microshock	310
12.4	Intentional Application of Electricity to Human Body	310
12.4.1	Defibrillator	311
12.4.2	Pacemaker	312
12.4.3	Diathermy	313

	12.4.4 Transcutaneous Electrical Nerve Stimulation	315
	12.4.5 Implantable Smart Medical Devices	318
12.5	Additional Simple Bio-instruments	318
	12.5.1 Electronic Stethoscopes	319
	12.5.2 Plethysmograph	319
12.6	Gamma Scintillation Counting	319
	12.6.1 Using a Single-Channel Analyzer	320
	12.6.2 Using the Multi-Channel Analyzer	322
References		323
Problems		323

Appendix A: More on Circuit Analysis ... 325

A.1	Nodal Analysis of a Circuit	325
A.2	Mesh Analysis	327
A.3	Linearity, Superposition, Source Transformation	329
A.4	Thévenin's and Norton's Theorems	332

Appendix B: Fourier Transform ... 335

B.1	Sinusoidal Excitation	335
B.2	Fourier Series	335
B.3	Fourier Transform	336
	Example B.M.1	337
B.4	Fourier Transform and the R, L, C Elements	338
B.5	Use of Phasors in the Analysis of a Linear Circuit	339
References		342

Appendix C: Laplace Transform ... 343

C.1	Complex Excitation	343
C.2	Complex Frequency and Laplace Transform	345
	Example C.M.1	348

Appendix D: MATLAB and Simulink Tutorial ... 351

D.1	Computer Aided Analysis	351
D.2	Basics of MATLAB	353
	D.2.1 Matrices and Arrays	354
	D.2.2 Graphics	356
	D.2.3 Programming	358
	D.2.4 Useful Functions	359
D.3	Basics of Simulink	362
	D.3.1 Building a Model	363
	D.3.2 Useful Blocks	364
References		366

Index ... 369

PART I

DEVICES AND CIRCUITS

Chapter 1

Circuit Fundamentals

1.1 Introduction

Electronics deals with signals, i.e. electrical voltages or currents that carry information. The current in the telephone receiver, in the loudspeakers of the radio or television sets, in the computer processing unit etc are examples of signals. Also signals are the currents inside the human body that carry information related to the operation of various organs of the body. These are generated by electro-chemical activities in the living cell and are generally called *bio-electrical signals* or *biosignals* for short. Electronics is required to sense or detect the biosignal and extract the information content from it.

In this chapter, initially, we briefly review basic definitions of certain electrical quantities and proceed to define and explain the representation of signals and their characteristics in physical as well as mathematical terms. Then, basic electrical circuit components, such as voltage and current sources, resistance, capacitance and inductance are introduced as they are used in electronics. Kirchhoff's current and voltage laws are briefly discussed and used in studying the concepts of differentiation and integration. The operation of the integrator circuit as a low-pass filter is also demonstrated. Finally, the concept of circuit analysis and indeed the computer-aided analysis is discussed and the program MATLAB is introduced.

1.2 Current, Voltage, Circuit

An *electrical current* or simply current in one direction in a conductor is defined as the instantaneous rate of net positive charge passing through a cross section of the conductor in the given direction. Its symbol is I or i and is mathematically defined as

$$i = \frac{dq}{dt} \tag{1.1}$$

where dq is the net positive charge passing through the cross section of the conductor in the time dt. It is measured in Amperes (A). Electrical current in a conductor is movement of electrons, but in other conducting materials, like electrolytes or semiconductors, both positive and negative ions are the electrical carriers. For example, the bioelectrical current is due to movement of ions like sodium (Na^+), potassium (K^+), chlorine (Cl^-) and other ions.

Voltage or potential difference across the terminals of a circuit element is defined as the ratio of the energy required for transferring a positive electric charge from one terminal to the other divided by this charge. The symbol of the voltage is V or v and is given in Volts (V) with $1V = 1J/C$.

A *circuit element* is the mathematical model of a physical (electrical) element that has two terminals (*two-terminal element*) (Fig. 1.1). The circuit element may be characterized completely by the relationship between the current i through it and the voltage v across its terminals and it cannot be split into any other two-terminal elements.

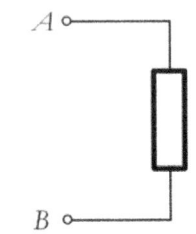

Figure 1.1. A two-terminal element

In general, the model is based on experimental measurements. Terminal A is said to be positive with respect to (w.r.t.) B, if the current enters in terminal A and goes out from terminal B and it is an external energy source that supplies this current. Then v_{AB} and i are both positive and the element dissipates power

$$p = v_{AB} \cdot i \tag{1.2}$$

In this case the element is said to be *passive*. On the other hand, if v_{AB} and i are of opposite signs, i.e. the current enters in B and goes out from A, while A is positive w.r.t. B, the power p will be negative and the element supplies energy, i.e. it is a voltage or current source. In this case the element is said to be *active*. The power is measured in Watts (W), with $1W = 1V \cdot 1A$.

Two or more circuit elements can be connected by perfect conductors to form a *network*. The point where two or more of them are connected together is a *node*. If there is a closed path in a network, this is called a *loop* and the network is said to be a *circuit*. However, in the literature the terms network and circuit are used independently of whether the network contains any loops or not. In the *circuit diagram*, the elements are represented by their symbols and are connected to each-other by pieces of conductors the length, diameter and in general their presence does not affect the operation of the circuit, which depends only on the type of elements and the way they are connected in the circuit.

An alternative name of the circuit element is the *component*. Both these terms will be used alternately in this book, with the second term used mostly for passive elements like resistors, capacitors and inductors.

1.3 Electric Signals

A voltage or a current in general has energy content and carries some type of information. In cases of heavy currents used in an electric stove or to run a motor the energy content is the most important. However in the current in the loudspeakers of our radio or TV set the information content is of utmost importance. The latter or the associated voltage is an *electric signal*.

The information in the signal is contained as a specific type of fluctuation of the amplitude, the frequency or phase of the current or voltage involved. An electric signal is generally represented by a voltage or current being a function of time, say $v(t)$ or $i(t)$.

1.3.1 Signal Representation

A voltage or current is characterized by the time dependence of its value. In mathematical terms we may write

$$v = v(t) \qquad i = i(t)$$

denoting that the voltage v and the current i are functions of time t. We may distinguish some specific cases, which occur in practice, the following:

a. *Direct current (dc) and voltage*

In this case the current and the voltage are constant and we use capital letters as their symbols

I or I_{dc} for the *dc* current

V or V_{dc} for the *dc* voltage

b. *Alternating current (ac) and voltage*

$$v = V_m \sin(\omega t + \theta) \qquad i = I_m \sin(\omega t + \varphi)$$

In these V_m is the *amplitude* of the voltage v and I_m the *amplitude* of the current i, while ω is the *frequency* of these quantities given in *radian/s (rad/s)*. Finally, θ and φ is the phase of the voltage and the current respectively, expressed in *radians* or *degrees* in agreement with the ωt expression. Plotting υ or i against ωt or t gives their *waveform*, which in this case is *sinusoidal*. Depending on the magnitude and sign of θ and φ, υ or i may advance or lag each other.

An *ac* voltage or current is a *periodic function* of the angle ωt or of time t, because the values of υ or i are repeated every 2π radians or every time interval $T = 2\pi/\omega$ seconds. T is the *period* of the sinusoid, while $f = 1/T$ is the frequency in *Hz*.

c.   *Line spectrum*

In general, the waveform of a voltage or current may be periodic, but not sinusoidal, when it will contain a number of sinusoidal components, which can be obtained by analyzing the pertinent waveform in a Fourier series (see Appendix B). If we plot the amplitudes of these individual sinusoidal components against frequency the plot will consist of discrete vertical lines (perpendicular to the frequency axis) the length of which will be proportional to the amplitude of the corresponding frequency component. This is called a *line spectrum*.

d.   *Continuous spectrum*

A general non-periodic waveform is characterized by a *continuous spectrum*, which can be studied using the Fourier transform (see Appendix B).

Any voltage or current is equivalent to a *dc* voltage or current respectively producing the same heating effect. This is its *root-mean-square*, or *rms* value given by the equation

$$V_{rms} = \sqrt{\frac{1}{T}\int_0^T \upsilon^2 dt} \qquad I_{rms} = \sqrt{\frac{1}{T}\int_0^T i^2 dt} \qquad (1.3)$$

where T is the period of the periodic waveform (or the time up to which the *rms* value is calculated for non-periodical signals).

It can be easily shown using Eq. (1.3), that for the *dc* current I_{dc} or the *dc* voltage V_{dc} the *rms* value is I_{dc} or V_{dc} respectively, while for the *ac* sinusoidal voltage or current we have

$$V_{rms} = \frac{V_m}{\sqrt{2}} \qquad I_{rms} = \frac{I_m}{\sqrt{2}} \qquad (1.4)$$

The *bandwidth* of a waveform is the difference between the highest and the lowest frequencies in the spectrum, or the highest frequency

Circuit Fundamentals 7

contained in its spectrum, if its lowest frequency is zero. Clearly, the bandwidth of a non-periodic waveform is infinite.

Example 1.M.1

Using MATLAB plot the voltage $v(t)=10\sin\omega t$ (*S.I.*) and the current $i(t)=20\cos(\omega t+30°)$ (*S.I.*) with period $T=50ms$ for $t=0$ to $2T$.

Code

```
>> t=0:0.001:0.1;              % define time range (in sec)
>> w=2*pi/0.05;                % define frequency (in rad/sec)
>> phi=30*pi/180;              % convert degrees to radians
>> v=10*sin(w*t);
>> i=20*cos(w*t+phi);
```

% the plots below are derived with the following procedure (the reader is advised first to consult the MATLAB tutorial in Appendix *D*)

```
>> subplot(2,1,1); plot(t,v); axis([0 0.1 -15 15]); grid; xlabel('time (sec)'); ylabel('v(t) (volts)')
>> subplot(2,1,2); plot(t,i); axis([0 0.1 -25 25]); grid; xlabel('time (sec)'); ylabel('i(t) (amperes)')
```

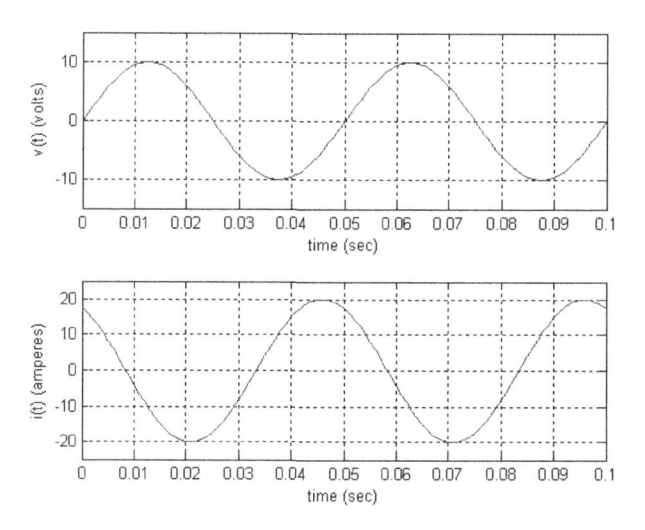

Figure 1.M.1. Voltage $v(t)$ and current $i(t)$

Example 1.M.2

Using MATLAB find the *rms* values of voltage $v(t)$ and current $i(t)$ of example 1.M.1 according to Eqs. (1.3). Then use Eqs. (1.4) to verify the results.

Code

```
>> T=0.05;
```

```
>> vsquare_integral=quad('vsquare',0,T);    % use m-file vsquare.m and take the
                                             integral
>> isquare_integral=quad('isquare',0,T);    % use m-file isquare.m and take the
                                             integral
>> vrms=sqrt(vsquare_integral/T)
>> irms=sqrt(isquare_integral/T)

             vrms =                                    irms =

             7.0711                                    14.1421

>> vrms1=10/sqrt(2)                         % verification with Eqs. (1.4)
>> irms1=20/sqrt(2)                         % verification with Eqs. (1.4)

             vrms1 =                                   irms1 =

             7.0711                                    14.1421
```

The m-files vsquare.m and isquare.m are:

```
function v2=vsquare(t)              function i2=isquare(t)
v=10*sin((2*pi/0.05)*t);            i=20*cos((2*pi/0.05)*t+30*pi/180);
v2=v.^2;                            i2=i.^2;
end                                 end
```

1.4   Resistance, Resistor

The *resistance* is a circuit element defined by the relationship between the voltage v and the current i in it, which is (Ohm's law)

$$R = \frac{v}{i} \tag{1.5}$$

It is a *linear* element measured in Ohms (Ω) with $1\Omega=1V/1A$. In the resistance, v and i have the same sign, therefore the power dissipated is

$$p = v \cdot i = i^2 R = \frac{v^2}{R} \tag{1.6}$$

by virtue of Eq.(1.5). Also according to Eq. (1.6) it is a passive element. The physical element is the *resistor*. Its symbol is shown in Fig. 1.2.

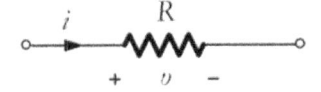

Figure 1.2. Symbol of resistance

The *conductance* G of a resistor is defined as

$$G = \frac{1}{R} \tag{1.7}$$

or

$$G = \frac{i}{v} \tag{1.8}$$

and is given in S (Siemens) (earlier in the past it was given in *mho* or Ω^{-1}) with $1S=1\Omega^{-1}$.

The resistance of a uniform conductor as in Fig. 1.3 of length l and cross section area A is given by

Figure 1.3. Uniform conductor

$$R = \rho \frac{l}{A} \tag{1.9}$$

where ρ is the *resistivity* given in $\Omega \cdot m$. The resistivity characterizes the material out of which the conductor is made. In Table 1.1 the resistivity of some materials is given.

Table 1.1. Resistivity of some materials (at 20°C)

Material	ρ ($\Omega \cdot m$)
silver	1.6×10^{-8}
copper	1.7×10^{-8}
constantan	50×10^{-8}
silicon	~ 1000
human skeleton	5

A material of theoretically infinite resistivity is called an *insulator*. The *conductivity* σ of the material is the inverse of the resistivity i.e.

$$\sigma = \frac{1}{\rho} \tag{1.10}$$

and is measured in $S \cdot m^{-1}$.

Example 1.M.3

Using MATLAB complete the following table with voltage and power dissipation values for the resistance $R=20\Omega$.

i (amperes)	0	1	2	3	4	5	6	7	8	9
v (volts)										
p (watt)										

Code

```
>> R=20;
>> i=0:1:9;                          % define current values
>> v=i.*R;
>> p=(i.^2)*R;
>> table=[i v p]                     % present the table
table =
   Columns 1 through 10
       0      1      2      3      4      5      6      7      8      9
   Columns 11 through 20
       0     20     40     60     80    100    120    140    160    180
   Columns 21 through 30
       0     20     80    180    320    500    720    980   1280   1620
```

1.5 Ideal Voltage and Current Sources

The *ideal voltage source* (*IVS*) is a two terminal source of energy, in which the voltage across its terminals is independent of the current in it. If its voltage does not depend on the voltage or the current in any other element in the circuit, in which it is embedded (connected), it is called an *independent IVS*.

The symbol of an *IVS* is one of those shown in Fig. 1.4(a), (b) and (c), depending on whether its voltage is constant with time (*dc*), alternating (*ac*), or of general form respectively.

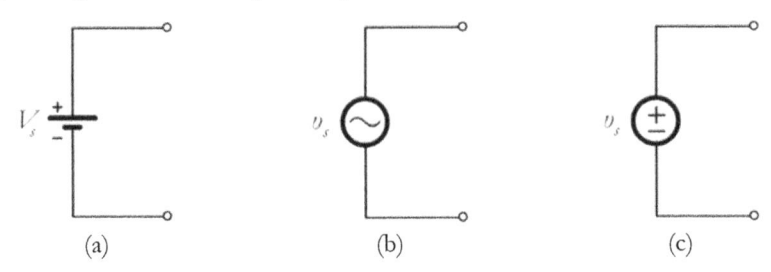

Figure 1.4. Symbols of independent *IVS*: (a) *dc*, (b) *ac* with $v_s = V_m \cos\omega t$ and (c) general symbol

In accordance with these definitions an *ideal current source* (*ICS*) is a two-terminal energy source, in which the current that it supplies to the circuit it is connected to, is independent of the voltage across its terminals. If it is also independent of the current or voltage in any other element in the circuit, then it is called an *independent ICS*. In Fig. 1.5(a), (b) and (c) the most common symbols of ideal current sources are shown.

The terminals of any voltage or current source constitute its output. Both, voltage or current sources are active elements, since the current comes out of the positive terminal and gets into the negative terminal.

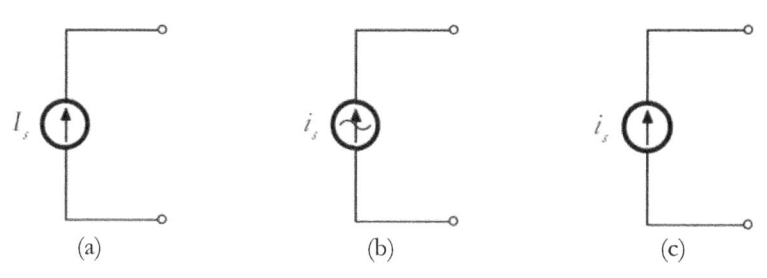

Figure 1.5. Symbols of independent current sources: (a) dc ICS, (b) ac ICS, (c) general ICS

1.6 Kirchhoff's Laws

Kirchhoff's Voltage Law or KVL states that the algebraic sum of voltages in a loop equals zero i.e.

$$\sum_{j=1}^{k} v_j = 0 \tag{1.11}$$

According to this law the sum of voltages in the loop in Fig. 1.6(a) is

$$v_1 - v_2 + v_3 + v_4 - v_5 = 0 \tag{1.12a}$$

or

$$-v_1 + v_2 - v_3 - v_4 + v_5 = 0 \tag{1.12b}$$

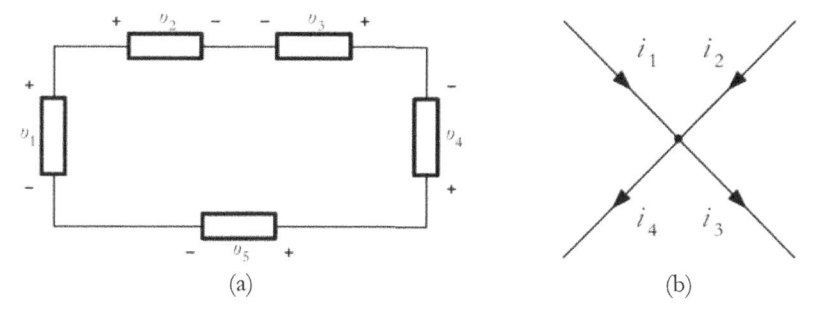

Figure 1.6. Applying (a) KVL and (b) KCL

In applying KVL in a loop as the one above, one may write as positive the voltage in a circuit element, when the current flowing in the clockwise direction enters the negative terminal and negative, when it enters the positive terminal, as written in Eq. (1.12a). Alternatively, one may write as positive the voltage, when the current running in the clockwise direction meets first the positive terminal and write it as negative, when the same current meets first the negative terminal of a circuit element, as it is written in Eq. (1.12b). To avoid mistakes when applying KVL in different loops of the same circuit, one has to stick to one's choice of convention.

Kirchhoff's Current Law or *KCL* states that, the algebraic sum of currents in a node equals zero. Taking as positive the currents leaving the node and negative those entering the node in Fig. 1.6(b) we may write

$$-i_1 - i_2 + i_3 + i_4 = 0 \qquad (1.13a)$$

On the other hand this equation can be written as

$$i_1 + i_2 - i_3 - i_4 = 0 \qquad (1.13b)$$

if the choice of positive currents is for those entering the node and negative for those leaving the node. Again one has to stick to one's choice to avoid mistakes when applying *KCL* for different nodes in the same circuit.

Applying *KCL*, *KVL* and Ohm's law one can easily show the following:

a. When two or more resistances are connected *in series*, the same current will pass through all of them. Then, the total resistance will be the sum of the individual resistances.

b. Alternately, when two or more conductances are connected *in parallel*, the same voltage will be applied to them. Then, the total conductance will be the sum of the individual conductances.

Example 1.1

Consider the circuit in the following Fig. E.1.1 where $v_s = 10\sin\omega t$ (*V*). Determine the power dissipated in the resistance $R_L = 1k\Omega$.

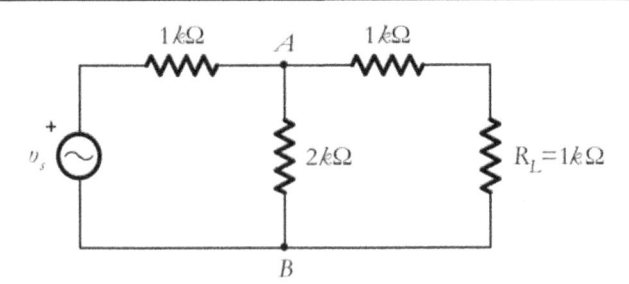

Figure E.1.1.

Solution

The power dissipated in R_L is

$$P_L = I_L^2 R_L$$

where I_L is the *rms* value of the current in R_L. Let i_s be the current coming out of the source, as shown in Fig. E.1.2. Clearly,

$$i_s = \frac{v_s}{1k\Omega + R_{AB}}$$

where R_{AB} is the total resistance between nodes A and B. Here R_{AB} is the parallel combination of $2k\Omega$ and $(1+R_L)\ k\Omega$ and since $R_L = 1k\Omega$

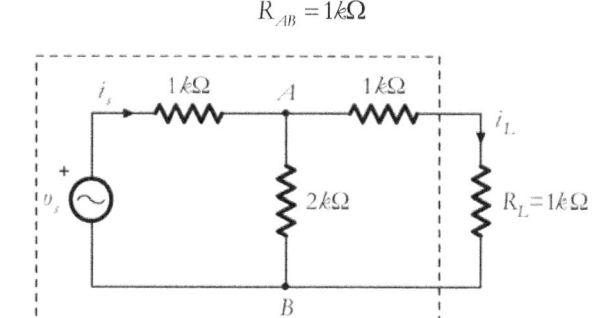

$$R_{AB} = 1k\Omega$$

Figure E.1.2.

Therefore,

$$i_s = \frac{v_s}{(1+R_{AB})k\Omega} = \frac{10\sin\omega t(V)}{2k\Omega} = 5\sin\omega t\,(mA)$$

This is divided by 2 to give the current i_L, which is found to be

$$i_L = \frac{1}{2}i_s = 2.5\sin\omega t\,(mA)$$

Its *rms* value I_L is

$$I_L = \frac{2.5}{\sqrt{2}}mA$$

as i_L is sinusoidal.

Therefore, the power dissipated in R_L will be

$$P_L = I_L^2 R_L = \frac{2.5^2 \times 10^{-6}}{2} \times 10^3 W = 3.125\,mW$$

The current i_s could have been found easier or at least quicker, if we had used Thevenin's Theorem (see Appendix A.4). According to this theorem, the part of the circuit in the box within the broken lines in Fig. E.1.2 can be substituted by an equivalent non-ideal voltage source v_T with internal resistance R_T, as shown in Fig. E.1.3, where

$$v_T = \frac{2}{1+2}v_s = \frac{2}{3}v_s$$

and

$$R_T = \frac{1 \times 2}{1+2} + 1 = \frac{2}{3} + 1 = 1.67 k\Omega$$

Then

$$i_L = \frac{0.67 \times 10 \sin \omega t}{1.67 + R_L} (mA) = \frac{6.7}{2.67} \sin \omega t \, (mA) = 2.5 \sin \omega t \, (mA)$$

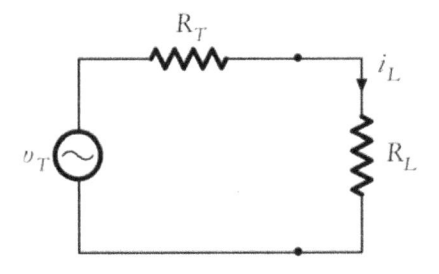

Figure E.1.3.

Example 1.M.4

Using MATLAB find the voltages at nodes 1, 2 and 3 for the circuit in Fig. A.1(a) in Appendix A.

Code

Applying nodal analysis in the circuit of Fig. A.1(a) (see Section A.1) the following set of equations is derived:

$$12v_1 - 4v_2 - 8v_3 = 2$$
$$-12v_1 + 10v_2 + 10v_3 = 0$$
$$v_2 - v_3 = 4$$

which can be written in matrix form as:

$$\begin{bmatrix} 12 & -4 & -8 \\ -12 & 10 & 10 \\ 0 & 1 & -1 \end{bmatrix} \begin{bmatrix} v_1 \\ v_2 \\ v_3 \end{bmatrix} = \begin{bmatrix} 2 \\ 0 \\ 4 \end{bmatrix}$$

```
>> g=[12 -4 -8; -12 10 10; 0 1 -1];
>> i=[2; 0; 4];
>> v=inv(g)*i

v =

   -1.2500
    1.2500
   -2.7500
```

1.7 Non-ideal Voltage and Current Sources

In a *non-ideal voltage source* the voltage across its terminals depends to some extent on its current and this is modeled by placing a resistance in series with the symbol of the voltage source (*IVS*), as shown in Fig. 1.7(a). This is called the *internal resistance* of the non-ideal *VS*. If the current of the current source is not completely independent of the voltage across its terminals, the *CS* is non-ideal and its symbol is as shown in Fig. 1.7(b), where a resistance R_s is connected in parallel to it. This is its internal resistance.

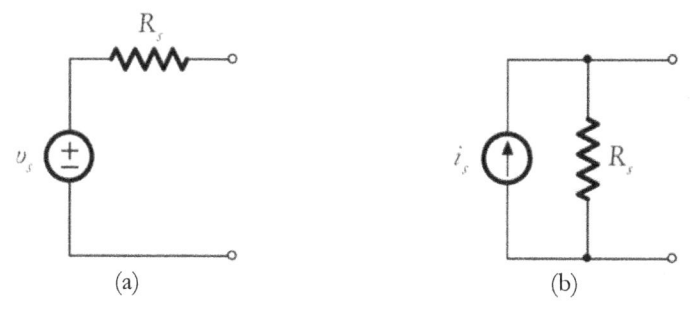

Figure 1.7. Non-ideal (a) voltage and (b) current source

It can be shown (Appendix *A*.3), that a non-ideal voltage or current source can behave as a non-ideal current or voltage source respectively depending on their load, i.e. the resistance connected to it. Then a non-ideal voltage source and a non-ideal current source are called equivalent, if they supply with the same voltage and same current the same load. In this case they must have the same internal resistance R_s. The equivalence is shown in Fig. 1.8.

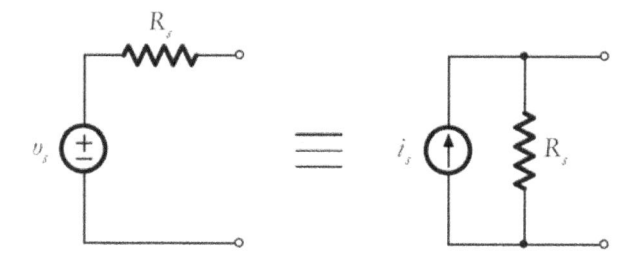

Figure 1.8. Equivalent sources when $v_s = i_s R_s$

Clearly, the equivalence can be demonstrated with the load resistance being infinite, as in Fig. 1.8. Then, the open-circuit voltage across the output of the voltage source is $v_{oc} = v_s$ and that across the output of the current source $v_{oc} = i_s R_s$. Since both supply zero current to their infinite load, it follows, that for equivalence their output voltages should also be equal i.e.

$$v_s = i_s R_s \qquad (1.14)$$

Eq. (1.14) can be used to transform a non-ideal voltage source to a non-ideal current source and vice-versa. Consider now a non-ideal voltage source feeding a load resistance R_L, as shown in Fig. 1.9(a).

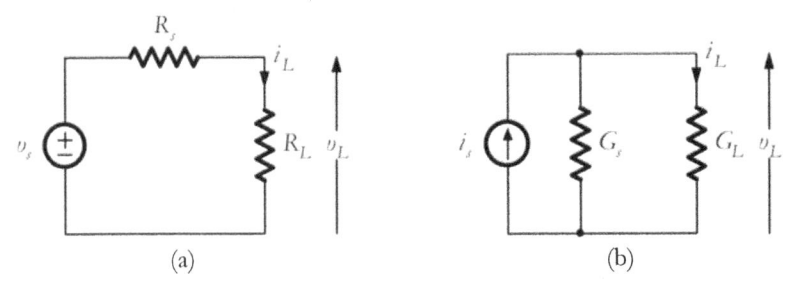

Figure 1.9. Loaded non-ideal (a) voltage and (b) current sources

The current i_L will be

$$i_L = \frac{v_s}{R_s + R_L}$$

and thus, the voltage v_L

$$v_L = i_L R_L = \frac{R_L}{R_s + R_L} v_s \quad (1.15)$$

In general, this equation represents voltage division and, since it appears very often in circuit analysis, it would be useful to be remembered by heart. By similar reasoning, it is easy to show that the current i_L in Fig. 1.9(b) is given by

$$i_L = \frac{G_L}{G_s + G_L} i_s = \frac{R_s}{R_s + R_L} i_s \quad (1.16)$$

Eq. (1.16) expresses the division of the source current i_s in the conductances G_L and G_s and it is also useful to be remembered by heart.

One more point is proper here: It can be shown (see problem 1.3) that the maximum power that can be transferred from a non-ideal voltage source or from a non-ideal current source to the load resistance R_L occurs when this is equal to the internal resistance of the source, i.e. $R_L = R_s$. This maximum power is

$$P_{max} = \frac{v_s^2}{4R_s} \quad (1.17)$$

for the non-ideal voltage source and

Circuit Fundamentals 17

$$P_{max} = \frac{i_s^2 R_s}{4} \qquad (1.18)$$

for the non-ideal current source (*theorem of maximum power transfer*). In Eqs. (1.17) and (1.18) v_s and i_s are *dc* or *rms* values.

Example 1.M.5

Consider the circuit in Fig. 1.9(a) with $v_s=1V$ and $R_s=20k\Omega$. R_L varies from 0 to $80k\Omega$. Using MATLAB plot the power dissipated by R_L and ascertain that the maximum power that can be transferred from the source to R_L occurs when $R_L=R_s=20k\Omega$.

Code

```
>> vs=1;
>> Rs=20000;
>> RL=0:1:80000;
>> n=length(RL);
>> for i=1:n
pL(i)=((vs/(Rs+RL(i)))^2)*RL(i);
end
>> plot(RL,pL); grid; xlabel('RL (ohms)'); ylabel('pL (watts)')
```

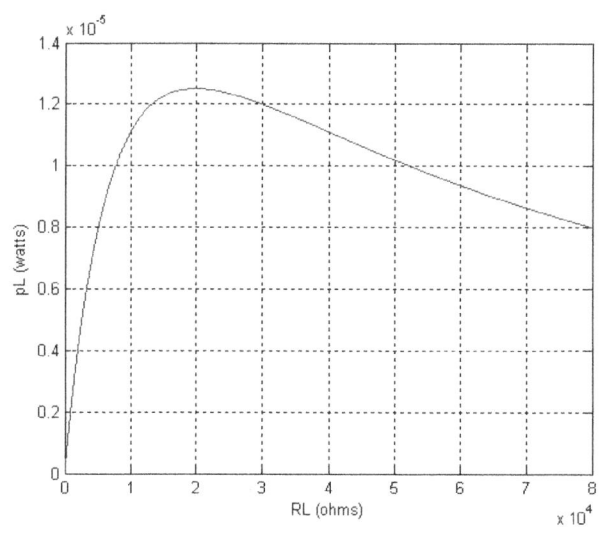

Figure 1.M.2. Power dissipated by R_L

1.7.1 Potentiometers

A *potentiometer* is a variable resistor with three terminals, as shown in Fig. 1.10(a). The moving terminal is the wiper, which can follow a linear or rotary movement. The nominal (or total) resistance of the potentiometer is R_p. As shown in Fig. 1.10, a potentiometer can be used as a variable resistance, Fig. 1.10(b), or as a potential divider, as shown in Fig. 1.10(c).

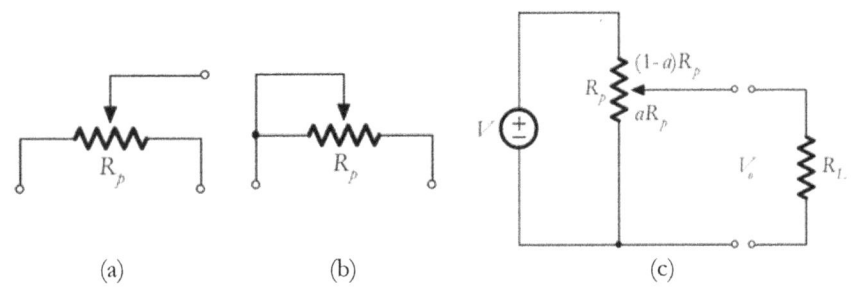

Figure 1.10. (a) The symbol of a potentiometer and its use (b) as a variable resistance and (c) as a potential divider

As a variable resistance it can take any value between zero and R_p. As a potential divider, Fig. 1.10(c), it can provide R_L with any value of voltage between V and zero. However, one should remember, that the voltage V_o obtained with R_L absent will not be the same when R_L is present. To show this, suppose that the potentiometer wiper is set to a position that the resistance to the common terminal is αR_p, when the rest of the potentiometer resistance will be $(1-\alpha)R_p$, where $0 \leq \alpha \leq 1$. With α different from 0 and 1, the voltage V_o with R_L absent will be

$$V_o = \frac{\alpha R_p}{\alpha R_p + (1-\alpha)R_p} V = \frac{\alpha R_p}{R_p} V = \alpha V$$

Now, when R_L is present, this will be in parallel with αR_p and thus, their combination will give

$$R'_L = \frac{\alpha R_p R_L}{\alpha R_p + R_L}$$

Then the new output voltage, V'_o say, will be

$$V'_o = \frac{R'_L}{R'_L + (1-\alpha)R_p} V = \frac{\alpha R_p R_L}{\alpha R_p R_L + (1-\alpha)R_p(\alpha R_p + R_L)} V =$$

$$= \frac{\alpha R_p R_L}{\alpha R_p^2 + R_p R_L - \alpha^2 R_p^2} V = \frac{\alpha R_L}{\alpha(1-\alpha)R_p + R_L} V$$

which is different from αV when R_L is absent. It is said that R_L is loading the potentiometer. Subtracting V'_o from V_o one can find the pertinent error. However, if $R_L \gg R_p$ and with $\alpha < 1$, V'_o can be approximately

Circuit Fundamentals

$$V'_o = \frac{\alpha R_L}{\alpha(1-\alpha)R_p + R_L} V \approx \alpha V = V_o$$

that is approximately equal to V_o. In practice, to avoid loading the potentiometer, R_L (or R_p) is chosen such that

$$R_L \geq 10 R_p$$

1.8 Capacitance, Capacitor

The *capacitor* is a passive element that can store energy in the form of an electric field. Its symbol is shown in Fig.1.11. The *capacitance* of the capacitor is defined by means of the relationship between the voltage and current in it, which is as follows:

Figure 1.11. Symbol of the capacitor

$$i = C \frac{dv}{dt} \qquad (1.19)$$

Capacitance unit is the *farad* (F), defined as $1C/V$ (coulomb/volt). The capacitor is made up of two parallel conducting plates on which electric charges can be stored. A dielectric layer is placed between the plates. The capacitance of the capacitor is given by

$$C = \varepsilon \frac{A}{l} \qquad (1.20)$$

where ε is the permittivity of the dielectric, A is the area of the plates and ℓ the distance between them. This relation is valid when the dimensions of the plates are much larger than the distance ℓ.

From the defining Eq. (1.19), it is clear, that there is a current in the capacitor only when its voltage changes with time. Thus, there is no current in the capacitor, if it is charged to a *dc* voltage. From the same equation we conclude, that the voltage in the capacitor cannot change suddenly, as this would mean infinite current and there is no source in nature that would provide the required infinite energy.

From Eq. (1.19) we may obtain the voltage in the capacitor as

$$v(t) = \frac{1}{C}\int_{t_0}^{t} i\,dt + v(t_0) \qquad (1.21)$$

where $v(t_0)$ is the capacitor voltage at $t=t_0$. Assuming that $v(t_0)$ is zero, the integral in Eq. (1.21) gives the total charge q stored in the capacitor and then

$$q = Cv \qquad (1.22)$$

On the other hand, the power dissipated in the capacitor is

$$p = iv = Cv\frac{dv}{dt} \qquad (1.23)$$

and, since v and dv/dt do not always have the same sign, it can be positive or negative. Clearly, for an ac voltage the total power stored each period is zero. So an ideal capacitor is a lossless element and, as this, it is considered to be passive. However, the practical capacitor is not exactly lossless as there is always some leakage current through its dielectric.

The main characteristics of the capacitor depend on their dielectric from which they take their name like polystyrene, paper, mica, electrolytic, etc.

The *KCL* and *KVL* are valid in circuits containing capacitors. It can then be easily shown, that the total capacitance of two or more capacitors connected in parallel is the sum of the capacitances of the individual capacitors. In a similar way it can be proved, that the total capacitance C_t of two capacitors with capacitances C_1 and C_2 connected in series is

$$C_t = \frac{C_1 C_2}{C_1 + C_2} \qquad (1.24)$$

1.9 Differentiation and Integration of Signals

Two simple circuits, very useful in practice, consisting of the elements we have met so far, are shown in Fig. 1.12.

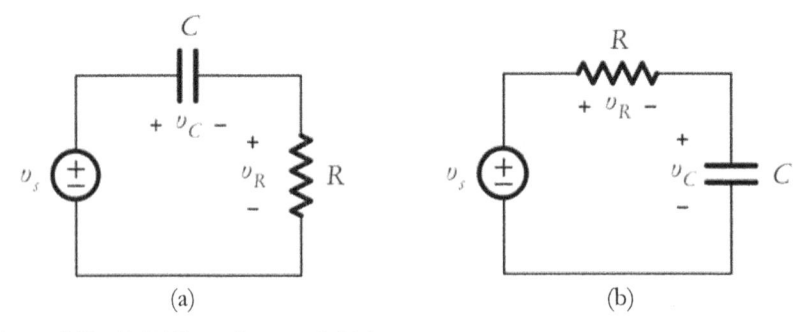

Figure 1.12. (a) Differentiator and (b) integrator

Consider Fig. 1.12(a) first. Applying *KVL* we get

$$v_s - v_C - v_R = 0 \quad \text{or} \quad v_s = v_C + v_R \tag{1.25}$$

and on differentiating w.r.t time *t*

$$\frac{dv_s}{dt} = \frac{dv_C}{dt} + \frac{dv_R}{dt} \tag{1.26}$$

The current *i* according to Eq. (1.19) is

$$i = C\frac{dv_C}{dt}$$

which makes v_R be

$$v_R = iR = RC\frac{dv_C}{dt} \tag{1.27}$$

From this we get

$$\frac{dv_C}{dt} = \frac{1}{RC}v_R \tag{1.28}$$

Substituting for dv_C/dt in Eq. (1.26) gives

$$\frac{1}{RC}v_R = \frac{dv_s}{dt} - \frac{dv_R}{dt}$$

which can be written as

$$v_R = RC\frac{d}{dt}(v_s - v_R) \tag{1.29}$$

If $v_s \gg v_R$ we will have

$$v_R \cong RC\frac{dv_s}{dt} \tag{1.30}$$

Thus, v_R is proportional to the rate of change of v_s, of course approximately, and the circuit acts as a *differentiator*.

Coming now to the circuit in Fig. 1.12(b), we may write again that

$$v_s = v_R + v_C \tag{1.31}$$

and on differentiating

$$\frac{dv_s}{dt} = \frac{dv_R}{dt} + \frac{dv_C}{dt}$$

Again

$$v_R = RC\frac{dv_C}{dt}$$

Substituting in Eq. (1.31) we get

$$v_s = RC\frac{dv_C}{dt} + v_C$$

or

$$\frac{dv_C}{dt} = \frac{1}{RC}(v_s - v_C)$$

If $v_C \ll v_s$, then

$$\frac{dv_C}{dt} \cong \frac{1}{RC}v_s \tag{1.32}$$

and on integrating

$$v_C(t) \cong \frac{1}{RC}\int_{t_0}^{t} v_s dt + v_C(t_0) \tag{1.33}$$

Thus, the circuit in Fig. 1.12(b) acts as an *integrator*.

In both Eqs.(1.30) and (1.33) the product RC appears. The unit for RC is 1s (1 second) and the RC product is called the *time constant* of the circuit. The role of the time constant τ is shown in Fig. 1.13, where the two circuits are used (a) to differentiate and (b) to integrate continuous square pulses. Two different time constants are used in each circuit.

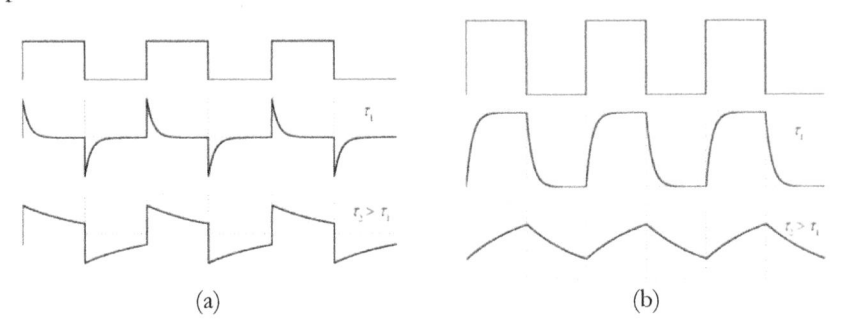

(a)   (b)

Figure 1.13. (a) Differentiation and (b) integration of square pulses

The integrator circuit in Fig. 1.12(b) is of great importance in practice, because it simulates the behaviour of the input of any practical system in that, it acts as a simple low-pass filter (see Chapter 4). Consider the case when the voltage v_s is sinusoidal i.e.

$$v_s(t) = V_m \cos \omega t \tag{1.34}$$

Circuit Fundamentals

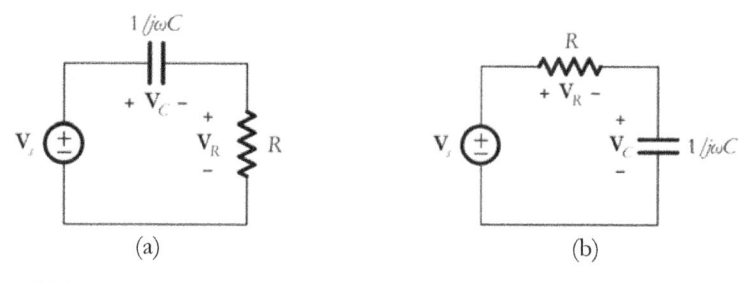

Figure 1.14.

In the real frequency domain, this is written as the phasor **V** (see Appendix B), which has an amplitude V_m and zero phase. In this case the capacitance C has an impedance $1/j\omega C$ and the circuits of differentiator and integrator are transformed into those in Fig. 1.14(a) and (b) respectively. Then, we may apply KVL to get

$$\mathbf{V}_s = \mathbf{V}_R + \mathbf{V}_C \tag{1.35}$$

with

$$\mathbf{V}_C = \frac{1}{j\omega C}\mathbf{I} \tag{1.36}$$

where **I** is the phasor of the current. Substituting in Eq. (1.35) gives

$$\mathbf{I}R + \frac{1}{j\omega C}\mathbf{I} = \mathbf{V}_s$$

and

$$\mathbf{I} = \frac{j\omega C}{1 + j\omega CR}\mathbf{V}_s \tag{1.37}$$

Subsequently, substituting the value of **I** in Eq. (1.36) gives

$$\mathbf{V}_C = \frac{1}{1 + j\omega CR}\mathbf{V}_s \tag{1.38}$$

Clearly $\mathbf{V}_C$, which is considered to be the output voltage of the integrator, see Fig. 1.14(b), is a function of the components and the frequency ω. Its magnitude is

$$|\mathbf{V}_C| = \frac{1}{|1 + j\omega CR|}|\mathbf{V}_s| = \frac{1}{\sqrt{1 + \omega^2 C^2 R^2}}|\mathbf{V}_s| \tag{1.39}$$

and its phase

$$\arg \mathbf{V}_C = -\tan^{-1}\omega CR \tag{1.40}$$

The ratio $\mathbf{V}_C/\mathbf{V}_s$ is called the *voltage transfer function* of the integrator. It is interesting to plot $|\mathbf{V}_C/\mathbf{V}_s|$ and $\arg(\mathbf{V}_C/\mathbf{V}_s)$ against ω, when we obtain the frequency response of the integrator (or *Bode* plots). These plots are shown in Fig. 1.15 with the $|\mathbf{V}_C/\mathbf{V}_s|$ and $\arg(\mathbf{V}_C/\mathbf{V}_s)$ called the magnitude (or amplitude) response and the phase response of the integrator. It can be seen that the amplitude of $\mathbf{V}_C$ is continuously decreasing as the frequency increases, i.e. the circuit rejects the higher frequencies and lets the lower frequencies pass. This is why it is called a low-pass filter. Also this is the reason why the response pulses in Fig. 1.13(b) appear curved at their rising and falling edges. (More about filtering will be said in Chapter 4).

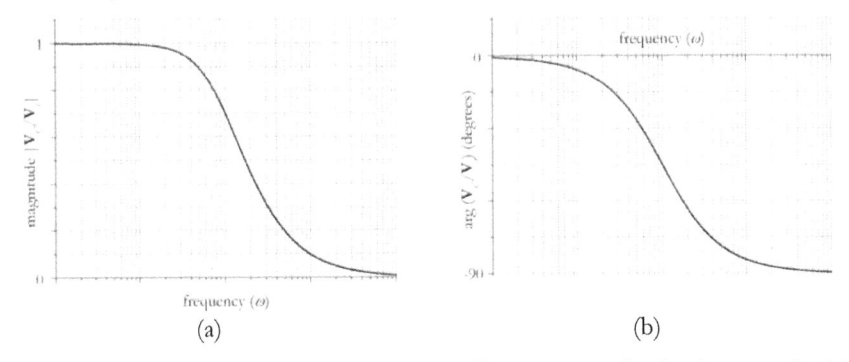

Figure 1.15. Frequency response of the filter: (a) amplitude (or magnitude) response and (b) phase response

On the other hand, the circuit in Fig. 1.14(a) also operates as a simple filter but as high-pass this time. To show this, consider the voltage across R, i.e. $\mathbf{V}_R$, which is easily obtained by virtue of Eq. (1.37) as

$$\mathbf{V}_R = \mathbf{I}R = \frac{j\omega CR}{1 + j\omega CR}\mathbf{V}_s \qquad (1.41)$$

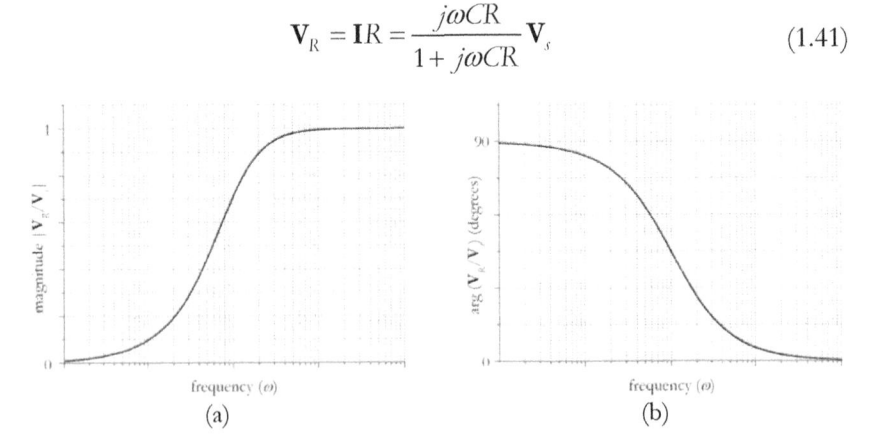

Figure 1.16. (a) Magnitude and (b) phase plots of the simple high-pass filter in Fig. 1.13(a)

The plots of the magnitude and phase of the voltage ratio $\mathbf{V}_R/\mathbf{V}_s$ against ω are as shown in Fig. 1.16. It can be seen, that at low frequencies,

Circuit Fundamentals

the magnitude of V_R/V_s is low and increases as the frequency increases, i.e. the circuit "attenuates" the lower frequencies and allows the higher frequencies to pass to the output.

Example 1.2

The values of R and C in the circuit of Fig. 1.12(b) are $1 k\Omega$ and $0.2 \mu F$. Determine the cutoff frequency of the amplitude of the ratio V_C/V_s and its phase at this frequency.

Solution

According to Eq. (1.38)

$$\frac{V_C}{V_s} = \frac{1}{1+j\omega CR}$$

The cutoff frequency is defined as the frequency ω_c at which

$$\left|\frac{V_C}{V_s}\right| = \frac{1}{\sqrt{1+\omega_c^2 C^2 R^2}} = \frac{1}{\sqrt{2}}$$

or

$$20 \log\left|\frac{V_C}{V_s}\right| = -3 dB$$

From the first of these two equations we get

$$1+\omega_c^2 C^2 R^2 = 2$$

or

$$\omega_c^2 = \frac{1}{C^2 R^2}$$

or

$$\omega_c = \frac{1}{CR}$$

Substituting values gives

$$\omega_c = \frac{1}{2\times 10^{-7} \times 10^3} = \frac{10^4}{2} = 5\times 10^3 \, rad/s = 5 krad/s$$

or in Hz

$$f_c = \frac{\omega_c}{2\pi} = \frac{5\times 10^3}{2\pi} = 796 Hz$$

The phase of $\mathbf{V}_C/\mathbf{V}_s$ at $\omega=5krad/s$ is

$$\varphi = \arg\frac{\mathbf{V}_C}{\mathbf{V}_s} = \arg\frac{1}{1+j\omega_c CR} = -\tan^{-1}\omega_c CR = \tan^{-1}1 = 45°$$

Example 1.M.6

Consider the integrator circuit in Fig. 1.12(b) with $R=40k\Omega$ and $C=4\mu F$. υ_s is a rectangular pulse with amplitude $2V$ and width $1s$. Plot $\upsilon_C(t)$ for a pulse period.

Code

Analysis of the circuit in Fig. 1.12(b) has already given

$$\frac{d\upsilon_C}{dt} = \frac{1}{RC}(\upsilon_s - \upsilon_C)$$

Solving the above differential equation we get

$$\upsilon_C(t) = \upsilon_s\left(1 - e^{\frac{-t}{RC}}\right)$$

for charging the capacitor with $\upsilon_C(t)=0$ for $t=0$ and

$$\upsilon_C(t) = \upsilon_{C_0} e^{\frac{-t}{RC}}$$

for discharging the capacitor ($\upsilon_s=0$) with υ_{C_0} being its initial voltage.

```
>> R=40000;
>> C=4e-6;
>> for i=1:100                          % charging the capacitor
t(i)=i/100;
vc(i)=2*(1-exp(-t(i)/(R*C)));
end
>> vco=vc(100);
>> for i=101:200                        % discharging the capacitor
t(i)=i/100;
vc(i)=vco*exp(-t(i-100)/(R*C));
end
>> plot(t,vc); axis([0 2 0 2.2]); grid; xlabel('time (sec)'); ylabel('vc(t) (volts)')
```

Example 1.M.7

Consider the integrator circuit in Fig. 1.12(b) with $R=40k\Omega$, $C=4\mu F$ and $\upsilon_s=2V$. Plot $\upsilon_C(t)$ using analytical solution as well as numerical solution of differential equation.

Circuit Fundamentals 27

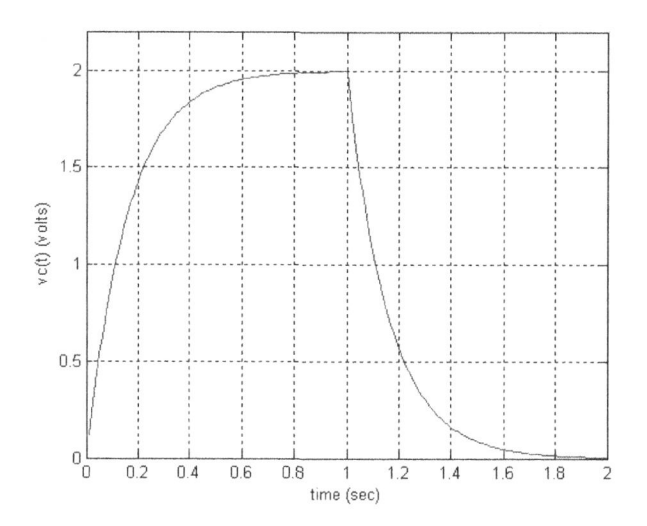

Figure 1.M.3. Integration of square pulse (example 1.M.6)

Code

Analysis of the circuit in Fig. 1.12(b) has already given

$$\frac{dv_C}{dt} = \frac{1}{RC}(v_s - v_C)$$

which becomes

$$\frac{dv_C}{dt} = 12.5 - 6.25 v_C$$

We have already seen that the analytical solution is

$$v_C(t) = v_s\left(1 - e^{\frac{-t}{RC}}\right)$$

considering $v_C(t)=0$ for $t=0$.

```
>> [t,vc1]=ode23('analytical',[0 2],0);% numerical solution using m-file analytical.m
                                        % till t=2s considering v_C(t)=0 for t=0
>> C=4e-6;
>> R=40000;
>> vc2=2*(1-exp(-t/(R*C)));                         % analytical solution
>> subplot(2,1,1); plot(t,vc1); axis([0 2 0 2.2]); grid; xlabel('time (sec)'); ylabel('vc(t) (volts)')
>> title('numerical and analytical solution for vc(t)')
>> subplot(2,1,2); plot(t,vc2); axis([0 2 0 2.2]); grid; xlabel('time (sec)'); ylabel('vc(t) (volts)')
```

The m-file analytical.m is:

```
function dx=analytical(t,x)
dx=12.5-6.25*x;
end
```

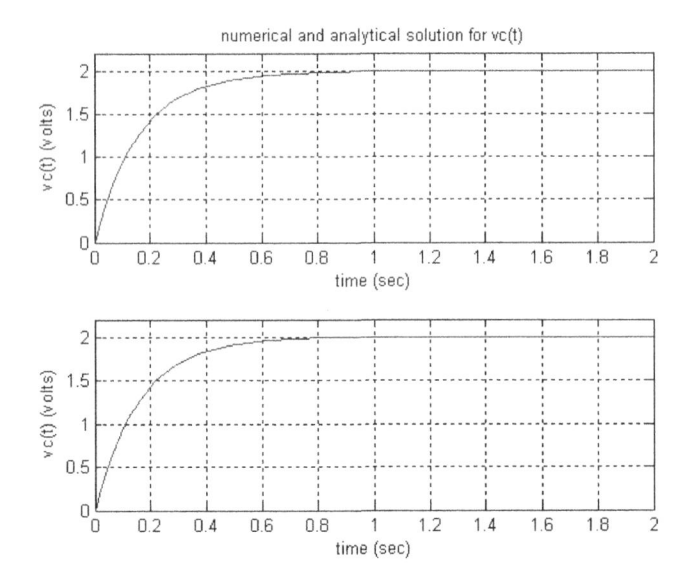

Figure 1.M.4. Numerical and analytical solution for finding $v_c(t)$

1.10 Inductance, Inductor

The inductor is a passive element capable of storing energy in the form of a magnetic field. Its symbol is shown in Fig.1.17. Its action is defined through the following relationship between the current and voltage in it:

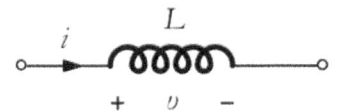

Figure 1.17. Symbol of the inductor

$$v = L\frac{di}{dt} \qquad (1.42)$$

L is the *inductance* of the inductor. The inductance unit is $1H$ (Henry). The physical element is known as the coil. It is constructed by wounding a wire around a cylindrical surface. Clearly, from the above defining equation a voltage appears between the terminals of the inductor only when i changes. Therefore, for a *dc* current the inductor acts like a *short circuit*. Also from Eq. (1.42), the current in the inductor can be found by integrating it i.e.

$$i(t) = \frac{1}{L}\int_{t_0}^{t} \upsilon dt + i(t_0) \tag{1.43}$$

where $i(t_0)$ is the current in the inductor at $t=t_0$. The power absorbed by this element is

$$p = i\upsilon = Li\frac{di}{dt} \tag{1.44}$$

and the energy stored in the form of magnetic field in the inductor from time instant t_0 to t is

$$w(t) - w(t_0) = \int_{t_0}^{t} pdt = L\int_{t_0}^{t} i\frac{di}{dt}dt = \frac{L}{2}\left[i^2(t) - i^2(t_0)\right]$$

If at t_0

$$i(t_0) = 0 \qquad w(t_0) = 0$$

then

$$w(t) = \frac{L}{2}i^2 \tag{1.45}$$

This total energy can be returned to the circuit by the ideal inductor. However the coil is not an ideal inductor having a small resistance, which is modeled to be in series with the inductance. Also the inductance is not perfectly linear and it changes with frequency. For these reasons and others (volume, weight), the use of inductors in modern electronics, produced in microelectronic form, is usually avoided.

It can be easily shown that, if two or more inductors are connected in series, the total inductance will be the sum of the inductances of the individual inductors. On the other hand, if inductors L_1 and L_2 are connected in parallel, the total inductance L_t will be

$$L_t = \frac{L_1 L_2}{L_1 + L_2} \tag{1.46}$$

In such combinations of inductors, it is important to observe, that they are placed in the circuit in such a way that *coupling* among these components is avoided. By virtue of Eq. (1.44) the ideal inductor is lossless.

1.11 Circuit Analysis

By circuit analysis we mean the determination of the node voltages and the branch currents in the circuit. This is achieved by the systematic application of *KCL* or *KVL* and Ohm's Law in the nodes or loops and branches. As is explained in Appendix *A*, there are two methods of analysis, namely, *nodal analysis* and *mesh analysis*. Nodal analysis is usually preferred to mesh analysis, because only this can be easily applied to *non-planar* circuits. For planar circuits their diagrams can be drawn on a sheet of paper without any crossings of branches which is not possible for non-planar circuits.

Node voltages and branch currents in any circuit exist only if there exists at least one current or voltage source in the circuit. The presence of such a source in the circuit results in the *excitation* of the circuit, while the subsequent creation of the node voltages and the branch currents constitute the *response* of the circuit.

Circuit analysis entails writing down a number of algebraic equations the solution of which becomes tedious when the circuit is complicated, i.e. if the circuit contains a lot of loops, branches and sources as well as nonlinear elements. For such cases in the study and design of the large integrated electronic circuits of today's technology, the analysis is carried out by special computer programs such as SPICE, MATLAB, etc. These programs can give the response of the circuit under consideration to various excitations as functions of frequency or time. The only requirement is the designer to introduce to the computer the description of the circuit diagram as well as the type and values of components, and, of course, the availability of such an analysis program in his computer.

Computer aided circuit analysis is further explained in Appendix *D* where MATLAB is introduced to some detail. The reader is advised to study this, because it is used in the text for the solution of worked examples.

References

[1]. Hayt W. H. Jr & Kemmerly J. E., *Engineering Circuit Analysis*, McGraw-Hill, New York, 1971.
[2]. Skilling N. H., *Electrical Engineering Circuits*, Wiley, 1968.
[3]. Smith R. J., *Circuits Devices and Circuits*, Wiley, New York, 1976.
[4]. Attia J. O., *PSPICE and MATLAB for Electronics: An Integrated Approach*, CRC Press, 2002.

MATLAB Problems

❖ *The reader is advised first to consult the MATLAB tutorial in Appendix D*

1.M.1. Using MATLAB find the charge q for the current $i(t)$ of example 1.M.1 at $t=10ms$ given that $q=0$ at $t=0$ according to Eq. (1.1).

1.M.2. Using MATLAB plot the power for the voltage $v(t)$ and the current $i(t)$ as given in example 1.M.1 and for $t=0$ to $2T$.

1.M.3. Using MATLAB find the equivalent resistance of the series connected resistances 10Ω, 20Ω, 30Ω, 40Ω, 50Ω and 60Ω.

1.M.4. Using MATLAB find the equivalent capacitance of the series connected capacitances 10F, 20F, 30F, 40F, 50F and 60F.

1.M.5. Using MATLAB solve problem 1.6.

1.M.6. Using MATLAB solve problem 1.7.

1.M.7. Consider the integrator circuit in Fig. 1.12(b) with $C=2\mu F$ and $v_s=2V$. Plot $v_C(t)$ for $R=5k\Omega$, $R=50k\Omega$ and $R=500k\Omega$ considering $v_C(t)=0$ for $t=0$. Use analytical as well as numerical solution.

Problems

1.1. Show that the total resistance of two or more resistances connected in series is equal to their sum, while the equivalent of two or more conductances connected in parallel is again equal to their sum.

1.2. Calculate the current in the resistance in the circuit in Fig. P.1.1, when the switch is in position 1. With the switch in position 2, what will be the new value of the current? Both voltage sources are ideal.

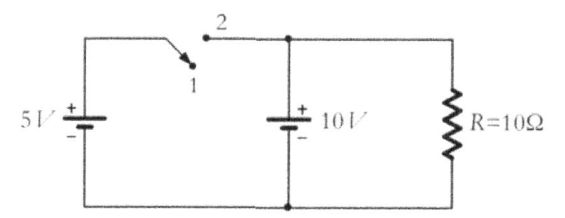

Figure P.1.1.

(Answers: a. 1A, b. not valid)

1.3. Calculate the voltage in the resistance in Fig. P.1.2 with the switch in position 1. Then turn the switch to position 2 and calculate the new value of the voltage in R. Both current sources are ideal.

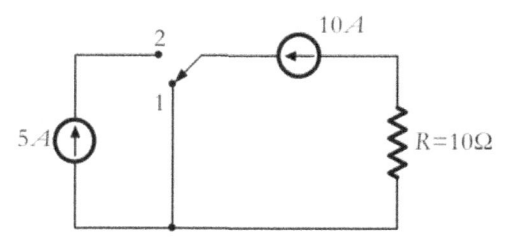

Figure P.1.2.

(Answers: a. 100V, b. not valid)

1.4. Show the validity of Eqs. (1.17) and (1.18).

1.5. Explain why the lights of your car dim when you try to start the engine with the lights on.

1.6. In the circuit in Fig. P.1.3 determine the current in the resistor of 2Ω.

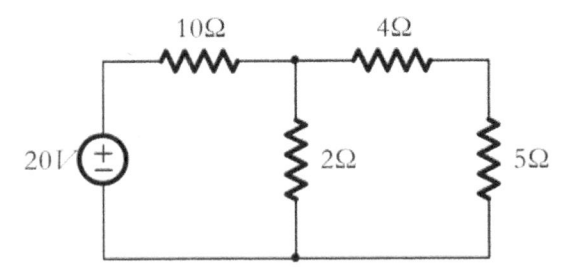

Figure P.1.3.

1.7. In the circuit in Fig. P.1.4 determine the voltage across the resistance of 4Ω.

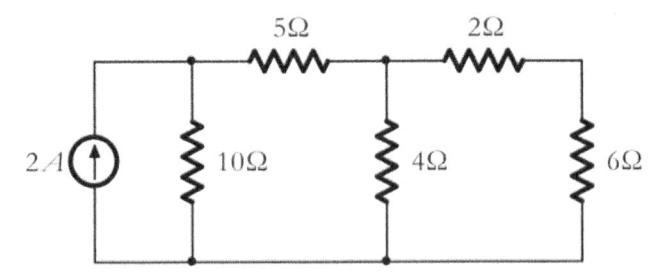

Figure P.1.4.

1.8. The switch in the circuit in Fig. P.1.5 closes at $t=0$. Determine the time it takes for the voltage across the capacitor to reach half of its final value, if $v_C(t)=0$ at $t=0$.

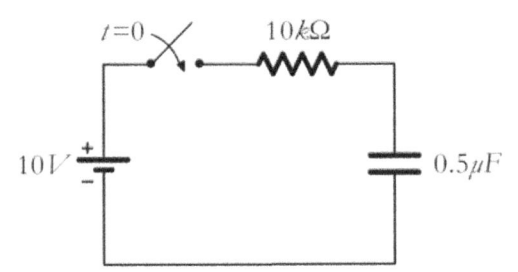

Figure P.1.5.

1.9. The switch in the circuit in Fig. P.1.6 closes at $t=0$. Determine an expression for the current i_s and plot it against time. What will be the final value of the voltage across the capacitor?

Circuit Fundamentals

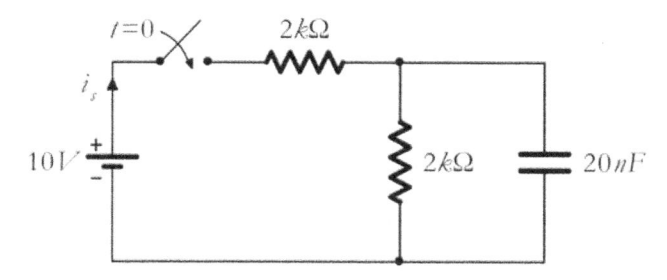

Figure P.1.6.

1.10. For the circuit in Fig. P.1.7 determine the a) frequency at which v_s and v_o are in phase and b) the magnitude of v_o/v_s at this frequency.

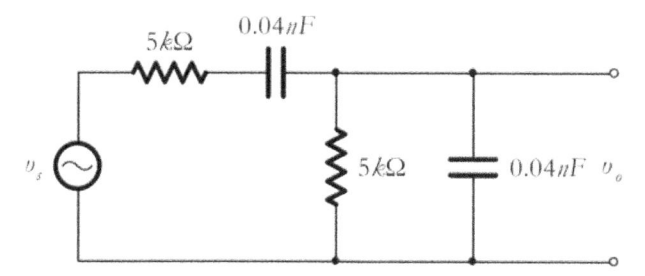

Figure P.1.7.

1.11. For the circuit in Fig. P.1.8 plot the magnitude and phase of the voltage ratio v_o/v_s against frequency ω.

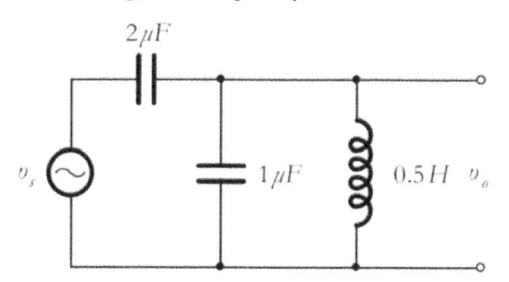

Figure P.1.8.

Chapter 2

Electronic Components – Devices

2.1 Introduction

The electric circuit components introduced in the previous chapter are also used in the electronic circuits. However, an electronic circuit, to be characterized as such, must employ additionally one or more electronic components. Electronic components are two-, three- or more terminal circuit elements that possess very useful characteristics and, mostly, are non linear. Nowadays they are all made of semiconductor material and withstand high currents and voltages. Also, their physical dimensions and weight are small. Great numbers of these can be placed together with the proper interconnections on a chip, a small piece of semiconductor, to produce an electronic integrated circuit. The important difference between an electronic circuit or system and an electric circuit or system is that the former processes information carried by a voltage or current in it, while in the latter the emphasis is given on the power content of the voltage or current.

This chapter deals first with the simplest and oldest electronic component, the diode. It is introduced through its ideal I-V characteristic and the physical operation of the semiconductor diode is briefly explained. Next two of its important applications, namely, as a rectifier and as a clipper, are explained with the diode being treated as ideal. Then the most important electronic component, the transistor, is introduced in two of its versions, namely, the *Bipolar Junction Transistor (BJT)* and the *Field-Effect Transistor (FET)*, and their physical operation is briefly explained. Also, the concept of the *Integrated Circuit (IC)* is introduced and briefly discussed. Finally, the important optoelectronic devices are introduced and their operation briefly explained.

2.2 Ideal Diode

The ideal diode is a two terminal nonlinear device, which is defined by its *I-V* characteristic shown in Fig. 2.1(a). The symbol of the diode is shown in Fig. 2.1(b). According to this *I-V* characteristic, when terminal *A*, the *anode*, is positive w.r.t. terminal *C*, the *cathode*, the diode is forward biased and current passes through the diode unprevented. On the other hand, if the anode *A* is negative w.r.t. cathode *C*, the diode is reverse biased and the current is zero. The arrow in the symbol indicates the direction of the current flow in the diode. According to its definition, the ideal diode acts as a short circuit or closed switch, when it is forward biased and as an open circuit or opened switch on reverse bias. This equivalence is useful, when one wants to explain qualitatively the operation of circuits to which diodes are embedded.

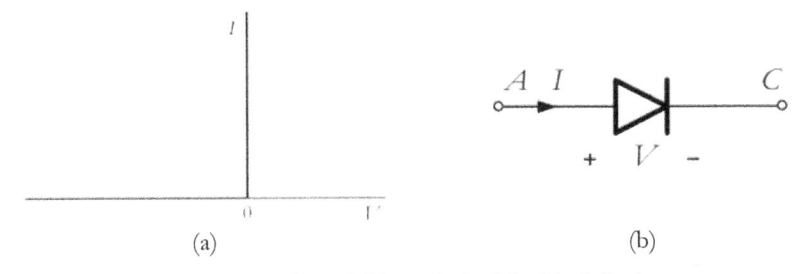

Figure 2.1. (a) *I-V* characteristic and (b) symbol of the ideal diode

Practical diodes, mostly semiconductor diodes today, differ from the ideal diode in the following:

a. When forward biased, the diode presents a small but nonlinear resistance, which is temperature dependent.

b. When reverse biased, the current is not exactly zero i.e. its resistance is not infinite albeit very high.

c. On the application of high reverse voltages to a semiconductor diode the reverse current at a certain voltage increases abruptly and if, it is not taken care of the current not to exceed a certain value, the diode will be destroyed. This is called the *breakdown* of the diode. However, technology has taken advantage of this phenomenon and adjusts the breakdown voltage to occur down to very low values, smaller even than two volts depending on the requirements of the application. Such special diodes are called Zener diodes and their use is explained in Section 2.3.

From among the numerous applications of the diode we site and explain here only two, which are useful in instrumentation. These are the use of diodes in *rectification* and in waveform *clipping*. In all cases the diode is considered ideal.

2.2.1 Rectification of *ac* Voltages

Rectification is the conversion of an *ac* voltage to a voltage of one polarity from which a current flowing in one direction only is obtained. Such a current can be used for charging a battery or for electrolysis purposes. The simplest rectifying circuit is shown in Fig. 2.2(a). The applied *ac* voltage is $e = E_m \sin\omega t$ with its waveform shown in Fig. 2.2(b). During the first half of the period ($0 < \omega t \leq \pi$) the diode D is forward biased, acts as a closed switch (short circuit) and the current i is e/R, following the waveform of e, i.e.

$$i = I_m \sin \omega t \quad \text{for} \quad 0 < \omega t \leq \pi$$

where

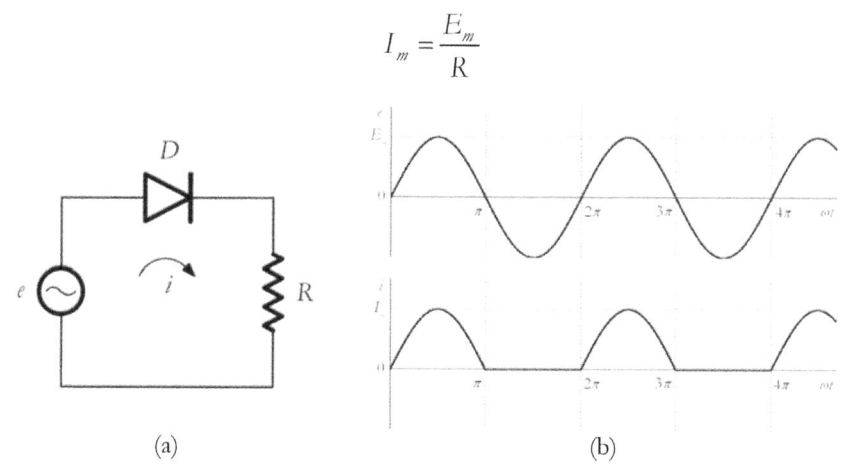

$$I_m = \frac{E_m}{R}$$

Figure 2.2. Rectification of *ac* voltage, (a) half-wave rectifying circuit and (b) waveforms

During the second half of the period the diode is reverse biased, it acts as an open switch and the current i is zero i.e.

$$i = 0 \quad \text{for} \quad \pi < \omega t \leq 2\pi$$

Thus the current flows in one direction in the load R. Similar to the current waveform is the waveform of the voltage across R. Since we take advantage of only half of the applied waveform, the action is *half-wave rectification* and the circuit is a *half-wave rectifier*.

Clearly, the current consists of positive pulses only, but it is not a *dc* current. It has a component I_{dc} which coincides with the mean value of the i waveform. This can be found as

$$I_{dc} = \frac{1}{2\pi}\int_0^{2\pi} id(\omega t) = \frac{1}{2\pi}\int_0^{\pi} id(\omega t) = \frac{1}{2\pi}\int_0^{\pi} I_m \sin \omega t d(\omega t) = \frac{I_m}{\pi} \quad (2.1)$$

To get a measure of the efficiency of this conversion we have to calculate the *dc* power and compare it to the total power consumed. The useful *dc* power is

$$P_{dc} = I_{dc}^2 R = \frac{I_m^2}{\pi^2} R \qquad (2.2)$$

To calculate the total power consumed we should first find the *rms* (root-mean square) value I_{rms} of the current waveform. We have

$$I_{rms} = \sqrt{\frac{1}{2\pi} \int_0^\pi i^2 d(\omega t)} = \frac{I_m}{2} \qquad (2.3)$$

and the power consumed

$$P_{ac} = I_{rms}^2 R = \frac{I_m^2}{4} R \qquad (2.4)$$

Then the efficiency of the conversion η will be

$$\eta = \frac{P_{dc}}{P_{ac}} \times 100\% = \frac{4}{\pi^2} \times 100\% = 40.6\%$$

This is rather low but we can increase it, as we will show next.

Example 2.M.1

Using MATLAB plot the voltage $e(t)=10\sin\omega t$ (*S.I.*) and the current $i(t)$ for the half-wave rectifying circuit of Fig. 2.2(a) from $t=0$ to $3T$. Take 100 points per period and let $R=10k\Omega$ and $T=100ms$.

Code

```
>> R=10;                    % resistance in kΩ and current in mA
>> T=0.1;
>> for i=1:300              % 3 periods × 100 points per period
t(i)=(i-1)*T/100;
e(i)=10*sin((2*pi/T)*t(i));
        if e(i)>=0
        current(i)=e(i)/R;
        else
        current(i)=0;
        end
end
```
% the plots below are derived by the following procedure (the reader is advised first to consult the MATLAB tutorial in Appendix *D*)

Electronic Components – Devices 39

```
>> subplot(2,1,1); plot(t,e); axis([0 0.3 -15 15]); grid; xlabel('time (sec)'); ylabel('e(t) (volts)')
>> subplot(2,1,2); plot(t,current); axis([0 0.3 -1.5 1.5]); grid; xlabel('time (sec)'); ylabel('i(t) (mA)')
```

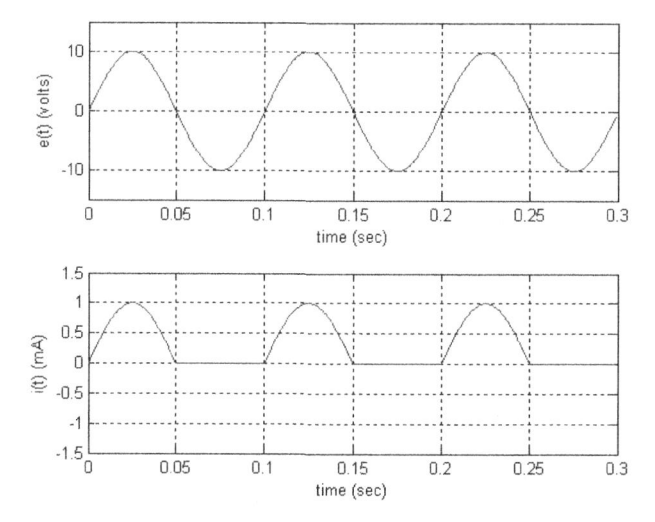

Figure 2.M.1. Voltage $e(t)$ and current $i(t)$ of the half-wave rectifying circuit

Example 2.M.2

Using MATLAB determine the I_{dc} and the *rms* values of the current $i(t)$ in example 2.M.1 according to Eqs. (2.1) and (2.3) by evaluating the integrals. Then check the results using the analytic outcome of the above equations.

Code

All currents in the example are in *mA*.

```
>> i_integral=quad('current',0,pi);   % use m-file current.m and take the integral
>> i2_integral=quad('current2',0,pi);  % use m-file current2.m and take the integral
>> idc=i_integral/(2*pi)
>> irms=sqrt(i2_integral/(2*pi))
```

```
       idc =                    irms =

          0.3183                   0.5000
```
```
>> idc1=1/pi                        % verification
>> irms1=1/2                        % verification

       idc1 =                   irms1 =

          0.3183                   0.5000
```

The m-files current.m and current2.m are:

```
function i=current(x)         function i2=current2(x)
i=1*sin(x);                   i=1*sin(x);
end                           i2=i.^2;
                              end
```

2.2.2 Full-Wave Rectifier

In order to improve the efficiency of rectification we should take advantage of the existence of the *e* waveform during the second half of the period i.e. for $\pi < \omega t \leq 2\pi$. This requires using a second diode and a specially made transformer. However, nowadays the use of such a transformer costs more than using a usual transformer and four diodes connected in a bridge as shown in Fig. 2.3(a).

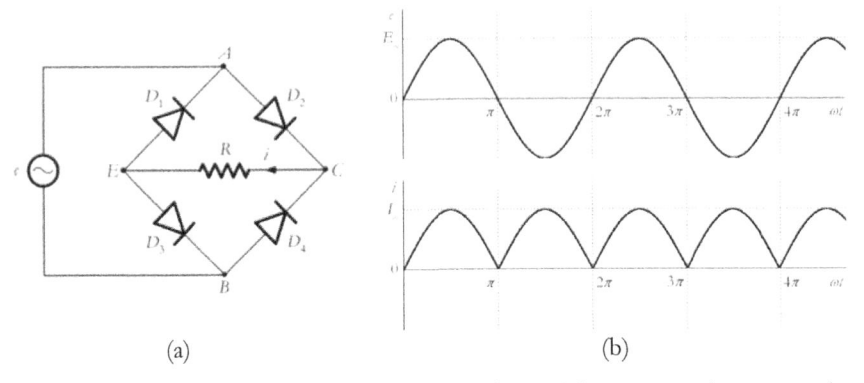

Figure 2.3. (a) Full-wave rectifier and (b) waveforms. The source voltage *e* may be obtained from the output of a usual transformer.

The operation of the circuit is as follows: During the first half of the period, node A is positive w.r.t. B. Then diode D_1 is reverse biased acting as an open switch but diode D_2 is forward biased, short-circuiting A to C. Thus node C is positive w.r.t. B and diode D_4 is reverse biased and therefore it does not conduct. But node E is positive w.r.t. B and thus diode D_3 is forward biased short-circuiting E to B. With only diodes D_2 and D_3 conducting the current flow follows the path *e* to AD_2CRED_3B and back to *e*. So the direction of *i* through R is from C to E following *e*.

Now during the second half of the period, A is negative w.r.t. B and diodes D_2 and D_3 are reverse biased, so they do not conduct. However, now with node B positive w.r.t. A, diodes D_4 and D_1 are forward biased and the current flow follows the path from the bottom terminal of *e* to BD_4CRED_1A to the upper terminal of source *e*. It can be seen that again the direction through R is from C to E, thus obtaining the same *i* waveform as in the first half of the period, as shown in Fig. 2.3(b).

Electronic Components – Devices 41

Clearly, since the waveform of i during the second half of the period is the same with that during the first half of the period, the I_{dc} value should be twice its value for the half-wave rectifier i.e.

$$I_{dc} = 2\frac{I_m}{\pi} \qquad (2.5)$$

To find the new efficiency we should determine the new value of I_{rms} as follows:

$$I_{rms} = \sqrt{\frac{1}{2\pi}\int_0^\pi 2i^2 d(\omega t)} = \frac{I_m}{\sqrt{2}} \qquad (2.6)$$

Therefore

$$P_{dc} = I_{dc}^2 R = \frac{4I_m^2}{\pi^2} R$$

and

$$P_{ac} = I_{rms}^2 R = \frac{I_m^2}{2} R$$

giving the efficiency as

$$\eta = \frac{P_{dc}}{P_{ac}} \times 100\% = \frac{8}{\pi^2} \times 100\% = 81.2\%$$

i.e. double that for the half-wave rectifier. The circuit in Fig. 2.3(a) is a full-wave rectifier and is the circuit we employ in practice almost always for rectification purposes in electronics.

In industrial applications, where high dc currents are required, other rectifying devices replacing diodes are used. Most often these devices belong to the *thyristor* family and play the most important role in *Power Electronics*. They are also semiconductor devices. Dealing with these devices and their applications is not included in this book, but any interested readers can consult one of the many excellent books in Power Electronics, for example ref. [13].

Example 2.1

In the full-wave rectification circuit of Fig. 2.3(a) each diode has a forward bias resistance of 10Ω otherwise being ideal. The source voltage is 220V *rms* and the load resistance 480Ω. Calculate

a. The dc current in the load
b. The power dissipated in the load

c. The power dissipated in the diodes
d. The total power taken from the source

Solution

a.   The dc current in the load is

$$I_{dc} = 2\frac{I_m}{\pi}$$

where

$$I_m = \frac{E_m}{R + 2r_d}$$

with $E_m = 220\sqrt{2}$ V and $r_d = 10\Omega$. Then

$$I_{dc} = \frac{2 \times 220\sqrt{2}}{\pi(480 + 2 \times 10)} = \frac{440\sqrt{2}}{500\pi} = 0.396\,A$$

b.   Dissipated power in the load

$$P_L = I_{rms}^2 R_L$$

where

$$I_{rms} = \frac{I_m}{\sqrt{2}}$$

Then

$$P_L = \left[\frac{E_m}{\sqrt{2}(R + 2r_d)}\right]^2 R_L$$

or

$$P_L = \left[\frac{220\sqrt{2}}{\sqrt{2}(480 + 20)}\right]^2 480 = 92.9W$$

c.   Power dissipated in the diodes

$$P_d = I_{rms}^2 \times 2r_d = \left(\frac{220}{500}\right)^2 \times 2 \times 10 = 3.87W$$

d.   Total power taken from the source

$$P_t = P_L + P_d = 92.9 + 3.87 = 96.77W$$

2.2.3 Waveform Clipping

Waveform *clipping* is another useful application of diodes. Consider the two circuits in Fig. 2.4(a) and (b) in which $e = E_m \sin \omega t$. During the first half of the period the diode in Fig. 2.4(a) conducts while in Fig. 2.4(b) the diode does not being reverse biased. Clearly, in both cases the voltage v_o follows e as shown in Fig. 2.4(c). However, during the second half of the period the diode in Fig. 2.4(a) does not conduct while the diode in Fig. 2.4(b) conducts being forward biased. The result is that in both cases v_o is zero as shown in Fig. 2.14(c). It can be seen that both circuits have clipped the negative part of e transmitting as v_o only the positive part. It can be similarly shown that reversing the diodes in both circuits the positive part of the e waveform will be clipped.

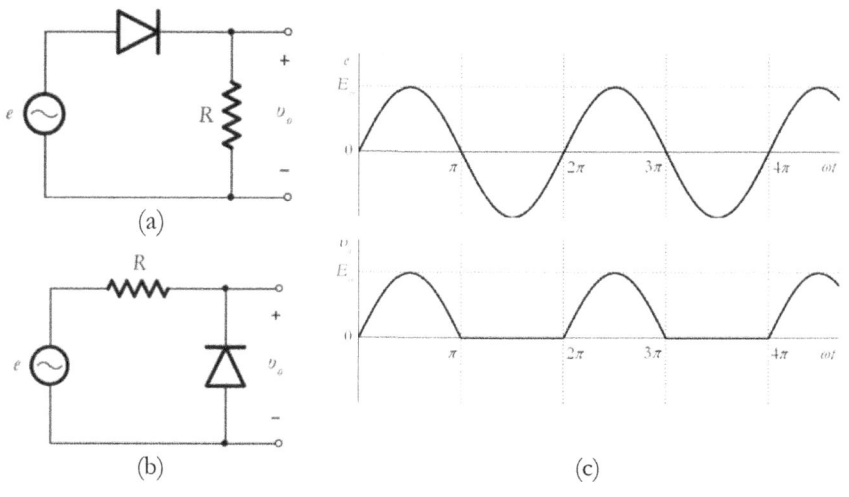

Figure 2.4. (a) and (b) Equivalent negative waveform clippers and (c) clipped waveform

Making use of a battery in the second of the above circuits we may clip part of the positive or negative part of a waveform. Consider the case in Fig. 2.5(a) in which the battery E reverse biases the diode. The latter does not conduct until it becomes forward biased by e. Till then v_o follows e, since there is not any voltage drop in R with the current being zero. This happens only when e becomes more negative than $-E$. When the diode conducts, acting as a closed switch, it causes v_o to take the value $-E$ remaining at this voltage as long as e is more negative than $-E$. Therefore, the waveform of v_o will be as shown in Fig. 2.5(b), in which part of the negative waveform is missing.

Following similar reasoning it can be easily shown that reversing the diode and the polarity of the battery E part of the positive e waveform can be clipped off.

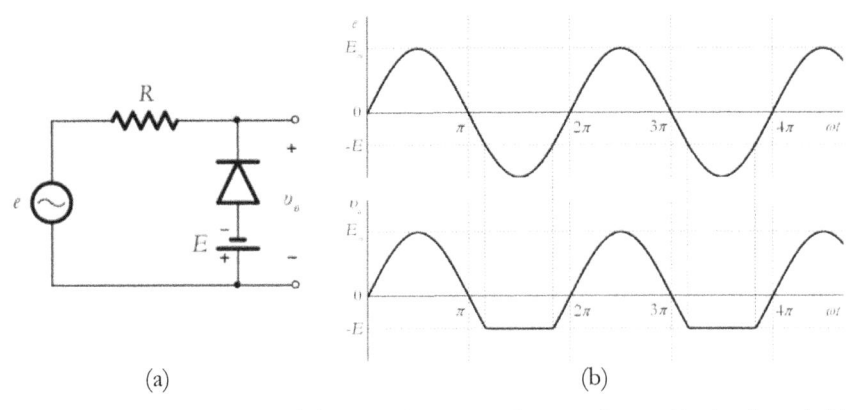

Figure 2.5. Clipping part of the negative part of a waveform, (a) circuit and (b) waveforms

Finally, by combining the two cases, as shown in Fig. 2.6(a), parts of the positive and negative e waveform can be clipped (see Fig. 2.6(b)).

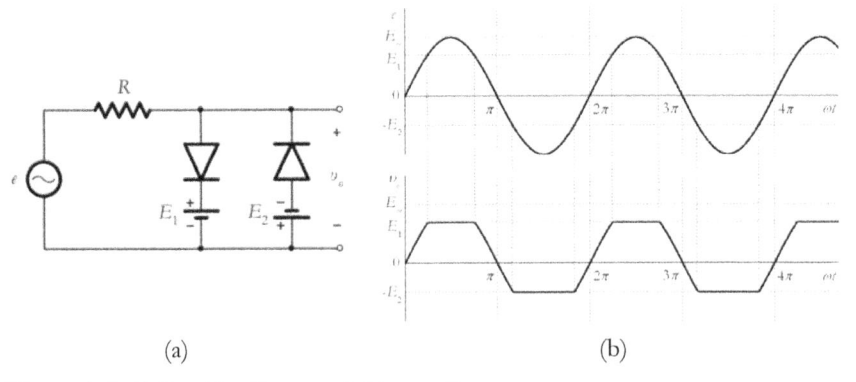

Figure 2.6. (a) Double clipper and (b) waveforms

Clippers are useful circuits for the protection of circuits in which the input signal should not exceed certain voltage levels.

Example 2.M.3

Using MATLAB plot the voltage $e(t)=10\sin\omega t$ (S.I.) and the voltage $v_o(t)$ for the double clipper of Fig. 2.6(a) from $t=0$ to $3T$ using 200 points per period. It is given that $E_1=4V$, $E_2=6V$ and $T=100ms$. Also, draw the transfer characteristic $v_o=f(e)$.

Code

```
>> T=0.1;
>> E1=4;
>> E2=6;
>> for i=1:600                    % 3 periods × 200 points per period
t(i)=(i-1)*T/200;
```

```
e(i)=10*sin((2*pi/T)*t(i));
        if e(i)>E1
        vo(i)=E1;
        elseif e(i)<-E2
        vo(i)=-E2;
        else
        vo(i)=e(i);
        end
end
```
\>> subplot(2,1,1); plot(t,e); axis([0 0.3 -12 12]); grid; xlabel('time (sec)'); ylabel('e(t) (volts)')
\>> subplot(2,1,2); plot(t,vo); axis([0 0.3 -12 12]); grid; xlabel('time (sec)'); ylabel('vo(t) (volts)')
\>> figure
\>> plot(e,vo); axis([-12 12 -12 12]); grid; xlabel('e (volts)'); ylabel('vo (volts)')
\>> title('transfer characteristic vo=f(e)')

Figure 2.M.2. Double clipper

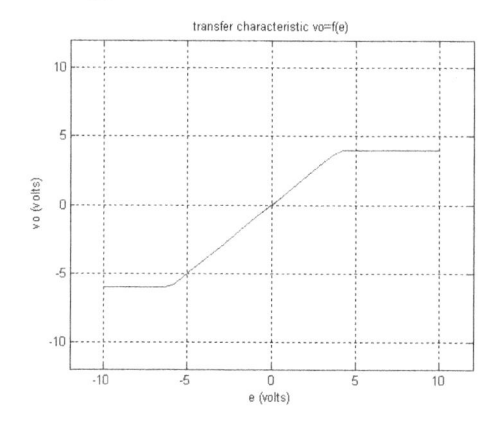

Figure 2.M.3. Transfer characteristic $v_o = f(e)$

Example 2.2

Draw the transfer characteristic $v_o = f(v_i)$ and $v_o = v_o(t)$, if $v_i = 10\sin\omega t$ (V), for the circuit shown in Fig. E.2.1. The diode is considered ideal.

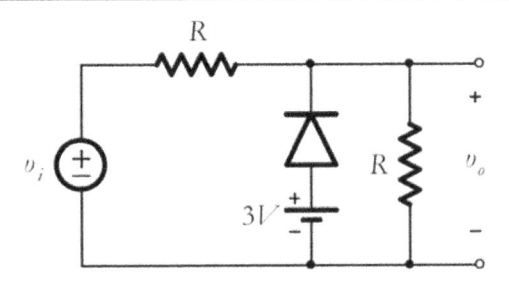

Figure E.2.1.

Solution

Clearly, when the diode is conducting the output voltage v_o will be $3V$ independently of the value of v_i. For values of v_i that the diode is reverse biased, v_o will be

$$v_o = \frac{R}{R+R} v_i = \frac{1}{2} v_i$$

The diode will not conduct when its cathode is positive w.r.t. its anode, i.e. when $v_o > 3V$. Then v_i will be $2v_o = 6V$. Therefore for $v_i > 6V$, v_o will be

$$v_o = \frac{1}{2} v_i$$

For $v_i < 6V$ the diode will be conducting and v_o will be constant at $3V$. Therefore, the v_o against v_i characteristic will be as shown in Fig. E.2.2(a).

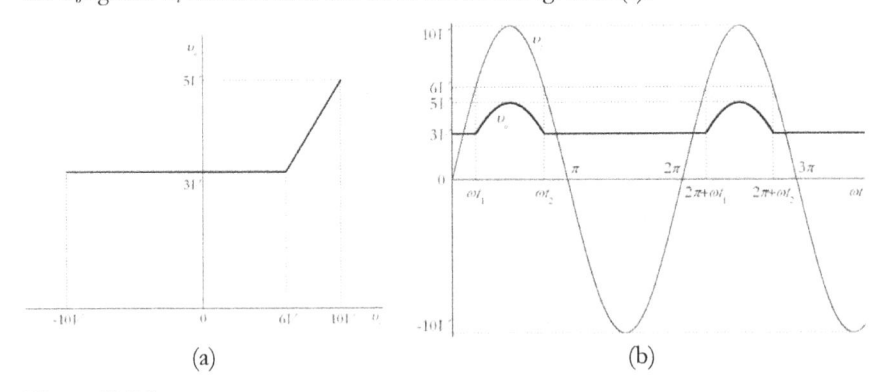

(a)   (b)

Figure E.2.2.

According to the above argument the v_o against ωt will be as shown in Fig. E.2.2(b). The values of ωt_1 and ωt_2 are found from the equation

$$6 = 10\sin\omega t$$

Electronic Components – Devices

which when solved for ωt gives

$$\omega t = \sin^{-1} \frac{6}{10} = 36.9°$$

Therefore

$$\omega t_1 = 36.9°$$
$$\omega t_2 = 143.1°$$

Example 2.M.4

Using MATLAB plot the voltage $v_i(t)=10\sin\omega t$ (*S.I.*) and the voltage $v_o(t)$ for the circuit in Fig. E.2.1 of Example 2.2 from $t=0$ to $2T$ evaluating 200 points per period. Let $T=100ms$. Also, draw the transfer characteristic $v_o=f(v_i)$ and determine the values of t_1 and t_2 in Fig. E.2.2(b) and the maximum value of $v_o(t)$.

Code

```
>> T=0.1;
>> for i=1:400                          % 2 periods × 200 points per period
t(i)=(i-1)*T/200;
v(i)=10*sin((2*pi/T)*t(i));
        if v(i)>6
        vo(i)=v(i)/2;
        else
        vo(i)=3;
        end
end
>> subplot(2,1,1); plot(t,v); axis([0 0.2 -12 12]); grid; xlabel('time (sec)'); ylabel('vi(t) (volts)')
>> subplot(2,1,2); plot(t,vo); axis([0 0.2 -12 12]); grid; xlabel('time (sec)'); ylabel('vo(t) (volts)')
>> figure
>> plot(v,vo); axis([-12 12 -12 12]); grid; xlabel('vi (volts)'); ylabel('vo (volts)')
>> title('transfer characteristic vo=f(vi)')
>> t1=(T/(2*pi)*asin(6/10))              % from 6=10sinωt
>> t2=(T/2)-t1

            t1 =                                    t2 =

              0.0102                                   0.0398

>> vomax=max(vo)

                    vomax =

                      5
```

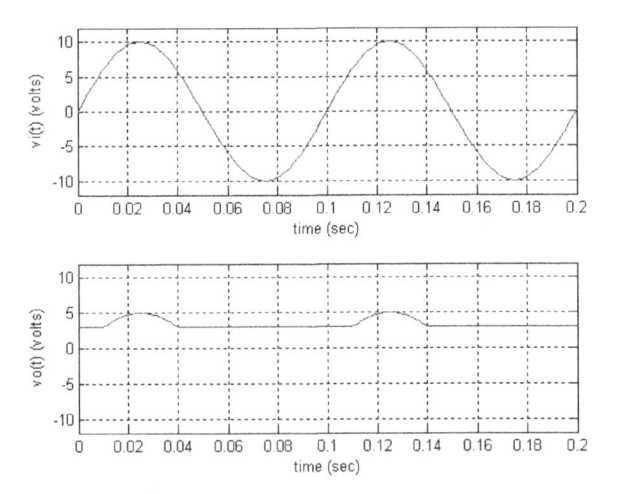

Figure 2.M.4. Input and output voltages

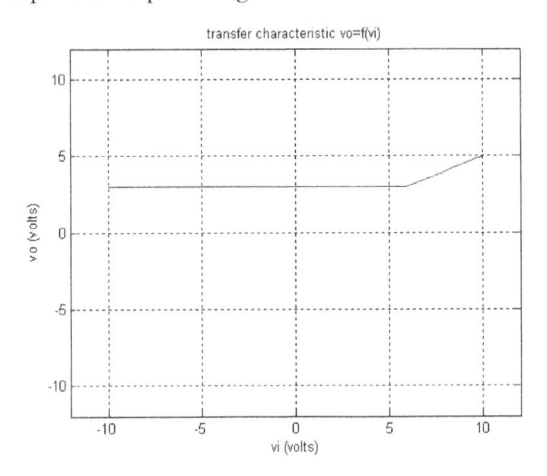

Figure 2.M.5. Transfer characteristic $v_o = f(v_i)$

2.3 Semiconductor Diodes

Real or practical diodes differ from the ideal in the following:

a. They require a small forward biasing voltage before they start conducting.

b. The relationship between current and voltage, when forward biased, is exponential.

c. There is a small current, when they are reverse biased, which increases with temperature.

These are the characteristics of crystalline semiconductor diodes, which are nearly exclusively in current use. A general purpose semiconductor diode is made of two types of semiconductor, namely, type-n and type-p in very close contact. A semiconductor has a low conductivity than a metal but not zero like an insulator. If it is pure, it behaves like an insulator at the $0K$ (zero degrees Kelvin). In a metal the electric charge carriers are free electrons, but in a semiconductor at temperatures above $0K$ the charge carriers are of two types: negative, which are free electrons and positive, which are called *holes*. A hole is a deficit of an electron i.e. an empty place in the valence band of an atom of the crystal. It has been shown theoretically that the hole can move from atom to atom inside the crystal, thus corresponding to a positive charge carrier which is equal but opposite to that of an electron and with mass nearly equal to that of the electron. In a pure semiconductor, like silicon *Si* for example, the concentrations of both electrons and holes are the same and increase with temperature always remaining equal. However, if certain impurities, like atoms of phosphor (P) or boron (B) are introduced into the pure semiconductor, they will create an imbalance in the carrier concentrations (extrinsic semiconductor). Thus, if the impurity atoms are of valence 5, like the phosphor, the higher concentration will be that of electrons, majority carriers, and the semiconductor will be called n-type. On the other hand, if the impurity atoms are of valence 3, like boron, the majority carriers will be holes and the semiconductor will be called p-type. It should be said that the product of the concentration of the majority carriers times the concentration of the minority carriers in each semiconductor type is equal to the product of the concentrations of electrons and holes in the pure (intrinsic) semiconductor at the same temperature. When the two types of semiconductor come to close contact, a potential barrier is established and in order to overcome this barrier and establish a current through the junction the type-p should be positively biased w.r.t the type-n. The junction is then said to be forward biased. If the type-p is negatively biased w.r.t. the type-n the height of the potential barrier increases and no significant current flows through the junction. Thus, the junction behaves like a diode having the I-V characteristic shown in Fig. 2.7.

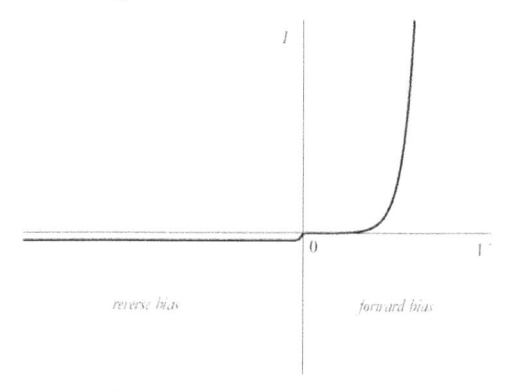

Figure 2.7. I-V characteristic of semiconductor diodes

General purpose semiconductor diodes are mostly made of silicon. Germanium was mostly used to make diodes in the earlier years of the semiconductor industry, but nowadays is used for special purpose applications.

There is a limit in the size of the reverse bias that can be applied to a semiconductor diode, because beyond this the junction breaks down and a heavy reverse current flows in it that can destroy the diode, if it is not kept low. Technology has taken advantage of this effect and produces diodes with this breakdown voltage very low, even less than $2V$. Such diodes are named *Zener diodes* with the breakdown voltage called *Zener voltage*. The I-V characteristic of a Zener diode is shown in Fig. 2.8. Zener diodes are used as voltage regulators in power supplies, clippers and for voltage reference.

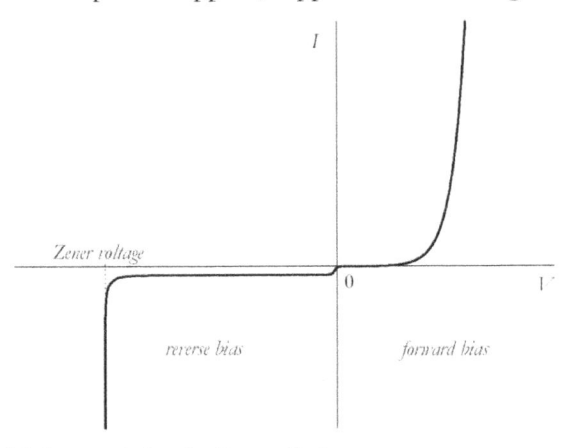

Figure 2.8. I-V characteristic of a Zener diode

Other type semiconductor diodes can be obtained using also different semiconductor materials, like GaAs, with different properties very useful in practice. Such diodes are the photodiodes (sensitive to light), the solar cells, the light-emitting diodes (*LED*), the laser diodes etc, which are used for special purpose applications and are briefly presented later in this chapter.

2.4 *dc* Power Supplies

A power supply is a nearly ideal *dc* voltage source that is an essential part of every electronic equipment. Its purpose is to bias all non-passive electronic components in the equipment to operate actively. Such electronic components are the various types of transistors and of course all integrated circuits, analog or digital, which however are built using transistors.

Since the currents in most electronic circuits are of the order of milliamps or less, a battery is very close to the ideal power supply. All portable electronic equipment employ batteries as their power supply.

However, when the demand for higher *dc* power is greater or when one wants to avoid using batteries in the laboratory or at home, usually a power supply is installed inside the equipment. Such a power supply converts *ac* power taken from the electrical installation (mains) to the required *dc* power. In general, such a power supply consists of the following parts shown in Fig. 2.9.

```
110 or 220V  ─[Transformer]─[Full Wave Rectifier]─[Filter]─[Voltage Regulator]─ dc voltage
    ac
```

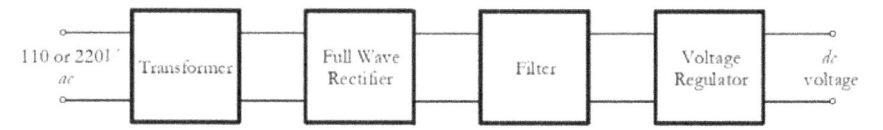

Figure 2.9. General construction of a power supply

The transformer is made up of two closely coupled coils. If a voltage is applied to the primary coil, an induced voltage appears in the secondary with its value being proportional to the ratio n of the number of turns of the secondary to that of the primary. In the ideal transformer there is no power loss in the transformation, therefore the power in the primary coil is equal to the power in the secondary. Since the power is the product of the current times the voltage in each coil, it follows, that the ratio of the currents is the inverse of the ratio of the corresponding voltages. The transformers used in power supplies have a laminated iron core shared by the coils in order to make the coupling between the primary and the secondary windings most effective and to reduce any power loss due to Eddy currents in the core.

The transformer is required to reduce the magnitude of the mains voltage (110 or 220V rms) down to a value suitable for the final *dc* voltage. The transformer feeds the *FWR*, usually that in Fig. 2.3(a). The rectified current waveform contains apart from the useful *dc* component $2I_m/\pi$ an infinite number of *ac* components which should be rejected. This is mainly achieved by the filter, which allows the dc component to pass, but prevents the *ac* components to be transmitted to the following stage, the regulator. The latter is nowadays an integrated circuit that clears out the final voltage from the remaining *ac* components and gives the power supply the characteristics of a nearly ideal *dc* voltage source.

2.5 Transistor

The transistor is a three-terminal semiconductor element, which can be active if properly biased. Two main kinds of transistor exist:

a. The *Bipolar Junction Transistor* or *BJT*

b. The *Field-Effect Transistor* which can be either a *Junction FET* or a *Metal-Oxide Semiconductor FET (MOSFET, MOST)*

The *BJT* is called bipolar, because in its operation both electrons and holes participate. On the other hand the operation of the *FET* is due only to one kind of electric carriers, either electrons or holes. This is the reason why the *FET* is also referred to as *unipolar transistor*.

We will briefly describe the operation of the common *BJT* and the common *MOSFET*, but the interested reader can consult any of the many specializing books in Electronics that exist in the literature.

2.5.1 The Bipolar Junction Transistor

The *BJT* is a sandwich of *n* and *p*-type semiconductor regions, either two regions of *n*-type with a *p*-type in between or two regions of *p*-type with a *n*-type region in between. We choose the *npn BJT* to describe its operation, but the operation of the *pnp* transistor is the same except that the role of the carriers is the opposite of that in the *npn* transistor.

The symbol of the *npn BJT* transistor is shown in Fig. 2.10. The letters in the terminals correspond *e* to the *emitter*, *b* to the *base* and *c* to the *collector*. For normal operation the base-emitter junction forming a diode is forward biased, whereas the collector-base junction is reverse biased. If a small current of holes is injected into the base, this causes a much larger current of electrons to be injected into the base region from the emitter. Part of this electron current from the emitter will be used to neutralize the incoming holes in the base and the rest, much larger, will cross the base-collector junction to form the collector current. Therefore, a small base current causes a much larger current in the collector, thus the base current is amplified β times, where β is the current amplification factor of the *BJT*. So the normal and main transistor action of the *BJT* is used in order to amplify signals i.e. to increase their power content.

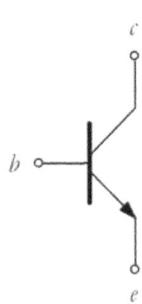

Figure 2.10. The symbol of the *npn BJT*

Another use of the *BJT* is as a switch in digital electronics. Consider the circuit in Fig. 2.11. This is the *common-emitter* configuration of the *BJT* use, since the emitter is a common terminal between input and output. In certain applications the *BJT* is also used in the common-base or common-collector configuration. If I_B, the current in the base is zero, there is no current I_C in the collector and V_o will be equal to V_{CC}. It is said that the transistor is OFF. On the other hand, if I_B is not zero, the collector current I_C will be βI_B and passing through R_C will cause a voltage drop in it so that the output voltage V_o will be less than V_{CC} i.e.

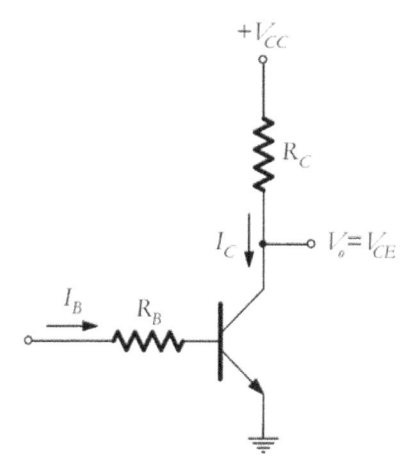

Figure 2.11. The BJT switch

$$V_o = V_{CE} = V_{CC} - I_C R_C \qquad (2.7)$$

If I_B is large enough, it will bring the BJT to saturation, i.e. I_C will not become larger than a value I_{Cs}. In this case

$$V_o = V_{CEs} = V_{CC} - I_{Cs} R_C \qquad (2.8)$$

with V_{CEs} being very small, of the order of $0.2V$. The transistor is now said that it is ON. The output voltage V_o may thus be either V_{CC}, transistor OFF, or nearly $0V$, transistor ON. Therefore the transistor acts as a switch which is controlled by the base current. This type of transistor operation is made use of in the bipolar digital electronics presented in Chapters 5 and 6.

For intermediate values of I_B the BJT works such that $\Delta I_C / \Delta I_B$ is nearly constant for small changes ΔI_B and thus, it can be used as an amplifying element. We will not endeavor more time and space here for the use of the BJT transistor in the design of amplifiers or switches as its use has been superseded nowadays by the so-called integrated circuits explained in section 2.6.

2.5.2 The MOSFET

It has been proved in practice that the most useful type of FET is the MOSFET and this is chosen to be described here. The MOSFET is constructed as shown in Fig. 2.12. In a p-type bar, called the substrate, two n-type regions are formed, which are more heavily doped than in a usual n-type semiconductor. For this reason they appear in the figure as n^+-type. These two regions are termed the *source* and the *drain* of the MOST. In one type of MOSFET, the *depletion* type, the regions of source and drain are connected by a very thin layer of n-type material which forms the channel. In the *enhancement* type there is not a built-in channel. Above the channel a conducting material is placed, forming the electrode of the *gate* of the MOSFET insulated from the channel by a thin layer of insulating material, usually silicon oxide.

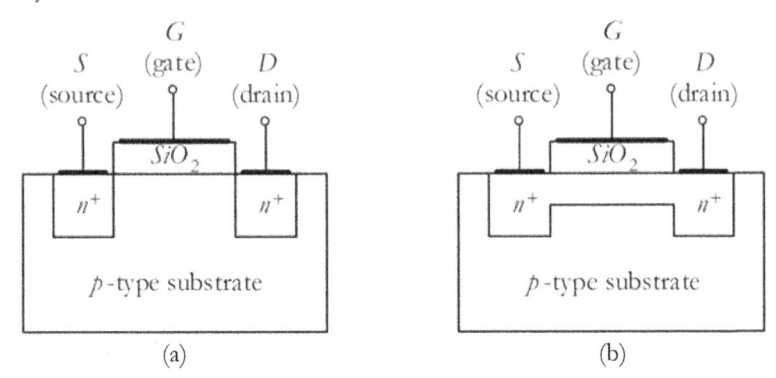

Figure 2.12. Structure of the (a) enhancement type and (b) depletion type MOSFET

The above structure applies to the n-channel MOST. A similar structure, but with n-type substrate and p-type source, drain and channel regions, gives the p-channel MOST. The operation of the n-channel depletion type MOSFET is briefly described as follows: With a positive voltage V_{DS} applied to the drain w.r.t. the source, the drain-source current I_D in the channel is affected by the electric field created by the gate-to-source voltage V_{GS}. The gate V_{GS} voltage, if negative, it leads to reduction of the channel current, because it affects the channel depth and finally, if negative enough, the channel current becomes zero. On the other hand, if V_{GS} is positive, the field attracts electrons from the n^+ source region to the channel thus increasing its conductivity. This last mode of operation is called channel enhancement. Clearly, the operation of the depletion-type MOST is affected by the field due to the gate-to-source voltage V_{GS} and justifies the name as field-effect transistor.

In the enhancement n-type MOST there is not initially built-in n-channel and this cannot operate in the depletion mode, i.e. with negative V_{GS}. It can only operate in the enhancement mode, when V_{GS} is positive.

The operation of a MOSFET is affected by the voltage between the substrate and the source. This voltage can either be zero or negative. Thus the MOSFET, when it comes as a discrete component has a forth terminal, which is the substrate terminal.

From the above description of the n-channel MOSFET, it is clear, that only one type of electric carriers, the electrons, is responsible for the formation of the drain current. Similarly in the p-channel MOSFET the role of electrons is taken up by the holes. In this case all externally applied voltages i.e. V_{GS}, V_{DS} and between source and substrate should be the opposite to the corresponding voltages in the n-channel MOSFET.

The MOSFET, as in fact any type FET device, is a voltage-controlled device in contrast to the BJT, which is a current-controlled device. Fig. 2.13 shows the usual common-source connection of the n-channel MOSFET. For an enhancement type MOSFET, if $V_s \leq 0$, the MOST is OFF and

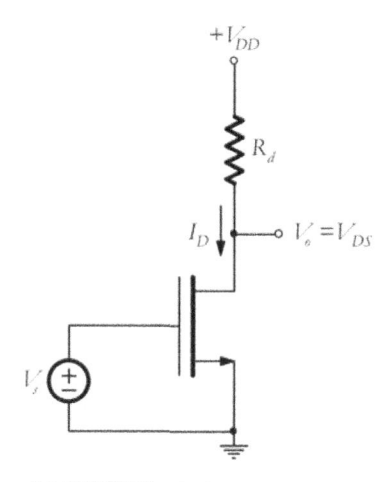

Figure 2.13. An n-channel MOSFET switch

$$V_o = V_{DS} = V_{DD} \tag{2.9}$$

while if $V_s > 0$, the MOSFET is ON. There is then a current, say I_D, in the channel, and the output voltage V_o will be

$$V_o = V_{DD} - I_D R_d \tag{2.10}$$

The channel current I_D is controlled by V_{GS} and, for small variations of V_{GS}, $\Delta I_D / \Delta V_{GS}$ will be nearly constant and given by g_M the *transconductance* of the MOSFET. In this case the MOST acts as an amplifying element.

For adequate positive $V_s = V_{GS}$ the channel current will come to saturation and V_o will take a small value V_{DSS}.

Thus the MOSFET can amplify signals, although not so effectively as the BJT, and also operate as a switch. However, its use is most effectively applied in the construction of digital integrated circuits, as discussed in the next section.

2.6 Integrated Circuits

The *BJT* and *MOSFET* transistors can be constructed in small dimensions and placed, lots of them, on a chip of silicon. With proper interconnections among the transistors, a whole circuit can be constructed on the chip. Passive components, like resistors and capacitors, can also be included in this construction thus placing a whole circuit on the chip of very small dimensions. This is the so-called *integrated circuit (IC)*. The usual material has been the silicon, although other, more expensive materials, like gallium-arsenide etc, are used in applications of high frequency electronics.

The integrated circuits in the past have been analog, digital or hybrid in that they may include both analog and digital subcircuits. Analog integrated circuits are usually amplifiers, filters and other analog functional blocks. Digital integrated circuits are functional blocks like gates, flip-flops, digital adders, multipliers, encoders, decoders, multiplexers, demultiplexers etc (see Section 6.11). On the other hand, hybrid integrated circuits include analog-to-digital and digital-to-analog converters, microprocessors etc.

However, nowadays the progress of the integrated circuit has been tremendous with the number of components (transistors) on the chip getting nearly doubled every 1.5 years (Moore's law). Silicon has been the main semiconductor material and *MOSFET*s predominate. The technology is characterized by the length of the *MOST* channel which has become as low as of the order of nanometers (10^{-9}m). Thus we are living in the age of nanoelectronics with the possibility of building whole electronic systems on a chip. It has also become possible to construct integrated circuits on a multilayer basis i.e. not only on the surface of the chip. Nanochips are already commercially available.

For the time being, silicon serves the purposes of contemporary electronics although in high frequency applications, of analog mainly electronics, gallium-arsenide proves very useful. It is foreseen that silicon will predominate at least up to 2020 when other materials will take its place in most applications. However, because of its high abundance in nature, consequently its low cost, its endurance to radiation and other useful properties the use of silicon will not easily come to a complete end.

Why integrated circuits?

Most important reasons for the development and use of integrated circuits are the following:

a. Low cost. Integrated circuits are very cheap, if they are produced in large quantities

b. Very small size and weight (hundreds of thousands of components in a very small space)

c. High speed of operation

d. Simultaneous fabrication of components with the same characteristics as they are made under the same conditions

e. Absence of problems due to faulty connections (soldering) of the components

After fabrication an *IC* is available to the customer as a package mainly in one of the types shown in Fig. 2.14. The various pins in each package are terminals for interconnecting the *IC* with other or similar components. Since it is an active circuit, two of the pins are connected to the *dc* power supply to provide the necessary bias for the transistors.

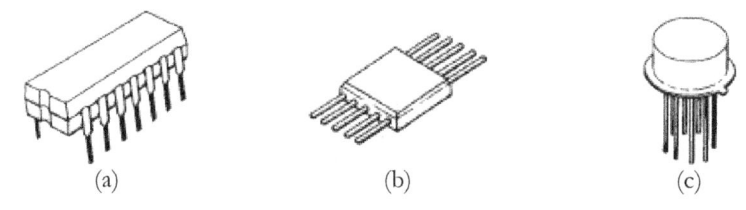

(a) (b) (c)

Figure 2.14. Main *IC* packages: (a) *DIP* (Dual-In-line-Package), (b) *SOP* (Small Outline Package) and (c) *TOP* (Transistor Outline Package)

2.7 Optoelectronics

When electromagnetic radiation falls on a semiconductor, it is possible to produce electron-hole pairs in excess of those that the temperature of the material dictates. Technology has taken advantage of this phenomenon to produce some optoelectronic devices, which are useful in practice. Such devices convert light signals to electric and vice versa. In what follows some of these devices are introduced briefly.

2.7.1 Photoconductors

The simplest optoelectronic device is the photoconductor. When light falls on a semiconductor, part of it is reflected, part absorbed and part passes through, if the crystal is very thin. Of interest here is the absorbed part, because, if the photons have enough energy, they can kick electrons out of their bonds and thus produce electron-hole pairs, which will increase the conductivity. If the semiconductor is extrinsic, this effect will affect mainly the concentration of the minority carries and the change of conductivity at the ambient temperature will be slight and of no practical interest. However, if the semiconductor is intrinsic and is kept at constant temperature, the change in conductivity due to the incident light will be significant and this is of practical interest. The intrinsic semiconductor can then act as a photodetector being called *photoconductor*. Suitable semiconductor materials for visible light detection are the PbSe (lead selenide), PbS (lead sulfide), Ge, Si, GaAs (gallium arsenide) and CdTe (cadmium telluride). Such a device takes the name *Light Dependent Resistor* (LDR). One advantage of these devices in some applications is that they respond to different wavelengths of light in a manner similar to the human eye. Unfortunately, their response is rather slow.

2.7.2 Photodiodes

Photodiodes are devices in which photoconductivity is exploited to change their I-V characteristics under the influence of electromagnetic radiation. When photons reach the junction region, if they are of suitable energy, they can cause the production of additional electron-hole pairs, which increase the minority carrier current of the diode. Clearly the effect is insignificant when the diode is forward biased. However, when the diode is reverse biased, the increase of its leakage (reverse) current is significant as shown in Fig. 2.15.

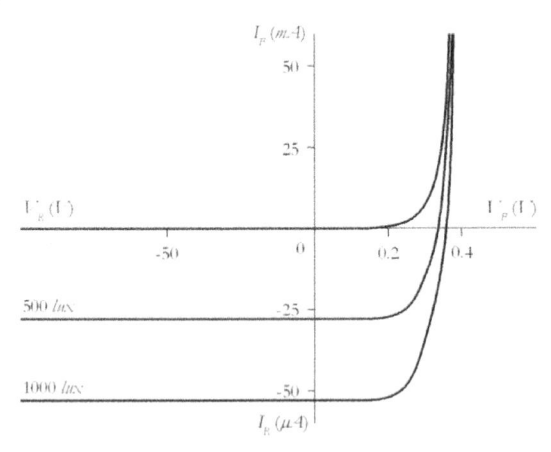

Figure 2.15. *I-V* characteristics of a photodiode for different light intensities

In a photodiode the region p or n in which the light falls is made very thin for the light to reach the pn junction region. The effect on the reverse bias I-V characteristics depends on the intensity, the wavelength and the angle of the incident light, and on the size of the area of the junction.

Photodiodes are also made by introducing a layer of intrinsic semiconductor between the regions p and n in the junction (p-i-n diodes). On the other hand heterojunction photodiodes, i.e. consisting of two different semiconductors, e.g. GaAs (n) and Ge (p), are made, which have shorter response times.

2.7.3 Phototransistor

The current of the photodiode in the photoconductive mode is low, and in order to obtain a larger one, a phototransistor can be used. The bipolar phototransistor is a *pnp* or *npn* transistor in which the *pn* (or *np*) collector-base junction is a photodiode. The physical operation of the phototransistor is the same as that of the usual transistor with the difference that the base current is a photocurrent. This is amplified by the transistor action and thus the collector current is many times larger than the photocurrent.

In the family of phototransistor belongs the photo-*JFET*, in which the photosensitive part is the gate-channel junction. In this case, the photocurrent passing through an external resistance changes the gate to source voltage thus affecting the current in the channel.

2.7.4 Photovoltaics

When light falls on an unbiased *pn* junction, the created electron-hole pairs are split by the electric field in the region with the holes moving to the *p* region and the electrons to the *n* region. Thus, excess charges are created in each of these regions and this leads to the appearance of an external potential difference between the *p* and *n* regions. This conversion of light energy into electrical energy is the photovoltaic effect, and it is exploited today to produce energy from the sunlight. The devices for this conversion are called *solar cells*. Apart from Si, other semiconductor materials exploited for this purpose are GaAs, CdS, CdTe, AlSb etc. However, Si seems to be most suitable due to conversion efficiency and endurance.

2.7.5 Photoemission

In a photoemissive device the incident radiation causes the emission of an electron from the surface of the absorbing material into the surrounding space. The emitted electrons from the *photocathode* can be collected by another electrode, the *anode*, which is at a higher potential than the cathode. Photoemission is used in vacuum phototubes and *photomultiplier* tubes. The latter are used, for example in the so called scintillation counters for nuclear radiation detection and measurements.

2.7.6 Light-Emitting Diodes (*LED*)

Light-Emitting Diodes or *LED* are semiconductor diodes, which emit light when current passes through them. They are formed from compound semiconductors such as GaAs, GaP etc. Under the passage of current holes and electrons crossing the junction change energy levels and thus light is emitted at a wavelength dependent by the difference between the energy levels. The emitted light may be in the infrared, red, yellow, green and blue regions of the electromagnetic spectrum. Different semiconductors produce different colors or wavelengths. They are useful in displays and in optical communications.

2.7.7 Optoisolators

An *optoisolator* is an optoelectronic device combining a light source with a phototransistor in a single package and is useful in providing electrical isolation between two parts of the same circuit, usually between input and output parts. The latter is particularly useful when the two parts operate at different ground levels. Fig. 2.16 shows the arrangement of the components in the isolator and an example of its use.

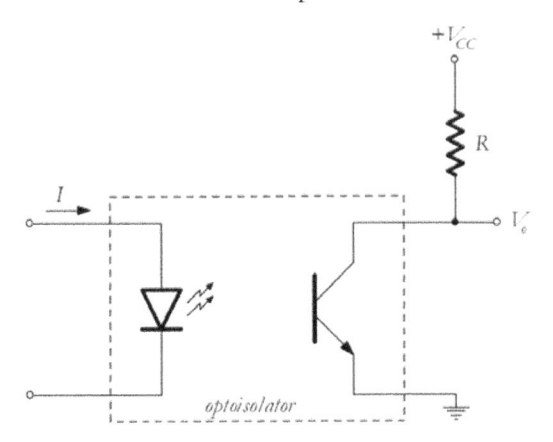

Figure 2.16. The optoisolator and its use as a switch

When a suitable current passes through the LED the emitted light falls on the base of the phototransistor and causes a collector current. This current passing in the resistor R produces a voltage drop that makes V_o to be low. On the other hand, if the current in the LED is not adequate, there is no light falling on the transistor base and therefore no current in R. This means that V_o will be equal to V_{CC}, i.e. a high value. Thus the optoisolator acts as a switch and the above arrangement as an inverter circuit (see Chapter 5).

References

[1]. Angelo E. J., *Electronics: BJTs, FETs and Microcircuits*, McGraw-Hill, 1969.
[2]. Ryder J. D., *Electronic Fundamentals and Applications*, Pitman, 1973.
[3]. Gray P. E. and Searle C. L., *Electronic Principles: Physics, Models and Circuits*, Wiley, 1969.
[4]. Millman J. and Halkias C., *Electronic Devices and Circuits*, McGraw-Hill, 1967.
[5]. Alley C. L. and Atwood K. W., *Semiconductor Devices and Circuits*, Wiley, 1971.
[6]. Alley C. L. and Atwood K. W., *Electronic Engineering*, Wiley, 1973.
[7]. Millman J. and Halkias C., *Integrated Electronics*, McGraw-Hill, 1972.

Electronic Components – Devices

[8]. Gibbons J. F., *Semiconductor Electronics*, McGraw-Hill, 1969.
[9]. Dance J. B., *Photoelectronic Devices*, Iliffe, 1969.
[10]. Bliss J., *Theory and Characteristics of Phototransistors*, Motorola AN-440.
[11]. Pallàs-Areny R. & Webster J. G., *Sensors and Signal Conditioning*, Wiley, 1991.
[12]. Smith R. J., *Electronics: Circuits and Devices*, Wiley, 1973.
[13]. Lander C. W., *Power Electronics*, 3rd edition, McGraw-Hill, 1993.

MATLAB Problems

❖ *The reader is advised first to consult the MATLAB tutorial in Appendix D*

2.M.1. Using MATLAB plot the voltage $e(t)=20\sin\omega t$ (S.I.) and the current $i(t)$ for the full-wave rectifying circuit of Fig. 2.3(a) from $t=0$ to $3T$ evaluating 100 points per period. It is given that $R=5k\Omega$ and $T=200ms$.

2.M.2. Using MATLAB determine the I_{dc} and the *rms* values of the current $i(t)$ of problem 2.M.1 according to Eqs. (2.5) and (2.6).

2.M.3. Using MATLAB plot the voltage $e(t)=10\sin\omega t$ (S.I.) and the voltage $v_o(t)$ for the clipper of Fig. 2.5(a) from $t=0$ to $3T$ evaluating 100 points per period. It is given that $E=8V$ and $T=100ms$. Also, draw the transfer characteristic $v_o=f(e)$.

2.M.4. Using MATLAB solve problem 2.4.

2.M.5. Using MATLAB solve problem 2.5.

2.M.6. Using MATLAB solve problem 2.8. Also, draw the transfer characteristic $v_o=f(v)$.

Problems

2.1. In a rectifying system the resistance of the load is $10k\Omega$, whereas each diode has a forward bias resistance $1k\Omega$. If the supplying voltage is sinusoidal, calculate the efficiency of rectification in the following two cases:

    a. The system is a half-wave rectifier
    b. The system is a bridge full-wave rectifier

(Answers: a. 36.8%, b. 67.6%)

2.2. Four diodes having a 10Ω forward bias resistance otherwise being ideal are used in a bridge full-wave rectifier to provide with $2A$ dc current an 80Ω load resistance from the $220V$ mains voltage. Draw the circuit and calculate the following:

a. The highest value of the reverse voltage in each diode
b. The power dissipated in each diode
c. The required turns ratio of the transformer

(Answers: a. 314V, b. 49.3W, c. 1:1.01)

2.3. A rectifying bridge circuit with ideal diodes is to supply a 50Ω resistive load with a 500mA dc current. The under rectification voltage is obtained from 110V mains. Draw the circuit and determine

a. The required turns ratio of the transformer
b. The rectification efficiency
c. The maximum reverse voltage in each diode

2.4. For the circuit in Fig. P.2.1(a) plot the v_o against v_i characteristic and v_o against ωt. The source voltage v_i is 10sinωt (V). The diodes are considered ideal.

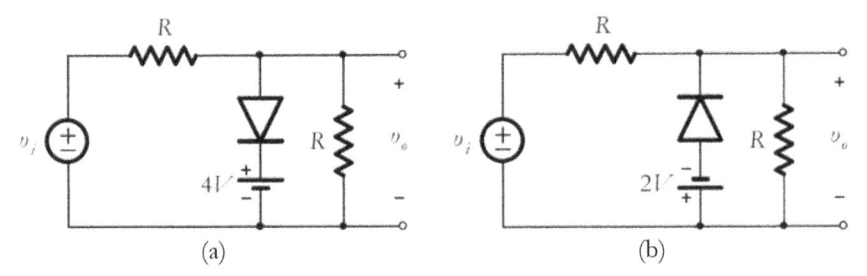

Figure P.2.1.

2.5. Repeat problem 2.4 for the circuit in Fig. P.2.1(b).

2.6. Using an ideal diode, resistors, and a dc voltage source design a circuit the output voltage v_o of which will vary with the input voltage v_i according to the characteristic shown in Fig. P.2.2. If v_i is equal to 12sinωt draw v_o against ωt.

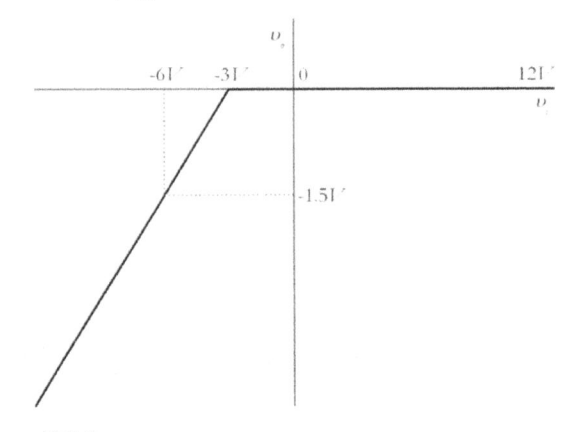

Figure P.2.2.

2.7. Repeat problem 2.5 if the circuit transfer characteristic v_o against v_i is to be as shown in Fig. P.2.3.

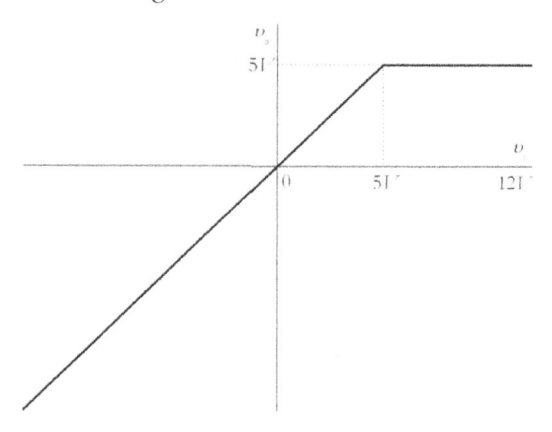

Figure P.2.3.

2.8. The diode in Fig. P.2.4. is an ideal Zener diode, i.e. it acts as an ideal diode when forward biased and has a sharp breakdown region at $5V$ when reverse biased, while the maximum power it can safely dissipate is $250mW$. The supply voltage v is $8\sin 318t$ (V).

a. Calculate the required value of the resistance R.
b. Plot the waveforms of v and v_o as functions of time.

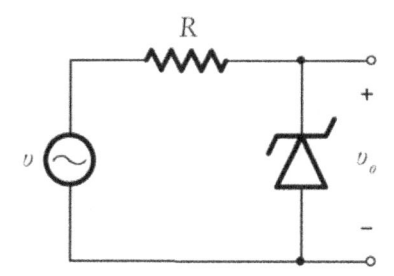

Figure P.2.4.

(Answers: a. 60Ω)

2.9. An ideal Zener diode with Zener voltage $10V$ is used to stabilize the voltage across a variable load resistor R_L, as shown in Fig. P.2.5. Voltage v varies between 13 and $16V$ while the current in the load varies between 13 and $85mA$. For satisfactory stabilization the current in the load should not fall below $15mA$.

a. Determine a suitable value for resistance R so that the voltage stabilization will be satisfactory under all the above conditions.
b. Calculate the maximum power dissipated in the diode for this value of R.

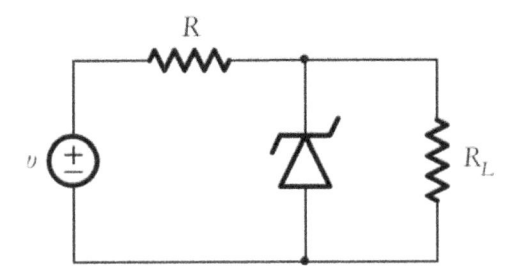

Figure P.2.5.

(Answers: a. 30Ω, b. 1.87W)

2.10. The current I in Fig. P.2.6 takes only two values, zero or high enough to bring the transistor into saturation, where V_{CEsat} is 0.2V. For the LED to be on (lit) a current of $16mA$ is required when the voltage across it will be 1.6V. Calculate the value of resistance R_C for the diode to emit light.

(Answers: 200Ω)

2.11. The voltage V_s in the circuit in Fig. P.2.7 can be either zero or high enough to bring the MOSFET into saturation, where V_{DS} is 0.2V.

   a. Describe the way the circuit operates.
   b. If the LED in order to emit light requires a current of $15mA$ when the voltage across will be 1.8V, calculate the required value of R_D.

(Answers: 200Ω)

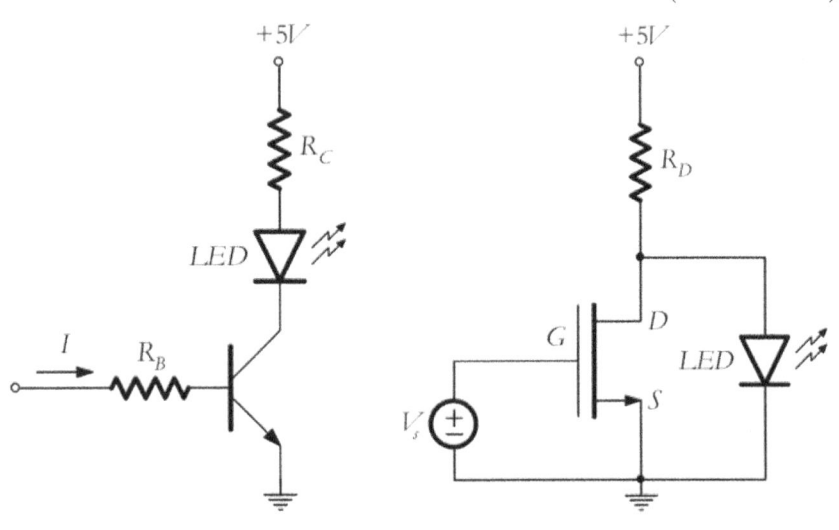

Figure P.2.6.

Figure P.2.7.

Chapter 3

Amplifiers

3.1 Introduction

The amplifier is an electronic device, that increases the energy content of the signal applied to its input. Depending on its structure, emphasis may be given on the amplification of the voltage, voltage amplifiers, or the current, current amplifiers. However, in general, it is a power amplifier, since in either case, increasing the voltage, while keeping the current unchanged or, increasing the current keeping unchanged the voltage, in fact, it is the power that is increased. The extra power given to the signal comes from the *dc* power supply that the amplifier needs for its operation. Therefore, the amplifier transforms *dc* power to signal power thus increasing the latter.

In the outset, the amplifier appears as a three- or four-terminal electronic circuit. Two terminals are required for applying the signal to be amplified and two for taking the amplified signal out. It is built up of *BJTs*, or *FETs*, or both transistor types. Extra terminals are provided for connecting the *dc* power supply to the amplifier. Contemporary technology provides the amplifiers as integrated circuits. The user does not have to know its internal structure, or how it has been designed, as long as he knows how to make it operational.

In this Chapter we introduce the various types of amplifiers as ideal devices. The integrated circuit operational amplifier is considered as such and is used to obtain other amplifiers as well as perform operations like signal addition, signal subtraction and signal integration. One important amplifier, the instrumentation amplifier, for the amplification of biosignals and other low amplitude signals is examined to some detail. Other useful amplifiers namely, the Operational Transconductance Amplifier (*OTA*) and the Amplifier with Negative Input Capacitance are presented, because of their use in some applications in bioinstrumentation.

3.2 The General Amplifier

An amplifier is a 3- or 4-terminal device, that increases the power level of a signal. Considering it as a black box, as shown in Fig. 3.1, it has a pair of terminals (1, 1') to which the signal to be amplified is applied and constitutes its input, and a pair of terminals (2, 2') from which the amplified signal is obtained and constitutes the output of the amplifier.

Figure 3.1. For the definition of the amplifier

The increase of the power level of the signal can be achieved either by increasing the current or the voltage keeping the other quantity (voltage or current respectively) constant, or amplifying both current and voltage. We can distinguish the amplifiers as current amplifiers, voltage amplifiers and power amplifiers respectively. However, in all cases it is the signal power that is amplified and the previous classification serves the purpose of better understanding and design.

For an amplifier to serve its purpose, it has to be supplied by dc power to operate actively and increase the power level of the signal. Thus the amplifier converts dc power taken from the power supply to signal power.

In the general case the application of the signal to the input of the amplifier causes a voltage difference V_{in} to develop across terminals 1, 1' and a current I_{in} flowing into it. If V_{in} and I_{in} are in phase, the amplifier at its input consumes an input signal power P_{in} given by

$$P_{in} = V_{in} I_{in} \tag{3.1}$$

and it behaves as a resistance of value

$$R_{in} = \frac{V_{in}}{I_{in}} \tag{3.2}$$

This is the amplifier input resistance, which in the case that V_{in} and I_{in} are not in phase, will be complex taking the name input impedance Z_{in}.

Coming now to the amplifier output, in delivering the signal to its load, it behaves like a voltage or current source, in general non-ideal, and, according to the discussion in Section 1.7, it has an output resistance R_o. Thus the equivalent circuit of an amplifier will be as shown in Fig. 3.2.

Amplifiers

Figure 3.2. Amplifier equivalent circuit with the output behaving as a voltage or current non-ideal source

Usually in practice, terminals 1' and 2' are short-circuited. However there is a class of amplifiers, the *isolation* amplifiers, in which this is not allowed. If the signal power delivered to the load is P_L and the input power is P_{in}, then there is a power gain A_P

$$A_P = \frac{P_L}{P_{in}} \qquad (3.3)$$

Because the power gain in practice may assume values in a very wide range, we usually express A_P in decibels (dB) as follows:

$$A_P(dB) = 10\log\frac{P_L}{P_{in}} \qquad (3.4)$$

It is also useful to define the voltage gain A_v and the current gain A_i of an amplifier as

$$A_v = \frac{V_L}{V_{in}} \qquad (3.5)$$

$$A_i = \frac{I_L}{I_{in}} \qquad (3.6)$$

where V_L and I_L are the voltage and the current delivered to the load respectively. It is also customary to express A_v and A_i in dB. Observing that the power is proportional to the square of the voltage and the square of the current, then

$$A_v(dB) = 20\log\left|\frac{V_L}{V_{in}}\right| \qquad (3.7)$$

$$A_i(dB) = 20\log\left|\frac{I_L}{I_{in}}\right| \qquad (3.8)$$

Absolute values of the ratios are required because the corresponding quantities, in general, will not be in phase. Also *rms* values of these quantities are considered.

Example 3.1

In the circuit shown in Fig. E.3.1(a) the voltage source V_s is ideal. When the switch S is open, the output voltage V_2 is $1V$, whereas when it is closed $V_2=0.975V$. Determine: a) The gain of the amplifier and b) its output resistance. Draw the equivalent circuit of the amplifier assuming that there is no feedback in it, i.e. the output signal does not affect the input signal.

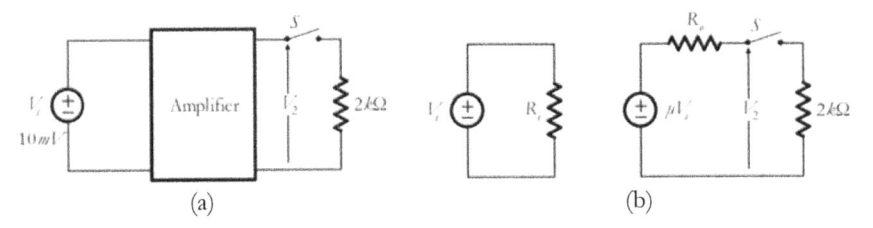

Figure E.3.1. (a) The circuit of the amplifier and (b) its equivalent circuit

Solution

Clearly the amplifier is non-ideal. It has an input resistance R_i and an output resistance R_o. So the equivalent circuit is as shown in Fig. E.3.1(b) with R_i playing no role since the signal source is ideal.

a) When the switch S is open, there is no current in the output resistance R_o and so V_2 is equal to μV_s, where μ is the amplifier gain. Therefore,

$$\mu = \frac{V_2}{V_s} = \frac{1V}{10mV} = 100$$

b) When the switch is closed, the voltage V_2 becomes $0.975V$ meaning that there is a voltage drop across R_o equal to

$$1V - 0.975V = 0.025V$$

which is due to the current I_o in it. This current is

$$I_o = \frac{V_2}{R_L} = \frac{0.975V}{2000\Omega} = 0.488mA$$

Therefore,

$$R_o = \frac{0.025V}{0.488mA} = 51\Omega$$

Example 3.M.1

Using MATLAB find the voltage gain A_v for the circuit of Fig. E.3.1 of Example 3.1 in dBs. Also plot the voltage gain versus the output resistance R_o for $R_o=0$ to $20k\Omega$ in order to demonstrate the effect of non zero R_o on the voltage gain.

Amplifiers

Code

```
>> Vin=0.01;
>> VL=0.975;
>> Av=VL/Vin;
>> AvdB=20*log10(Av)

          AvdB =

          39.7801

>> RL=2000;
>> for i=1:2000                    % take 2000 points
Ro(i)=i*10;                        % define range of R_o 0-20kΩ
VL(i)=1*RL/(RL+Ro(i));
Av(i)=VL(i)/Vin;
end
```

% the plots below are derived by the following procedure (the reader is advised first to consult the MATLAB tutorial in Appendix *D*)

```
>> plot(Ro,Av); axis([0 20000 0 105]); grid; xlabel('Ro (Ohms)'); ylabel('Av')
```

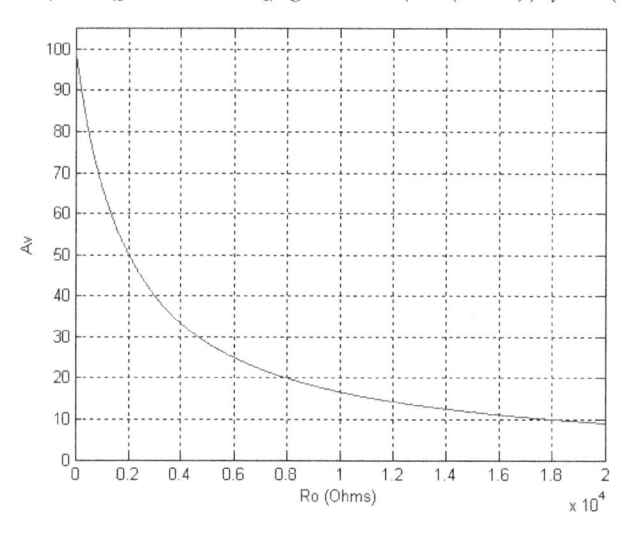

Figure 3.M.1. Voltage gain A_v versus output resistance R_o

Another approach would be:

```
>> Vin=0.01;
>> RL=2000;
>> Ro=1:10:20000;                  % take 2000 points
>> VL=1*RL./(RL+Ro);
>> Av=VL/Vin;
>> plot(Ro,Av); axis([0 20000 0 105]); grid; xlabel('Ro (Ohms)'); ylabel('Av')
```

3.2.1 Ideal Amplifiers

The characteristics of an ideal amplifier are the following:

a. Stable gain, voltage or current, independent of frequency.
b. Infinite input resistance for voltage amplifiers and zero for current amplifiers.
c. Zero output resistance for voltage amplifiers and infinite for current amplifiers.

Following this, we can classify the ideal amplifiers, considering them as ideal *controlled sources*, as shown in Table 3.1:

Table 3.1. Ideal amplifiers

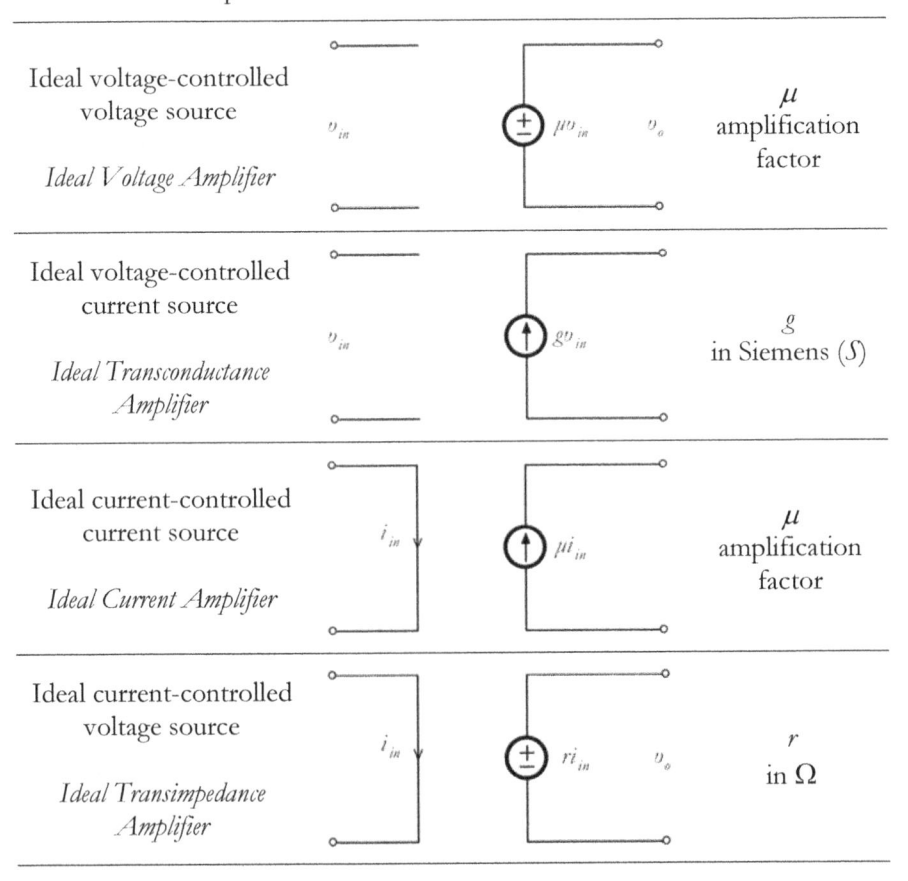

Ideal voltage-controlled voltage source v_{in} *Ideal Voltage Amplifier*	μv_{in} v_o	μ amplification factor
Ideal voltage-controlled current source v_{in} *Ideal Transconductance Amplifier*	gv_{in}	g in Siemens (S)
Ideal current-controlled current source i_{in} *Ideal Current Amplifier*	μi_{in}	μ amplification factor
Ideal current-controlled voltage source i_{in} *Ideal Transimpedance Amplifier*	ri_{in} v_o	r in Ω

Example 3.2

The amplifier in Fig. E.3.2 is an ideal voltage amplifier having a gain $A=-100$. Determine in *dB* the voltage ratio V_o/V_i, when $R_1=1k\Omega$, $R_g=10k\Omega$ and $R_f=100k\Omega$.

Amplifiers 71

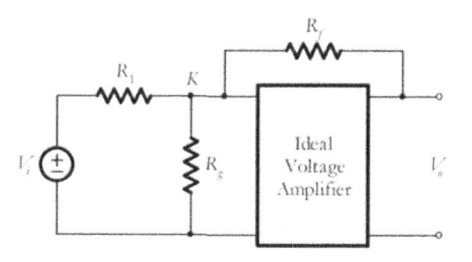

Figure E.3.2.

Solution

Let V be the voltage at node K. Since the voltage amplifier is ideal there is no current entering its input terminal. Then summing the currents at node K gives:

$$\frac{V_i - V}{R_1} + \frac{V_o - V}{R_f} - \frac{V}{R_g} = 0$$

or

$$\frac{V_i}{R_1} + \frac{V_o}{R_f} = \left(\frac{1}{R_1} + \frac{1}{R_f} + \frac{1}{R_g}\right) V \qquad \text{(E.3.1)}$$

But

$$V = \frac{V_o}{A}$$

Then substituting in Eq. (E.3.1) for V gives

$$V_o \left[\left(\frac{1}{R_1} + \frac{1}{R_f} + \frac{1}{R_g}\right)\frac{1}{A} - \frac{1}{R_f} \right] = \frac{V_i}{R_1}$$

and

$$\frac{V_o}{V_i} = \frac{\frac{A}{R_1}}{\frac{1}{R_1} + \frac{1}{R_g} + \frac{1}{R_f} - \frac{A}{R_f}} = \frac{A}{1 + \frac{R_1}{R_g} + \frac{(1-A)R_1}{R_f}}$$

Substituting values we get

$$\frac{V_o}{V_i} = \frac{-100}{1 + 0.1 + \frac{101}{100}} = \frac{-100}{2.11} = -47.4$$

or

$$20 \log \left|\frac{V_o}{V_i}\right| = 33.5 dB$$

3.3 The Operational Amplifier, Opamp

The ideal *operational amplifier* or *opamp* is defined as a differential-input $VCVS$ which has

a. infinite voltage gain
b. infinite input resistance
c. zero output resistance

The term *differential-input* means that the opamp has two input terminals not connected to ground, one being the *inverting* (-) and the other the *non-inverting* (+) input terminal. The output voltage of the opamp is the difference of the signals connected to its inputs multiplied by the gain of the opamp. The symbol and the equivalent circuit of the ideal opamp are shown in Fig. 3.3.

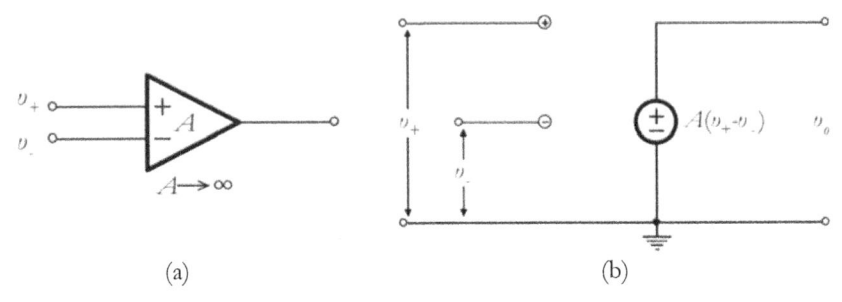

Figure 3.3. (a) Symbol and (b) equivalent circuit of the ideal opamp

The input voltages v_+ and v_- as well as the output voltage v_o are referred to the ground voltage, which is considered to be at zero volts.

The practical opamp is also a differential input but non-ideal $VCVS$ that has non infinite but high ($>10^5$) voltage gain, non infinite but high input resistance ($>150 k\Omega$) and non zero but low output resistance ($<100\Omega$). The input resistance is the resistance between the two input terminals, while the output resistance is the resistance in series with the output. Since the opamp is always used with an impedance of a few $k\Omega$ connected between the inverting input and the output terminal, the output resistance does not affect in practice the operation of the opamp output voltage source, provided that the output current does not exceed few mA. To be sure that these impedances will not affect the performance of the circuit using bipolar IC opamps, the impedance level of the associated circuit should be chosen greater than $1 k\Omega$ and smaller than $100 k\Omega$ with $10 k\Omega$ being the most appropriate choice. The upper limit is set by other imperfections of the opamp, which are explained below.

Referring to the gain of the practical opamp, we must stress that, it is a function of frequency having a frequency response as shown in Fig. 3.4.

This has been shaped in this form for stability reasons. A figure of merit for any opamp is the product Gain times Bandwidth, symbolized by f_T, which, in the case of Fig. 3.3, would be $10^5 \times 10 = 1 MHz$.

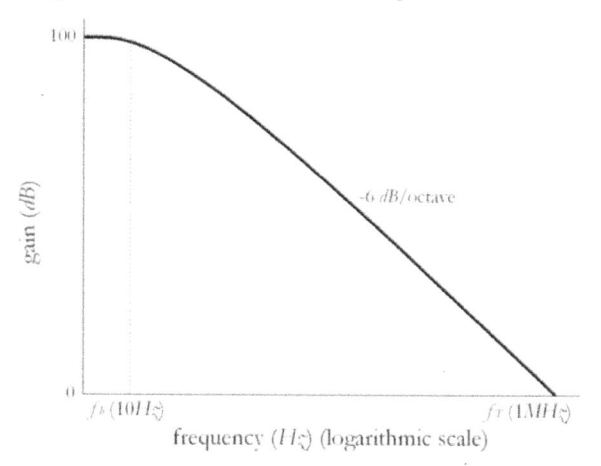

Figure 3.4. Frequency response of practical opamp (741 type)

3.4 Applications of the Opamp

In this section we introduce three simple but basic applications of the opamp, this being treated as ideal, to obtain

a. inverting or non inverting finite voltage gain,

b. sum of voltages and

c. voltage integration

Other opamp applications are given in following sections and chapters in this book. In all applications, in order to calculate the output voltage of an opamp, the following observation is applied:

The two voltages at the opamp inputs v_+ and v_- in Fig. 3.3(b) are equal. This is a consequence of the consideration that the gain A of the opamp is infinite. In fact, since

$$v_o = A(v_+ - v_-) \quad \text{or} \quad v_+ - v_- = \frac{v_o}{A}$$

if v_o is finite when $A \rightarrow \infty$, it requires that

$$v_+ - v_- \rightarrow 0 \quad \text{or} \quad v_+ \cong v_- \qquad (3.9)$$

This relationship should always be kept in mind when analyzing a circuit using opamps. Also the opamp should be treated as having infinite input resistance (i.e. no current is flowing into its input terminals) and zero output resistance (i.e. its output voltage remains constant for any load impedance).

3.4.1 Inverting and Non Inverting Finite Voltage Gain

Consider the circuit in Fig. 3.5(a). Since the non-inverting input terminal is connected to ground, it is $v_+=0$ and by virtue of Eq. (3.9) v_- should be zero too. The node voltage V at the inverting input is usually said to be a *virtual earth voltage*. Thus $V=0$. Also there is no current flowing into the opamp inverting input, because the input resistance is considered to be infinite. Then applying Kirchhoff's Current Law (KCL), we get

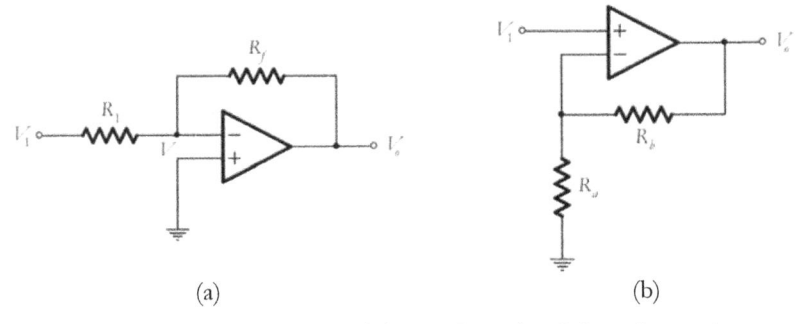

Figure 3.5. To obtain (a) inverting and (b) non inverting finite voltage gain

$$\frac{V_1}{R_1} + \frac{V_o}{R_f} = 0$$

or

$$V_o = -\frac{R_f}{R_1} V_1 \qquad (3.10)$$

Thus the circuit in Fig. 3.5(a) has a voltage gain G_I given by

$$G_I = -\frac{R_f}{R_1} \qquad (3.11)$$

which is finite and inverting. The negative sign has the meaning that V_1 and V_o are 180° out of phase.

If V_1/R_1 represents a current I obtained from a device that behaves as a current source, then Eq. (3.2) gives

$$V_o = -R_f I$$

meaning that the circuit in Fig. 3.5(a) behaves as a *current-to-voltage converter*.

Coming now to Fig. 3.5(b), it is observed that $v_+=V_1$ whereas

$$v_- = \frac{R_a}{R_a + R_b} V_o$$

and by virtue of Eq. (3.9)

$$V_1 = \frac{R_a}{R_a + R_b} V_o$$

or

$$V_o = \frac{R_a + R_b}{R_a} V_1 = \left(1 + \frac{R_b}{R_a}\right) V_1$$

Thus this circuit has a gain G_N given by

$$G_N = 1 + \frac{R_b}{R_a} \qquad (3.12)$$

which is finite and positive i.e. V_1 and V_o are in phase.

If R_a is removed from the circuit, when $R_a=\infty$, Eq. (3.12) gives $G_N=1$, i.e. V_o and V_1 are identical. Independently of whether V_1 is obtained from an ideal or non ideal voltage source the relationship $V_1=v_+$ will always hold. Thus the voltage V_o ($=V_1$) will be obtained from an ideal voltage source. It can also be shown that, even if the opamp input resistance is considered to be of the order of $200k\Omega$ (741 type) the input resistance of the circuit becomes $\sim 2G\Omega$. Therefore this circuit can be used as a *buffer stage* between the preceding stage giving the voltage V_1 and the following stage to which V_o will be applied. In practice this is a very useful circuit for such applications.

It can be shown [1] that the bandwidth of the non inverting amplifier is higher than that of the inverting one, meaning that the former can operate more accurately than the latter beyond the high frequency edge of the frequency response of the inverting amplifier.

Example 3.3

The opamp in Fig. 3.5 is ideal. If $R_1=R_a=10k\Omega$ and $R_f=R_b=100k\Omega$ determine the gain for each of the two circuits in the same figure.

Solution

a. Inverting amplifier:

$$G_I = \frac{V_o}{V_1} = -\frac{R_f}{R_1} = -\frac{100}{10} = -10$$

b.   Non inverting amplifier:

$$G_N = \frac{V_o}{V_1} = 1 + \frac{R_b}{R_a} = 11$$

So same amplifier, same resistances, different configurations, different gain.

Example 3.M.2

Consider the non inverting amplifier of Fig. 3.5(b) with $R_a=10k\Omega$, $R_b=100k\Omega$ and open-loop gain A. Using MATLAB plot the closed-loop voltage gain G_N against the open-loop gain A, for A equal to 10^1 up to 10^6 in order to demonstrate the effect of finite open-loop gain on the closed-loop gain.

Code

Analyzing the circuit we have

$$V_o = A(V_1 - V) \quad \text{with} \quad V = \frac{R_a}{R_a + R_b} V_o$$

Inserting the value of V in the first equation we get

$$G_N = \left(1 + \frac{R_b}{R_a}\right) \cdot \frac{AR_a}{AR_a + R_a + R_b}$$

```
>> Ra=10;
>> Rb=100;
>> A=logspace(1,6,2000);              % 2000 points between 10^1 and 10^6
>> G=(1+(Rb/Ra))*Ra*A./(A*Ra+Ra+Rb);
>> semilogx(A,G); axis([10 1e6 4 12]); grid; xlabel('Open-loop gain A');
ylabel('Closed-loop gain G')
```

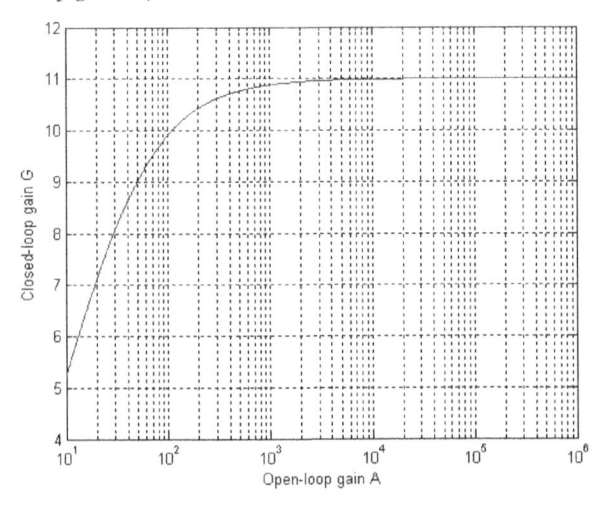

Figure 3.M.2. Closed-loop gain against open-loop gain

Amplifiers

3.4.2 Summation of Voltages

Consider the circuit in Fig. 3.6. Summing the currents at the virtual earth node gives

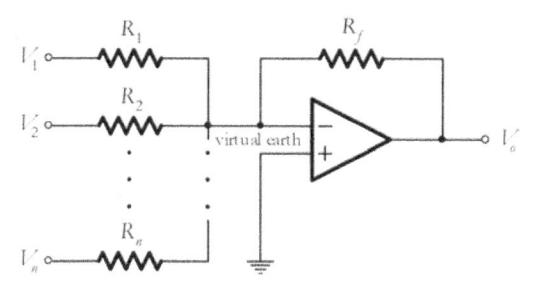

Figure 3.6. Voltage summer

$$\frac{V_1}{R_1} + \frac{V_2}{R_2} + \cdots + \frac{V_n}{R_n} + \frac{V_o}{R_f} = 0$$

Then solving for V_o

$$V_o = -R_f \left(\frac{V_1}{R_1} + \frac{V_2}{R_2} + \cdots + \frac{V_n}{R_n} \right) \quad (3.13)$$

If all resistances are equal, i.e. if $R_1 = R_2 = \ldots = R_n = R_f = R$, Eq. (3.13) gives

$$V_o = -(V_1 + V_2 + \cdots + V_n) \quad (3.14)$$

This is the negative sum of the voltages V_i, $i = 1, 2, \ldots, n$. The fact that this sum is of opposite sign to the sign of the sum of V_i is not a serious problem, because this sign can be reversed, if the summer is followed by the circuit in Fig. 3.5(a) with $R_1 = R_f$.

Evidently the summer can give the difference of two voltages $V_1 - V_2$, if this circuit is employed to add the voltages $-V_1$ and V_2. Of course if V_1 and V_2 are of the same sign, to get their difference by a summer would require the use of another opamp to reverse the sign of V_1 first before performing the summation. However, this can be avoided, if the circuit of a differential amplifier examined in Section 3.6 is employed.

3.4.3 Integration of a Voltage

Integration of a voltage υ_i can be achieved by the circuit in Fig. 3.7. The two currents added at the virtual earth point are υ_i/R and $Cd\upsilon_c/dt$ according to Eq. (1.19). Thus, since $\upsilon_c = \upsilon_o$

$$\frac{\upsilon_i}{R} + C\frac{d\upsilon_o}{dt} = 0 \quad (3.15)$$

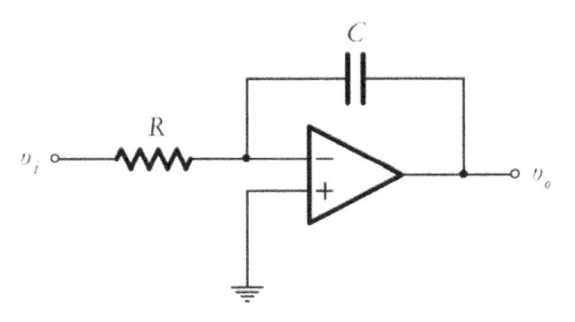

Figure 3.7. Analog integrator

On integrating and solving for v_o gives

$$v_o = -\frac{1}{RC}\int_0^t v_i dt + v_o(0) \qquad (3.16)$$

If v_o at $t=0$ is zero, then Eq. (3.16) simplifies to

$$v_o = -\frac{1}{RC}\int_0^t v_i dt \qquad (3.17)$$

Note that, if $v_i=V$, constant, with $v_o(0)=0$, Eq. (3.17) will give

$$v_o = -\frac{V}{RC}t \qquad (3.18)$$

i.e. the magnitude of the output of the integrator increases with time linearly tending to infinity as t tends to become infinite. Of course this is not physically possible and the practical opamp will reach the saturation state after a certain time determined by the time constant RC, which also determines the integrator gain $-1/RC$.

3.4.4 Constant Current Source

In some cases a load resistance R_L requires to be driven by a constant current rather than by a constant voltage. This situation arises for example, when temperature is measured using a thermistor, a temperature sensor (see Chapter 9). An easy way to have a constant current (approximately) through R_L is to connect a voltage source in series with a resistance R_s with $R_s \gg R_L$. Then, see Fig. 3.8

$$I_L = \frac{V}{R_s + R_L} \cong \frac{V}{R_s}$$

for $R_s \gg R_L$. With this condition any change in R_L leaves the current in it approximately constant.

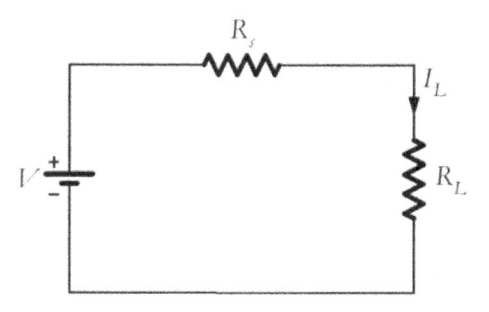

Figure 3.8. An approximate current source

However, the use of an opamp can produce an exact constant current through R_L. Consider the circuit in Fig. 3.9(a). The current in R_L is V_o/R_L, which equals $-V_i/R$. Keeping V_i and R constant makes the current in R_L to be constant when R_L changes. This is so, because decreasing or increasing R_L results in decreasing or increasing V_o respectively, so that the ratio of the two remains equal to $-V_i/R$.

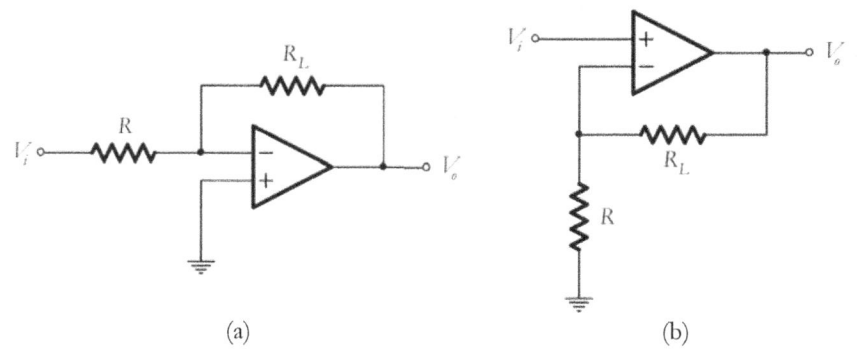

(a)   (b)

Figure 3.9. Two ways for passing a constant current through R_L.

In the case of Fig. 3.9(b), since the voltages v_+ and v_- at the opamp inputs are equal, the current in R_L is V_i/R, which remains constant independently of any change in R_L. This last circuit has the advantage over the previous one that no current is drawn from the voltage source V_i.

3.5 Additional Characteristics of Practical Opamps

As was stated in Section 3.3, practical opamps have characteristics, which differ from those of the ideal element we used in previous sections. Apart from their open-loop voltage gain, which is non-infinite, their input impedance and output admittance that are not infinite either, there are also some additional parameters associated with the operation of the practical opamp [1,2], which degrade its performance and the designer should always keep in mind.

Input Offset Voltage V_{IO}

If both inputs of the real opamp are grounded, the output voltage will not be zero in practice, as would be expected. This is a defect, which causes the output voltage to be offset with respect to ground potential. For large *ac* input signals the output voltage waveform will then be asymmetrically clipped, that is the opamp will display a different degree of nonlinear behaviour for positive and negative excursions of the input signals. The input offset voltage V_{IO} is that voltage which must be applied between the input terminals to balance the opamp. In many opamps this defect may be "trimmed" to zero by means of an external potentiometer connected to terminals provided for this reason.

Input Offset Current I_{IO}

This is defined as the difference between the currents entering the input terminals when the output voltage is zero. These currents are actually the base bias currents of the transistors at the input stage of the opamp (for bipolar opamps), and their effect is the appearance of an undesired *dc* voltage at the output. This defect of the opamp can be modelled by connecting two current generators at the input terminals of the ideal opamp. This is shown in Fig. 3.10 for the case of the circuit in Fig. 3.5(b), which is used to provide $1+R_f/R_1$ voltage gain. To reduce the effect of the input bias currents R_2 is inserted as shown below. Evidently this has no effect on the signal. If the input voltage V_s is zero, and assuming linear operation of the opamp, we may observe the following:

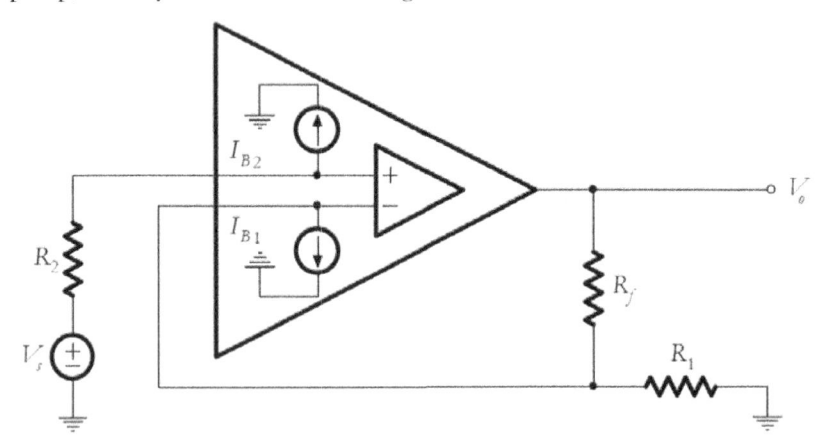

Figure 3.10. Current sources I_{B1}, I_{B2} represent the presence of input offset currents

The action of I_{B1}, assuming $I_{B2}=0$, causes the output voltage to be

$$V_{o1} = R_f I_{B1} \tag{3.19}$$

The action of I_{B2}, assuming $I_{B1}=0$, will result in the output voltage

$$V_{o2} = -\left(1 + \frac{R_f}{R_1}\right) I_{B2} R_2 \qquad (3.20)$$

Then applying superposition when both I_{B1} and I_{B2} are present, we get the output voltage

$$V_o = V_{o1} + V_{o2} = I_{B1} R_f - \left(1 + \frac{R_f}{R_1}\right) I_{B2} R_2 \qquad (3.21)$$

For this voltage to be zero when $I_{B1} = I_{B2}$, which is the optimistic case, the following relationship between the resistor values should hold:

$$R_2 = \frac{R_1 R_f}{R_1 + R_f} \qquad (3.22)$$

However, even under this condition, when $I_{B1} \neq I_{B2}$, the output voltage will be

$$V_o = (I_{B1} - I_{B2}) R_f = I_{IO} R_f \qquad (3.23)$$

i.e. non zero. Note though that without R_2, $V_o = I_{B1} R_f$, and, since $I_{IO} \ll I_{B1}$ in practice, the output voltage arising from the input bias currents is reduced by including R_2.

Input Voltage Range V_i

Assuming that the imperfections of the opamp due to Input Offset Voltage and Input Offset Current have been corrected, the voltage transfer characteristic of the amplifier will be as shown in Fig. 3.11, where V_i represents the differential input voltage. It can be seen that the opamp behaves linearly only in the range of V_i between $-V_2$ and V_1, i.e.

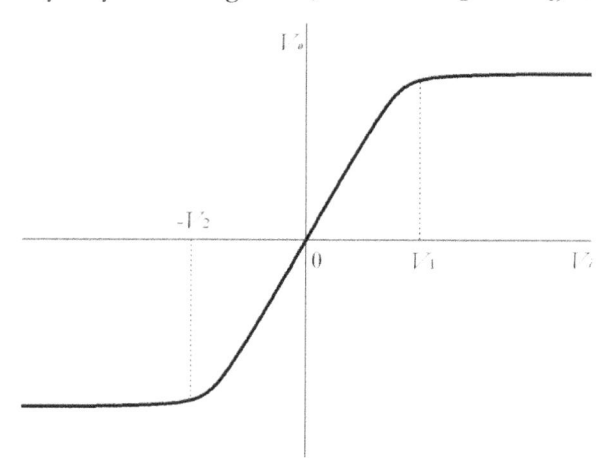

Figure 3.11. The saturation characteristic of the opamp

$$-V_2 \leq V_i \leq V_1$$

Therefore, only for this range of V_i one can get the benefit of the full voltage gain of the opamp. Beyond this voltage range the amplifier goes to saturation.

Although the nonlinear behaviour of the opamp is a cause of concern in the design of active RC filters, one can get advantage of the saturation characteristic to build analogue voltage comparators, which are very useful in practice (for example as zero crossing detectors).

Power Supply Sensitivity $\Delta V_{IO}/\Delta V_{GG}$

This is the ratio of the change of the input offset voltage ΔV_{IO} to the change in the power supply ΔV_{GG} that causes it. The change in the power supply is considered symmetrical.

Slew Rate *SR*

The rate of change of the output voltage cannot be infinite due to the various internal time constants of the opamp circuitry. The *Slew Rate* is defined as the maximum rate of change of the output voltage for a unit step input excitation. This is normally measured for unity gain at the zero voltage point of the output waveform.

The *SR* sets a serious limitation to the amplitude of the signal at high frequencies. This can be shown in the case of a sine wave as follows:

Let
$$v_o = V_m \sin \omega t$$

Then
$$\frac{dv_o}{dt} = V_m \omega \cos \omega t$$

which becomes maximal at the zero crossing points, i.e. when $\omega t = 0$, π, 2π, ...

Thus at $\omega t = 0$

$$\left. \frac{dv_o}{dt} \right|_{\omega t = 0} = V_m \omega \qquad (3.24)$$

Since this cannot be larger than the *SR* i.e.

$$SR \geq V_m \omega \qquad (3.25)$$

it is clear, that for linear operation at a high frequency ω, the amplitude of the output voltage cannot be greater than SR/ω. Thus, at high frequencies the opamp cannot work properly at its full input voltage swing as it does at low frequencies.

Short-Circuit Output Current

This denotes the maximum available output current from the opamp, when its output terminal is short-circuited with the ground or with one of its power supply rails.

Maximum Peak-to-Peak Output Voltage Swing V_{opp}

This is the maximum undistorted peak-to-peak output voltage, when the *dc* output voltage is zero.

Input Capacitance C_i

This is the capacitance between the input terminals with one of them grounded.

Common-Mode Rejection Ratio *CMRR*

Ideally the opamp should reject completely all common mode signals (i.e., the same signals applied to both inputs) and amplify the differential mode ones. However, for reasons of circuit imperfections, the amplifier gain is not exactly the same for both of its inputs. The result of this is that common mode signals are not rejected completely. A measure of this imperfection is the *Common-Mode Rejection Ratio* (CMRR). Expressed in *dB*, the *CMRR* is the ratio of open-loop differential gain to the corresponding common-mode gain of the opamp. Its value at low frequencies is typically better than 80*dB*, but it decreases at higher frequencies.

Total Power Dissipation

This is the total *dc* power that the opamp absorbs from its power supplies minus the power that the amplifier delivers to its load.

Rise Time t_r

This is the time required for the output voltage of the amplifier to increase from 10% to 90% of its final value for a step input voltage. It can be shown that $t_r \times f_t \cong 0.35$ for a first-order circuit.

Overshoot

This is the maximum deviation of the output voltage above its final value for a step input excitation.

In spite of all these imperfections though, the non ideal behaviour of the practical opamp does not prevent it from being the most versatile linear active element in use today.

Example 3.M.3

Plot the bandwidth against the maximum output voltage for an opamp having slew-rate $1V/\mu s$ with the output being free of distortion due to slew rate.

Code

```
>> SR=1e6;
>> vout=0:0.1:8;
>> BW=SR./(2*pi*vout);
>> plot(vout,BW); axis([0 8 0 4e5]); grid; xlabel('Maximum output voltage');
ylabel('Bandwidth')
```

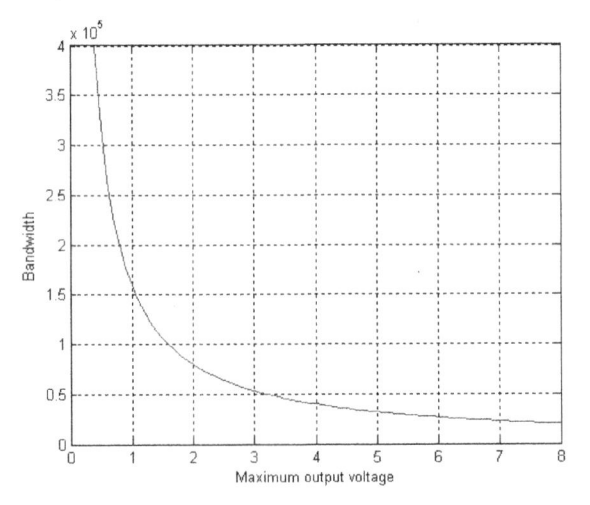

Figure 3.M.3. Bandwidth against maximum output voltage

3.6 The Differential (Difference) Amplifier

The *differential amplifier* presents at its output a voltage, which is dependent on the difference of the two voltages applied at its two inputs. The opamp is by nature a differential amplifier but, because of its very high gain, it is not used as such alone, except for one nonlinear application, i.e. when it is used as a comparator. However, the opamp combined with resistors, as shown in Fig. 3.12, gives the basic differential amplifier circuit, which is most useful in measurement applications.

Amplifiers

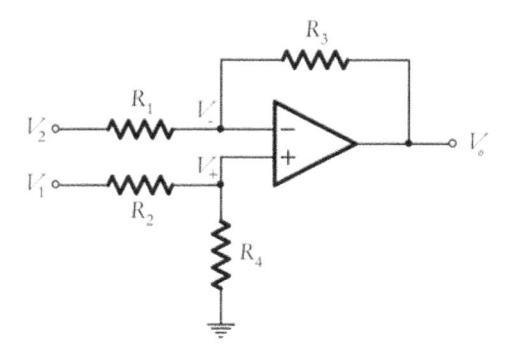

Figure 3.12. Differential amplifier

Considering the circuit in Fig. 3.12, it is clear, that neither V_+ nor V_- is zero. However as always $V_+ \cong V_-$. To analyze the circuit we can write for the currents at the input nodes the following:

$$\frac{V_2 - V_-}{R_1} + \frac{V_o - V_-}{R_3} = 0 \tag{3.26}$$

and

$$\frac{V_1 - V_+}{R_2} - \frac{V_+}{R_4} = 0 \tag{3.27}$$

Solving Eq. (3.26) for V_- gives

$$V_- = \frac{R_3 V_2 + R_1 V_o}{R_1 + R_3} \tag{3.28}$$

Similarly, solving Eq. (3.27) for V_+ gives

$$V_+ = \frac{R_4}{R_2 + R_4} V_1 \tag{3.29}$$

Then equating the expressions for V_+ and V_- and solving for V_o results in the following expression for V_o:

$$V_o = \frac{(R_1 + R_3)R_4}{R_1(R_2 + R_4)} V_1 - \frac{R_3}{R_1} V_2 \tag{3.30}$$

Eq. (3.30) gives the general dependence of V_o on V_1 and V_2, but in a rather complicated way. Of great interest are the following cases:

a. All resistances are equal i.e. $R_1 = R_2 = R_3 = R_4 = R$. Then Eq. (3.30) gives

$$V_o = V_1 - V_2 \tag{3.31}$$

Thus V_o is the difference of the two voltages V_1 and V_2 being zero if $V_1=V_2$.

b.   *Common mode (CM) operation:* $V_1=V_2$. Eq. (3.30) gives:

$$V_o = \frac{R_1R_4 - R_2R_3}{R_1(R_2 + R_4)} V_1 \tag{3.32}$$

Thus the common mode gain A_{CM} defined as

$$A_{CM} \equiv \frac{V_o}{V_1} \tag{3.33}$$

is found from Eq. (3.30) to be

$$A_{CM} = \frac{R_1R_4 - R_2R_3}{R_1(R_2 + R_4)} \tag{3.34}$$

Now if

$$R_1R_4 = R_2R_3 \tag{3.35}$$

A_{CM} becomes zero. Eq. (3.35) is thus the condition for zero *CM* gain and is of utmost importance for the differential amplifier, since all common mode signals, such as noise, common to both inputs are cancelled out and do not appear at the amplifier output.

c.   Of equal importance is also the differential gain defined for the case that $V_1=-V_2$. Then Eq. (3.30) gives

$$V_o = \frac{R_1R_4 + R_2R_3 + 2R_3R_4}{R_1(R_2 + R_4)} V_1 \tag{3.36}$$

Thus, the differential gain A_d is ($V_1=-V_2$)

$$A_d \equiv \frac{V_o}{V_1} = \frac{R_1R_4 + R_2R_3 + 2R_3R_4}{R_1(R_2 + R_4)} \tag{3.37}$$

which for $R_1R_4=R_2R_3$ is simplified to the following:

$$A_d = \frac{2R_3}{R_1} \tag{3.38}$$

Clearly, under the condition given by Eq. (3.38), the differential gain of the amplifier is determined by the ratio of resistances R_3 and R_1.

A figure of merit for a differential amplifier is the so-called *common mode rejection ratio*, which is discussed next.

3.7 Common-Mode Rejection Ratio, *CMRR*

This is a most important characteristic of any differential amplifier and is defined as follows: It is the ratio of the differential gain A_d of the differential amplifier divided by its common-mode gain A_{CM}, i.e.

$$CMRR = \frac{A_d}{A_{CM}} \tag{3.39}$$

According to this definition the *CMRR* of the amplifier in Fig. 3.12 is, in general,

$$CMRR = \frac{R_1 R_4 + R_2 R_3 + 2R_3 R_4}{R_1 R_4 - R_2 R_3} \tag{3.40}$$

When $R_1 R_4 = R_2 R_3$ this becomes infinite, which is the most important characteristic one seeks in a differential amplifier. The reason is that all noise signals (see Chapter 6) being common to both inputs of the amplifier will cancel out and will not affect the useful signal. The condition for zero *CMRR* is, in practice, better achieved when all resistances are equal. In this case they should be precisely matched and this can be best achieved, if the whole amplifier circuit is produced in microelectronic form. It should be said though that in this case the differential gain would take the value of only 2 according to Eq. (3.38).

Equally satisfactory from the *CMRR* point of view though is the case, when $R_1 = R_2$ and $R_3 = R_4$. The condition for infinite *CMRR* is satisfied while, unlike the equal resistance case, the differential gain can be greater than 2, if R_3 is higher than R_1.

The differential amplifier in Fig. 3.12 is employed in building up the circuit of the *instrumentation amplifier*, which we examine next.

3.8 Instrumentation Amplifier

The *Instrumentation Amplifier* (*IA*) is the most useful amplifying device employed in nearly all measuring equipment, where the transducer signal requires amplification (thermocouples, strain-gage bridges, biological probes, pre-amplification of small differential signals superimposed on high common-mode voltages etc as discussed in Chapter 9). The most important characteristics of the *IA* are the following:

a. High and stable gain
b. Ultra high input resistance
c. High *CMRR*

Additional, also important characteristics of an *IA*, are the following:

d. Low offset and drift
e. Low non linearity
f. Low effective output resistance

The differential amplifier in Fig. 3.12 lacks the first two characteristics, whereas it satisfies the third under, careful selection of the component values. However, by connecting extra circuitry in the front end, it can be made to satisfy all three characteristics of the IA. The whole circuit is shown in Fig. 3.13. The two opamps on the left are used in the non-inverting mode and are coupled by means of resistance R_c. Their outputs feed the two inputs of the differential amplifier of Fig. 3.12.

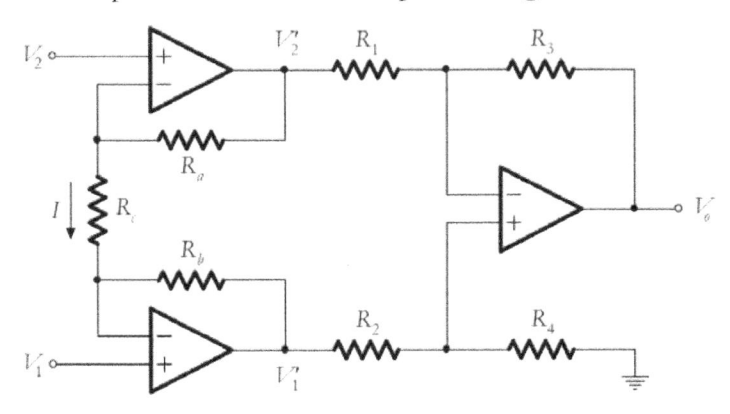

Figure 3.13. The instrumentation amplifier

Recalling that there is no input current into the opamp and that $v_+ = v_-$, one can write for the current I the following:

$$I = \frac{V_2' - V_2}{R_a} = \frac{V_2 - V_1}{R_c} = \frac{V_1 - V_1'}{R_b} \tag{3.41}$$

Solving for V'_1 and V'_2 gives

$$V_1' = \frac{R_c + R_b}{R_c} V_1 - \frac{R_b}{R_c} V_2 \tag{3.42}$$

and

$$V_2' = \frac{R_a + R_c}{R_c} V_2 - \frac{R_a}{R_c} V_1 \tag{3.43}$$

Voltages V'_1 and V'_2 are applied to the differential amplifier circuit. By virtue of Eq. (3.30) the output voltage V_o will be

$$V_o = \frac{(R_1 + R_3) R_4}{R_1 (R_2 + R_4)} V_1' - \frac{R_3}{R_1} V_2' \tag{3.44}$$

which in the case of equal resistances gives

$$V_o = V'_1 - V'_2$$

Substituting from Eq. (3.42) and (3.43) for V'_1 and V'_2 respectively gives for equal resistances

$$V_o = \left(\frac{R_b + R_c}{R_c}V_1 - \frac{R_b}{R_c}V_2\right) - \left(\frac{R_a + R_c}{R_c}V_2 - \frac{R_a}{R_c}V_1\right) \qquad (3.45)$$

To avoid the creation of different noise produced by R_a and R_b due to temperature variations (see Chapter 7) R_a should be selected equal to R_b. Then Eq. (3.45) gives for $R_a=R_b=R$

$$V_o = \frac{R+R_c}{R_c}V_1 - \frac{R}{R_c}V_2 - \frac{R+R_c}{R_c}V_2 + \frac{R}{R_c}V_1$$

or

$$V_o = \left(1 + \frac{2R}{R_c}\right)(V_1 - V_2) \qquad (3.46)$$

In case $R_3=R_4>R_1=R_2$, Eq. (3.44) becomes

$$V_o = \frac{R_3}{R_1}(V'_1 - V'_2)$$

Then, for $R_a=R_b=R$, Eq. (3.45) gives

$$V_o = \frac{R_3}{R_1}\left(1 + \frac{2R}{R_c}\right)(V_1 - V_2)$$

Thus the IA differential gain will be ($V_2=-V_1$)

$$A_d = \frac{V_o}{V_1} = \frac{2R_3}{R_1}\left(1 + \frac{2R}{R_c}\right) \qquad (3.47)$$

which for $R_3>R_1$ and $R>R_c$ can, theoretically, be as high as desired.

As an example consider $R=100k\Omega$, $R_c=4k\Omega$, $R_3=50k\Omega$ and $R_1=1k\Omega$. Then Eq. (3.47) gives

$$A_d = \frac{2\times 50}{1}\left(1 + \frac{2\times 100}{4}\right) = 5100$$

If $R_3=R_4=R_1=R_2$, Eq. (3.47) gives

$$A_d = 2\left(1 + \frac{2R}{R_c}\right) \qquad (3.48)$$

It can thus be seen that for constant R the differential gain can be varied by means of R_c.

3.8.1 Input Resistance of the *IA*

With the two opamps in the front end being used in the non-inverting mode, the input currents in the non-inverting inputs are theoretically zero. This implies that the input resistance at either input of the *IA* is theoretically infinite. In practice, in order to obtain input resistance in the order of $G\Omega$ the first input stage of the two opamps on the left of the *IA* circuit are built using *FETs*. This ultra high input resistance is necessary in bioinstrumentation, because some of the transducers have very high output resistance. Thus the circuit in Fig. (3.13) satisfies all three requirements to be a useful *IA*.

3.8.2 Integrated Circuit (*IC*) Instrumentation Amplifiers

A useful monolithic instrumentation amplifier is the AD524 of *Analog Devices* having *CMRR* 130dB, low non linearity, low offset voltage and pin programmable gains 1, 10, 100, 1000, while no external components are required for its operation.

Other also useful integrated circuit *IA* are the following:

AD620: Low Cost Low Power Instrumentation Amplifier (*Analog Devices*)

INA128/INA129: Precision, Low Power Instrumentation Amplifiers (*Burr-Brown Products from Texas instruments*)

MAX4194-MAX4197: Micropower, Single-Supply, Rail-to-Rail, Precision Instrumentation Amplifiers (*Maxim Integrated Products*)

3.9 Operational Transconductance Amplifier, *OTA*

The ideal *operational transconductance amplifier* (*OTA*) is a differential voltage to current converter having infinite input resistance and infinite output resistance. Its symbol and equivalent circuit are shown in Fig. 3.14(a) and (b) respectively. The activity of the *OTA* is characterized by its *transconductance* g_m, which connects the output current with the input voltages, i.e.

$$I_o = g_m(V_1 - V_2) \qquad (3.49)$$

The *OTA* transconductance, given usually in *mS* (milliSiemens), is controlled externally by a bias current. It is produced in a microelectronic

Amplifiers 91

form, which is quite simple and for this reason it has been proved to be a useful component in many more complicated analog microelectronic circuits. Eq. (3.49) describes the operation of practical OTAs only for voltage differences in the order of few tens of millivolts. Above these it operates non linearly.

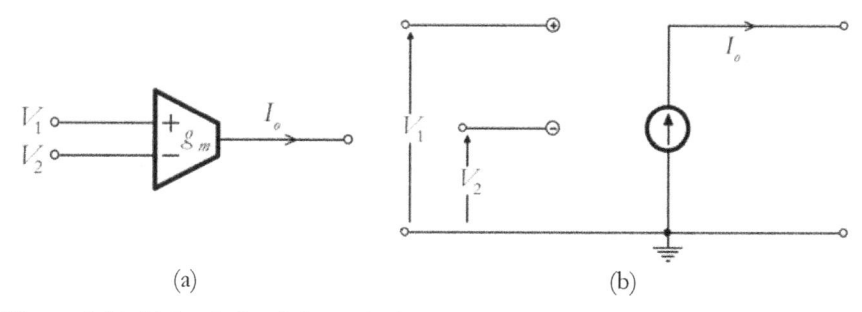

(a)   (b)

Figure 3.14. (a) Symbol and (b) equivalent circuit of the OTA

The OTA applications are quite numerous. If a resistance R_L is connected to its output, the voltage

$$V_o = R_L I_o \qquad (3.50)$$

appears across the load thus, resulting in a voltage gain G given by

$$G = \frac{R_L I_o}{V_1 - V_2} = g_m R_L \qquad (3.51)$$

This gain can be controlled by adjusting g_m for a certain R_L and can be positive or negative depending on the sign of V_1-V_2. The OTA operates normally even if V_1 or V_2 is zero. If, instead of R_L, a capacitance C_L is connected across its output, the circuit operates as an integrator.

One important application of the OTA, which is very useful in IC design, is its use as a variable resistance. To show this consider the set-up in Fig. 3.15. By definition,

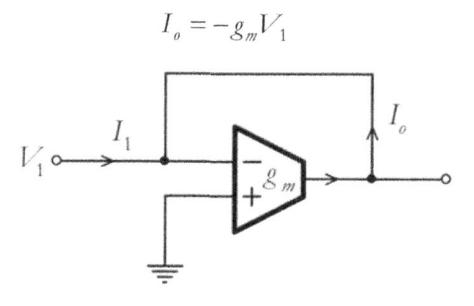

$$I_o = -g_m V_1$$

Figure 3.15. Use of OTA as a variable resistance

But as I_o=-I_1 (no current is flowing into the OTA input)

$$I_1 = g_m V_1$$

or

$$\frac{V_1}{I_1} = \frac{1}{g_m} \tag{3.52}$$

As g_m can be controlled externally, a variable resistance has been created. This resistance has one of its terminals earthed. However, a floating resistance can be obtained by employing an additional OTA (see Problem 3.8).

Other very useful applications of the OTA exist and the interested reader can consult more specialized books (see for example [1]).

3.10 Power Supply for an Opamp, an IA or an OTA

Any electronic amplifying device to operate properly, and thus be useful, requires the use of a *dc* power supply. This voltage supply biases the internal transistors of the opamps, the OTAs and the IAs to operate at the operating points of their characteristics, which have been chosen by the designers. Since these amplifiers should amplify both positive and negative signals, it is convenient to be biased both by positive and negative voltage. This assures that the output voltage of these amplifiers will be zero for zero input signals (all voltages are referred to ground potential, which is considered equal to zero). To achieve $+V_{cc}$ and $-V_{cc}$ voltages, two *dc* power supplies (two batteries for example) are connected in series with the common node earthed, as shown in Fig. 3.16. The common node is connected to the ground point to which the signal is referred.

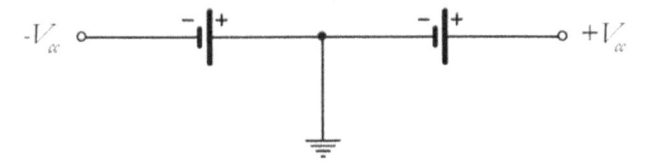

Figure 3.16.

3.11 Application of Feedback to Amplifiers

Feedback exists in a system, when the result affects the way in which the cause produces the result. There is feedback in an amplifier, when part of the output signal is added to (or subtracted from) the input signal. The feedback signal can be proportional to the amplifier output voltage or current, when it is called voltage or current feedback, and it can be applied in parallel or in series at the input of the amplifier. Depending on the type of feedback, this improves or impairs the characteristics of the amplifier. Feedback exists in all opamp circuits presented in the previous sections.

Consider for example Fig. 3.5(b), redrawn here for convenience (Fig. 3.17). The feedback voltage is

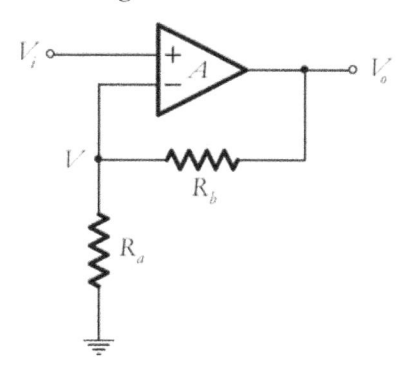

Figure 3.17.

$$V = \frac{R_a}{R_a + R_b} V_o \qquad (3.53)$$

applied in parallel with the input voltage V_i and, in effect, subtracted from it. Let β be the feedback ratio V/V_o i.e.

$$\beta \equiv \frac{V}{V_o} = \frac{R_a}{R_a + R_b} \qquad (3.54)$$

Also let A be the amplifier gain on open loop (i.e. without feedback). Then,

$$V_o = A(V_i - V) = A(V_i - \beta V_o) \qquad (3.55)$$

Solving Eq. (3.55) for V_o and dividing by V_i gives the gain A' with feedback, i.e.

$$A' \equiv \frac{V_o}{V_i} = \frac{A}{1 + \beta A} \qquad (3.56)$$

It is seen that the gain with feedback A' depends of the quantity $1+\beta A$, which here is always greater than 1 for an ideal opamp, and therefore $A'<A$. Clearly, the feedback here is voltage negative feedback. However, as was stated in Section 3.3, A for a practical opamp is a function of frequency, which for some opamps at higher frequencies can make the quantity βA negative, and, depending on its value, it can make $|1+\beta A|<1$, when from Eq. (3.56) $A'>A$. In such a case the feedback will be positive.

A similar procedure followed for Fig. 3.5(a) will lead to similar results.

3.12 Effects of Feedback on the Amplifier Characteristics

Eq. (3.56) is general, and is valid for any physical system, in which there is presence of feedback. In the case of any amplifier, application of feedback may cause important effects on its characteristics. As was already shown above, the amplifier gain is affected by the application of feedback. Some other important effects are given here below.

a. Effect on the Gain Variations

Consider the case, in which the amplifier open-loop gain varies for some reasons, for example due to variations in the power supply voltage. Differentiating Eq. (3.56) w.r.t. A gives

$$\frac{dA'}{dA} = \frac{1}{(1+\beta A)^2} \qquad (3.57)$$

and on dividing both sides by A' there results

$$\frac{dA'}{A'} = \frac{1}{1+\beta A}\frac{dA}{A} \qquad (3.58)$$

Clearly, the percentage variation of A' can be lower or higher than that of A depending on whether the feedback is negative or positive respectively.

It is worth mentioning here that the ratio

$$S \equiv \frac{dA'/A'}{dA/A} \qquad (3.59)$$

is called the *sensitivity* of the system, and it is always an important parameter in the design of any feedback system.

b. Effect on the Amplifier Bandwidth

Consider that the dependence on the frequency of the amplifier gain is described by the following equation:

$$A = \frac{A_o}{1+j\dfrac{f}{f_c}} \qquad (3.60)$$

where A_o is the open-loop amplifier gain at frequencies below f_c, with f_c being the amplifier cut-off frequency, which at the same time defines its bandwidth. Substituting this value of A into Eq. (3.56) gives

Amplifiers

$$A' = \frac{A}{1+\beta A} = \frac{A'_o}{1+j\dfrac{f}{f'_c}} \qquad (3.61)$$

where

$$A'_o = \frac{A_o}{1+\beta A_o} \qquad (3.62)$$

and

$$f'_c = f_c(1+\beta A_o) \qquad (3.63)$$

It can be seen that, apart from the effect of the feedback on the low frequency gain, the bandwidth has been changed, being wider for negative and more narrow for positive feedback respectively.

c. Effects on the Amplifier Input and Output Impedances

Similarly the following can be shown, which are stated here without proof:

- Application of voltage or current negative feedback in series with the input increases the input impedance, while if it is applied in parallel, it reduces it. The opposite happens, if the feedback is positive.

- Application of voltage negative feedback in series or in parallel with the input decreases the amplifier output impedance, whereas current negative feedback applied in series or in parallel with the input increases the amplifier output impedance. The opposite effects are caused when the current feedback is positive.

d. Effect on the amplifier distortion

It can also be shown that application of negative voltage or current feedback decreases the amplifier non-linear distortion, while if the feedback is positive the distortion increases.

Because of its high importance the effect of feedback on the amplifier stability is examined separately in Section 3.14.

Example 3.M.4

Using MATLAB plot the amplifier gain A' of Eq. (3.61) against frequency f for the feedback ratio $\beta=0$, 0.5 and 5 in order to see the effects of feedback. Consider a cut-off frequency without feedback $f_c=10Hz$ and open-loop gain $A_o=100$ at frequencies below f_c.

Code

```
>> Ao=100;
>> fc=10;
>> b=[0 0.5 5];                    % feedback values
>> w=0:1:2e5*pi;                   % define frequency range (in rad/sec)
>> for i=1:3
A1(i)=Ao/(1+b(i)*Ao);
f1(i)=fc*(1+b(i)*Ao);
num=[A1(i)];
den=[1/(2*pi*(f1(i))) 1];
A(i,:)=freqs(num,den,w);
end
>> f=w/(2*pi);
>> Ab_0=20*log10(abs(A(1,:)));
>> Ab_05=20*log10(abs(A(2,:)));
>> Ab_5=20*log10(abs(A(3,:)));
>>  semilogx(f,Ab_0,f,Ab_05,f,Ab_5);axis([1  1e5  -40  45]);grid;xlabel('frequency (Hz)');  ylabel('gain  (dB)');title('frequency  response  for  different  feedback values');text(15,38,'b=0'); text(15,8,'b=0.5');text(15,-12,'b=5')
```

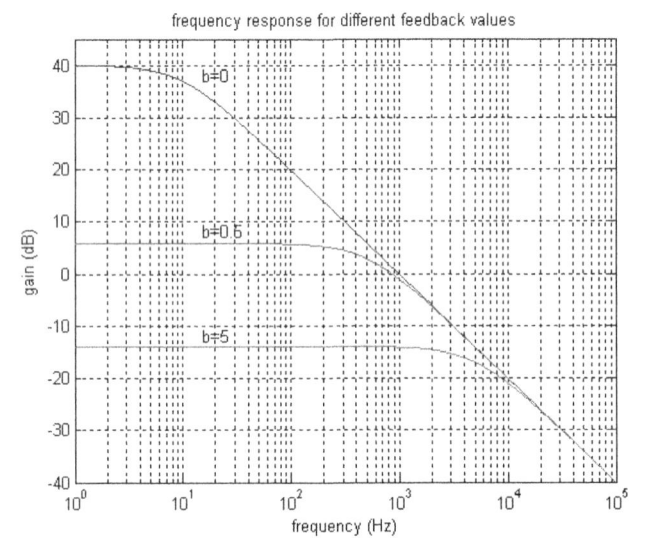

Figure 3.M.4. Amplifier gain against frequency for different feedback values

---

Example 3.M.5

Consider three amplifiers connected in cascade. Their individual gains are consistent with Eq. (3.60) with A_{o1}=100, A_{o2}=10, A_{o3}=2, f_{t1}=500Hz, f_{t2}=10kHz and f_{t3}=10Hz. Using MATLAB plot the total gain A of the cascade versus frequency as well as the individual amplifier gains versus frequency.

Code

```
>> Ao=[100 10 2];
>> fc=[500 10000 10];
>> w=0:1:2e5*pi;                        % define frequency range (in rad/sec)
>> f=w/(2*pi);
>> for i=1:3
num=[Ao(i)];
den=[1/(2*pi*(fc(i))) 1];
A(i,:)=freqs(num,den,w);
end
>> AdB1=20*log10(abs(A(1,:)));
>> AdB2=20*log10(abs(A(2,:)));
>> AdB3=20*log10(abs(A(3,:)));
>> subplot(3,1,1);semilogx(f,AdB1);axis([1  1e5  -20  45]);grid;xlabel('frequency (Hz)'); ylabel('gain (dB)');title('Ao1=100, fc1=500Hz');
>> subplot(3,1,2);semilogx(f,AdB2);axis([1  1e5  -20  45]);grid;xlabel('frequency (Hz)'); ylabel('gain (dB)');title('Ao2=10, fc2=10kHz');
>> subplot(3,1,3);semilogx(f,AdB3);axis([1  1e5  -20  45]);grid;xlabel('frequency (Hz)'); ylabel('gain (dB)');title('Ao3=2, fc3=10Hz');
>> num_total=[Ao(1)*Ao(2)*Ao(3)];
>> den3=(1/(2*pi*fc(1)))*(1/(2*pi*fc(2)))*(1/(2*pi*fc(3)));
>> den2=(1/(2*pi*fc(1)))*(1/(2*pi*fc(2)))+(1/(2*pi*fc(1)))*(1/(2*pi*fc(3)))+(1/(2*pi*fc(2)))*(1/(2*pi*fc(3)));
>> den1=(1/(2*pi*fc(1)))+(1/(2*pi*fc(2)))+(1/(2*pi*fc(3)));
>> den_total=[den3 den2 den1 1];
>> A_total=freqs(num_total,den_total,w);
>> AdB_total=20*log10(abs(A_total));
>> figure
>> semilogx(f,AdB_total);axis([1  1e5  -40  70]);grid;xlabel('frequency (Hz)'); ylabel('gain (dB)');
```

Figure 3.M.5. Amplifier gain versus frequency for each amplifier

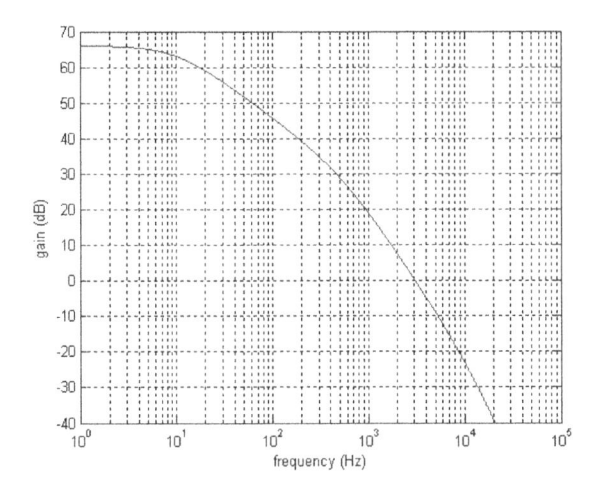

Figure 3.M.6. Amplifier gain versus frequency for the overall circuit

3.13 Amplifiers with Negative Input Capacitance

Any practical amplifier has an input impedance between its input terminals. This impedance is considered, in its simpler form, to consist of a resistance in parallel with a capacitance. The presence of the capacitance tends to reduce the high frequency performance of the amplifier. If the output impedance of the preceding stage has also a shunt capacitance the problem at high frequencies will be worse. In this case any signal pulse will be distorted.

To improve the situation one can use an amplifier with negative input capacitance, which will reduce the total capacitance between the output of the preceding the amplifier stage and the input of the amplifier. Care should be taken though that the size of the negative capacitance is not higher than a critical value, depending on the situation, because otherwise the whole circuit will become unstable, i.e. it will oscillate at frequencies different from those of the signal.

In practice, one way of obtaining negative capacitance at the input of an amplifier is as shown in Fig. 3.18.

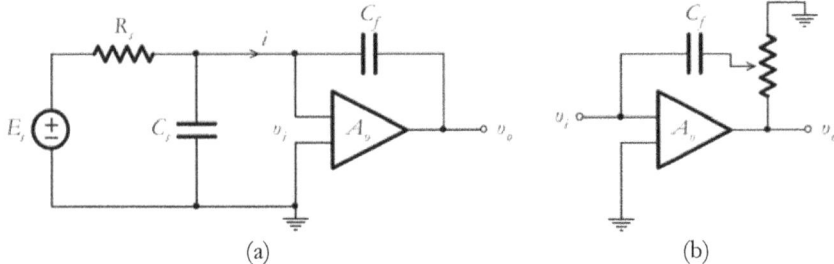

Figure 3.18. Arrangement for a negative capacitance appearing at the amplifier input

In Fig. 3.18(a) a capacitance C_f is connected between the input and the output terminals of the voltage amplifier of gain A_v. C_s represents the total capacitance from the output of the preceding the amplifier stage, including any wiring capacitance as well as the input capacitance of the amplifier itself. Let v_i be the voltage developed at the amplifier input, which being a voltage amplifier has an infinite input resistance. With no current flowing into the amplifier we may write, that the current i_f in C_f is

$$i_f = -C_f \frac{d(v_o - v_i)}{dt} \qquad (3.64)$$

Since $v_o = A_v v_i$, on substituting in Eq. (3.64) gives

$$i_f = +C_f \frac{d(v_i - A_v v_i)}{dt} = +(1 - A_v) C_f \frac{dv_i}{dt} \qquad (3.65)$$

Then integrating Eq. (3.65) gives

$$v_i = \frac{1}{(1 - A_v) C_f} \int i_f \, dt \qquad (3.66)$$

This can be interpreted as if the capacitance $C_f(1-A_v)$ appears at the amplifier input terminals. If $A_v > 1$, this capacitance is negative and can be subtracted from any capacitance C_s. Choosing the right C_f one can obtain

$$C_s + C_f (1 - A_v) = 0 \qquad (3.67)$$

obtained when

$$C_f = \frac{C_s}{A_v - 1} \qquad (3.68)$$

In practice, the neutralization of any capacitance C_s can be achieved by using only one value of C_f and adjusting the value of A_v in Eq. (3.68) by obtaining part only of the voltage v_o, as shown in Fig. 3.18(b). In this case the value of v_o in Eq. (3.64) will be changed to av_o, where a, the potentiometer setting, takes any required value between zero and one.

3.14 Stability

The effects of feedback on the amplifier characteristics were reviewed in previous sections using Eq. (3.56). One point though that has not been examined so far is the case when

$$1 + \beta A = 0 \qquad (3.69)$$

It can be seen that, under this condition, Eq. (3.56) results in A' becoming infinite. But such a condition would mean that the output voltage or current should be infinite, something that is unattainable in practice, because it requires infinite power from the power supply, which is not practically possible. So, what does this imply? A relatively simple answer is the following: Consider Eq. (3.56) written as follows:

$$V_o = \frac{A}{1+\beta A} V_i \qquad (3.70)$$

In the physical system, for V_o to be finite, when Eq. (3.69) holds, requires that V_i should be zero. This means that there is an output signal without an input signal. In this case the amplifier or the system in general, has become a generator otherwise called an *oscillator* (if the output voltage is not *dc*). Clearly this is an unstable system and the amplifier cannot be used to amplify any signal. For instability it is required from Eq. (3.69) that

$$\beta A = -1 \qquad (3.71)$$

which can be analyzed to

$$|\beta A| = 1 \quad \text{and} \quad \arg \beta A = 180° \qquad (3.72)$$

meaning that the magnitude of βA should be 1 and its phase 180°.

So, for a system with feedback to be stable, Eqs. (3.72) should not be both satisfied at any frequency. Therefore, the frequency response of βA i.e. both magnitude and phase responses of βA, should be shaped properly so that Eqs. (3.72) are not both satisfied at any frequency.

To have a measure of how stable is an amplifier the terms gain and phase margins are used. The *gain margin* is defined by how much $|\beta A|$ should increase until it becomes 1, when $\arg \beta A = 180°$. The *phase margin* is defined by how much the phase of βA should be changed to become 180°, when $|\beta A| = 1$. A useful stable practical amplifier should have a gain margin larger than 9dB or a phase margin at least 45°.

3.15 An *RC* Oscillator

The oscillator is an unstable electronic circuit. If its output signal is sinusoidal, it is called a *harmonic oscillator*. On the other hand, the output signal may contain a large number of harmonics, when it is called a *relaxation oscillator*. The oscillator gets power from its power supply and turns it to a periodic signal. For educational purposes one can distinguish the oscillators as oscillators using feedback and as negative resistance oscillators. The latter make use of elements like tunnel diodes or unijunction transistors whose part of their *I-V* characteristic presents negative slope. These oscillators are not of great interest nowadays and will not be considered any further here.

The operation of feedback oscillators is based on the condition described by Eq. (3.69). Usually they are built using an amplifier and a phase shifting network, which can consist of resistors and capacitors, RC oscillators, or of capacitors and inductors, LC oscillators. In physical terms the generation of oscillations can be explained as follows: If the condition (3.69) is satisfied, the gain of the amplifier increases tending to infinity and so does the output voltage of the amplifier. However, as the latter increases, the amplifier enters the region of its nonlinear characteristic, as can be seen for example from Fig. 3.11. In this case the gain decreases to such a value that the condition (3.69) is not satisfied any more. Then the output voltage becomes lower, the amplifier operation returns to the linear part of its I-V characteristic. Condition (3.69) will be satisfied again and the cycle will be repeated again and again. It is clear from this discussion that, because of the amplifier operation entering the nonlinear part of its transfer characteristic, the obtained periodic waveform will contain a lot of harmonics. However, these can be removed by connecting a suitable filter at the oscillator output.

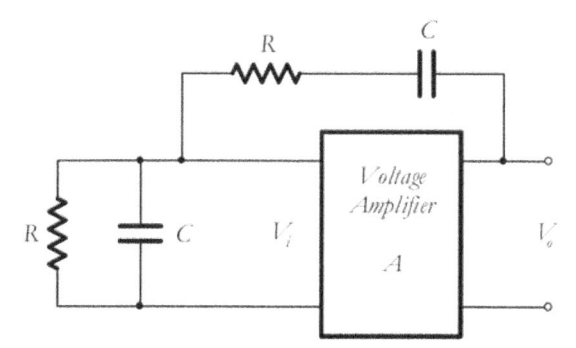

Figure 3.19. The Wien-Bridge oscillator

As an example, the Wien-Bridge oscillator is presented here, shown in Fig. 3.19. This is an RC oscillator as can be easily seen. Considering that the amplifier gain is A, analysis of the circuit gives that the condition (3.69) is satisfied if

$$\omega^2 C^2 R^2 = 1 \tag{3.73a}$$

and

$$\omega CR(3-A) = 0 \tag{3.73b}$$

Condition (3.73a) gives the frequency of oscillations, while condition (3.73b) gives the required gain from the amplifier, which here is 3. Such an amplifier can be easily obtained using an opamp.

In general RC oscillators using discrete components are suitable for operations at low frequencies. However, contemporary integrated circuit technology can push the oscillating frequencies to higher values. On the other hand, LC oscillators are more suitable for operation at high frequencies.

One problem with the oscillators using discrete components can be the stability of the oscillating frequency with time. If this is intolerable, a piezoelectric crystal can be used replacing part of the LC circuit that produces the required feedback signal. High frequency stability, as can be understood, is required, for example, in broadcasting radio or television.

One more point. In practice, starting the oscillations requires some type of excitation. This can be assured by making the required gain a little higher, about 5% more, than the critical value. In fact the application of the power supply or noise can trigger the circuit. Thus the output signal will get increasing from the initially zero value until the amplitude of the oscillations becomes constant due to the nonlinear characteristic of the amplifier.

References

[1]. Deliyannis T., Sun Y. & Fidler J. K., *Continuous-Time Active Filter Design*, CRC Press, 1999.
[2]. Millman J. & Grabel A., *Microelectronics*, 2nd Edition, McGraw-Hill, 1987.
[3]. Jackson A. S., *Analog Computation*, McGraw-Hill, New York, 1960.
[4]. Tobey G. E., Graeme J. G. & Huelsman L. P., *Operational Amplifiers – Design and Applications*, McGraw-Hill, New York, 1971.
[5]. Grabowski B., *Microélectronique Analogique*, Masson et Cie, Paris, 1971.
[6]. Moschytz G. S., *Linear Integrated Networks: Fundamentals*, Van Nostrand-Reinhold, New York, 1974.
[7]. Various Manufacturers, Specifications Books on Operational Amplifiers.
[8]. Storey N., *Electronics – A Systems Approach*, Addison-Wesley, 1992 (reprinted 1995).

MATLAB Problems

❖ *The reader is advised first to consult the MATLAB tutorial in Appendix D*

3.M.1. Using MATLAB find the current gain A_i and the power gain A_p for the circuit of Fig. E.3.1 of Example 3.1 in *dB*s, if the input resistance R_i is $R_i = 100k\Omega$.

3.M.2. Consider the inverting amplifier of Fig. 3.5(a) with $R_1 = 10k\Omega$, $R_f = 100k\Omega$ and open-loop gain A. Using MATLAB plot the closed-loop voltage gain G_l against the open-loop gain for $A = 10^1$ up to 10^6.

Amplifiers

3.M.3. Using MATLAB find in *dB*s the common mode gain A_{CM}, the differential gain A_d and the common-mode rejection ratio *CMRR* of the differential amplifier of Fig. 3.12 if $R_1=5k\Omega$, $R_2=2k\Omega$, $R_3=100k\Omega$ and $R_4=200k\Omega$.

3.M.4. Using MATLAB find in *dB*s the differential gain A_d of the instrumentation amplifier of Fig. 3.13, if $R_1=5k\Omega$, $R_2=10k\Omega$, $R_3=50k\Omega$, $R_4=100k\Omega$, $R_a=R_b=10k\Omega$ and $R_c=20k\Omega$.

3.M.5. Repeat example 3.M.5 for four amplifiers connected in cascade with $A_{o1}=50$, $A_{o2}=5$, $A_{o3}=1$, $A_{o4}=20$, $f_{c1}=200Hz$, $f_{c2}=2kHz$, $f_{c3}=100Hz$ and $f_{c4}=10kHz$.

Problems

3.1. The opamp in the circuit in Fig. P.3.1 has an open-loop gain of 5×10^4, input resistance $20k\Omega$ and output resistance 100Ω. Draw the equivalent circuit and determine the value of V_o when $V_s=1mV$.

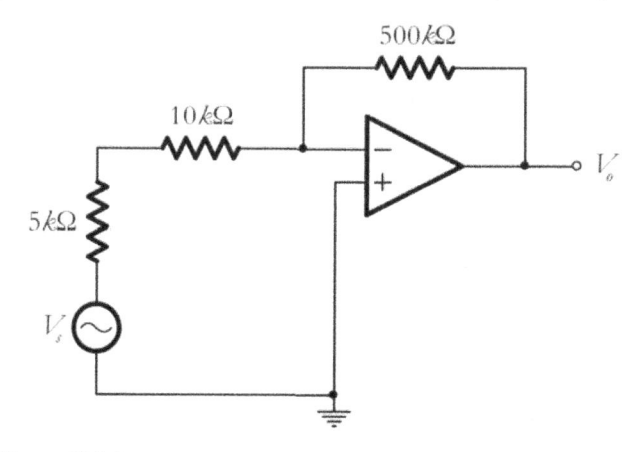

Figure P.3.1.

(Answers: $-33.3mV$)

3.2. The opamp in the circuit in Fig. P.3.2 has the following characteristics: open-loop voltage gain 10^3, input resistance (between its two inputs) $150k\Omega$ and output resistance 100Ω. Draw the equivalent circuit and calculate:

a. The voltage ratio V_o/V_s for $R_s=10k\Omega$
b. The equivalent resistance between the non-inverting input and ground and
c. The output resistance for $R_s=10k\Omega$ and for $R_s=0$

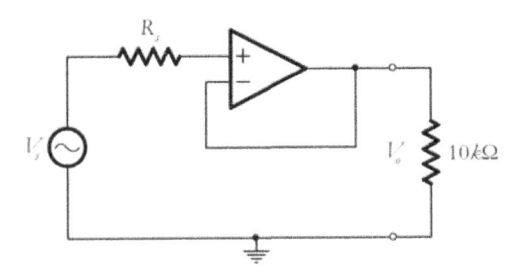

Figure P.3.2.

(Answers: a. 0.9989, b. 145.5$k\Omega$, c. 0.107Ω, 0.0999Ω)

3.3. The opamp in Fig. P.3.3 has the following characteristics: open-loop gain at low frequencies 10^5, input resistance (between its two inputs) 150$k\Omega$ and zero output resistance. Determine the input resistance between the non-inverting input and ground and the gain V_o/V_i of the circuit.

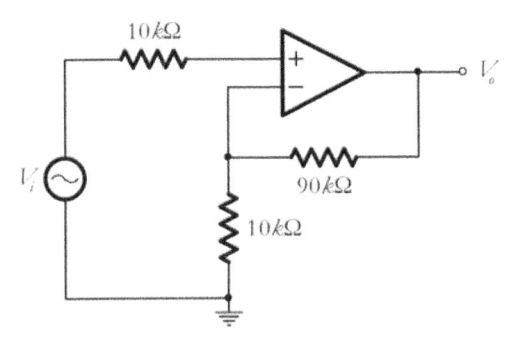

Figure P.3.3.

(Answers: 1.69×$10^6 k\Omega$, 10)

3.4. The amplifier in the circuit of Fig. P.3.4 has infinite input resistance, zero output resistance and gain −100. At $t=0$ the step voltage of 1V is applied at the input. Calculate 5s later the percent error of $v_o(t)$ from the corresponding value when the gain of the amplifier is -infinite.

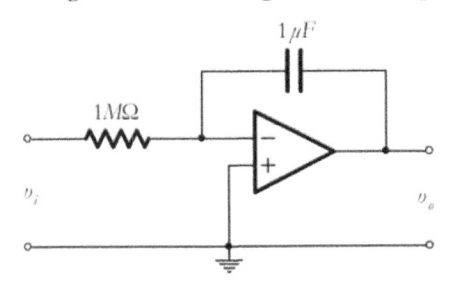

Figure P.3.4.

(Answers: 4%)

Amplifiers

3.5. In the integrator circuit in Fig. P.3.4, $R=100k\Omega$ and $C=0.5\mu F$, while the opamp is ideal. At $t=0$ a step voltage of $1V$ is applied.

  a. Draw the plot of the output voltage against time
  b. If the opamp saturates at $10V$, otherwise being ideal, determine the time it will take to reach saturation after the application of the input voltage

(Answers: b. 0.5s)

3.6. The opamp used in the inverting and non-inverting mode in Fig. 3.5 has a *dc* open-loop gain 10^5 and a cut-off frequency at $10Hz$. If $R_1=R_a=10k\Omega$ and $R_f=R_b=100k\Omega$ determine:

  a. The gain of each amplifier in Fig. 3.5 at *dc*
  b. The gain (magnitude and phase) of each amplifier in Fig. 3.5 at $1kHz$ considering that beyond cut-off the voltage gain of the opamp drops at the rate of $6dB/octave$ (Hint: Consider $A\approx 10^5/j\omega$ for $f>>10Hz$)
  c. The frequency of each circuit at which the magnitude of the gain becomes unity

(Answers: for inverting: a. -10, b. 8.23, $-34.6°$, c. $158kHz$,
for non-inverting: a. 11, b. 9.05, $-34.6°$, c. $174kHz$)

3.7. In the IA in Fig. 3.13 the opamps are ideal. If $R_1=R_2=R_3=R_4=40k\Omega$ and $R_a=R_b=25k\Omega$ while $R_i=10k\Omega$ determine:

  a. The differential gain A_d
  b. The common-mode gain A_{CM} and
  c. The CMRR

(Answers: a. 12, b. 10, c. ∞)

3.8. The OTAs in the following figure are ideal with $g_{m1}=g_{m2}=2mS$. Determine the resistance between terminals 1 and 2.

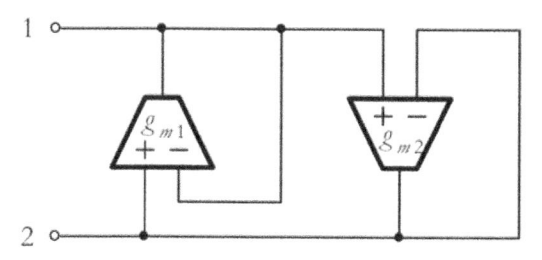

Figure P.3.5.

(Answers: 500Ω)

3.9. Determine the voltage gain V_o/V_1 for the two circuits below:

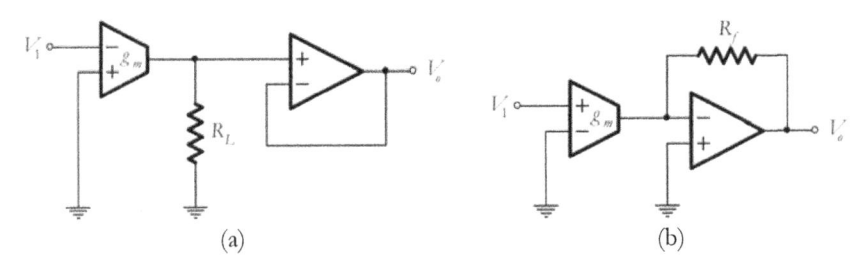

Figure P.3.6.

(Answers: a. $-g_m R_L$, b. $-g_m R_f$)

3.10. Derive the condition for the circuit in Fig. 3.19 to produce harmonic oscillations (sinusoidal) and the frequency of oscillations, when $R=10k\Omega$, $C=1nF$.

(Answers: 3, $1 Mrad/s$)

Chapter 4

Analog Filters

4.1 Introduction

An accepted definition of an analog filter is that, it is an electric/electronic circuit that has a predetermined response for a given input signal (excitation). There are various types of electronic filters and these can fall into two categories: *analog* and *digital*. Analog filters can be either *continuous-time* or *discrete-time* (sampled-data, switched-capacitor). In this Chapter we will concentrate on the continuous-time analog filters and in particular the *RC active filters*, as these can be easier understood by the users of this book, and easily designed and built in the laboratory. Digital filters will be dealt with in Chapter 7. We start by briefly explaining the general filter characteristics and proceed with the introduction of low-order *RC* active filters. The active element employed is the *operational amplifier*. The design of higher-order *RC* active filters is briefly discussed and demonstrated by means of two examples. The chapter finishes with MATLAB exercises and problems.

4.2 Filter Characteristics

Filters are usually characterized by their *frequency response* and therefore, are prescribed by means of their *passband(s)* and *stopband(s)*. Thus, they can be *low-pass* (LP), *high-pass* (HP), *band-pass* (BP), *band-reject* (BR) and *all-pass* (AP), depending on which frequency band is allowed to pass through them without any distortion of the amplitude of the signals at the corresponding frequencies. The *amplitude responses* of the ideal filters are shown in Fig. 4.1 (bold lines). Practical filters, for causality reasons, cannot possess the ideal responses but are similar to those shown with solid lines in the same figure. Their passband is defined by two frequencies ω_{pb} and ω_{pb} the upper and low *cut-off frequencies*, respectively, which denote the frequencies at which a certain passband error in *dB* is acceptable. Such an error can be for example 0.1, 0.5, 1 or 3dB. This is usually called the *maximum error in the passband* denoted as A_p. Similarly, a *minimum stopband attenuation* A_s is tolerated starting at frequencies ω_{sl} and ω_{sh} depending on the type of filter. Thus a low-pass filter has $\omega_{pl}=0$ and no ω_{sl}, while in a high-pass filter $\omega_{ph}=\infty$. Between the passband and the stopband the amplitude response of the filter is falling-off at a certain rate measured in *dB per octave* or *dB per decade* with the corresponding frequency band defined as *transition band*.

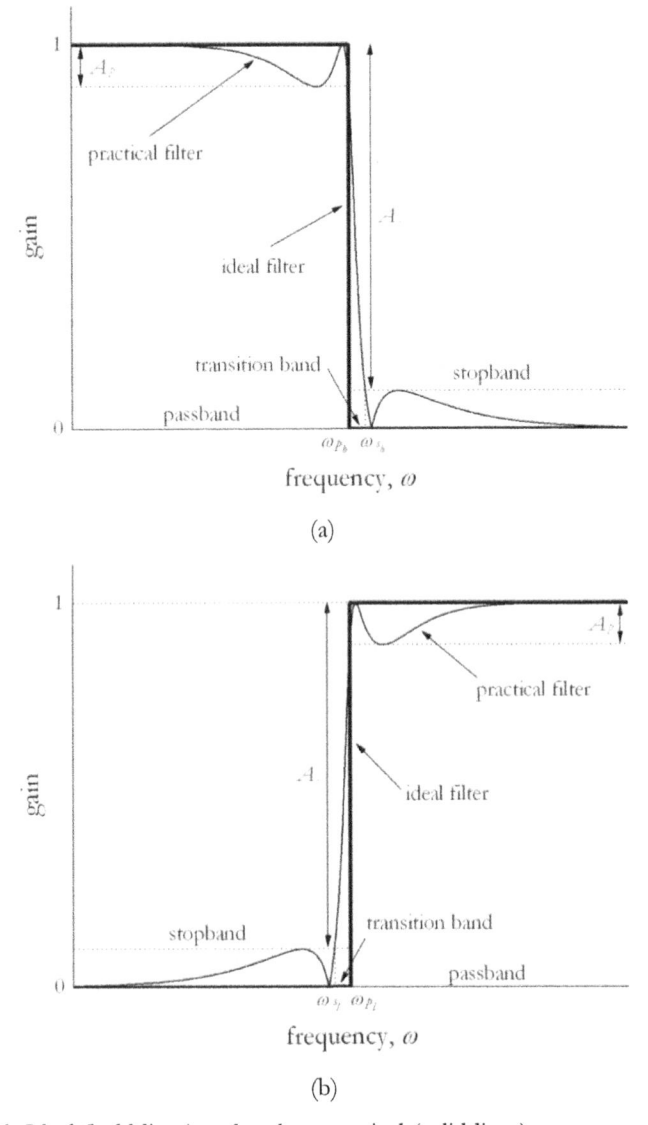

Figure 4.1. Ideal (bold lines) and real or practical (solid lines) responses of (a) low-pass filter, (b) high-pass filter

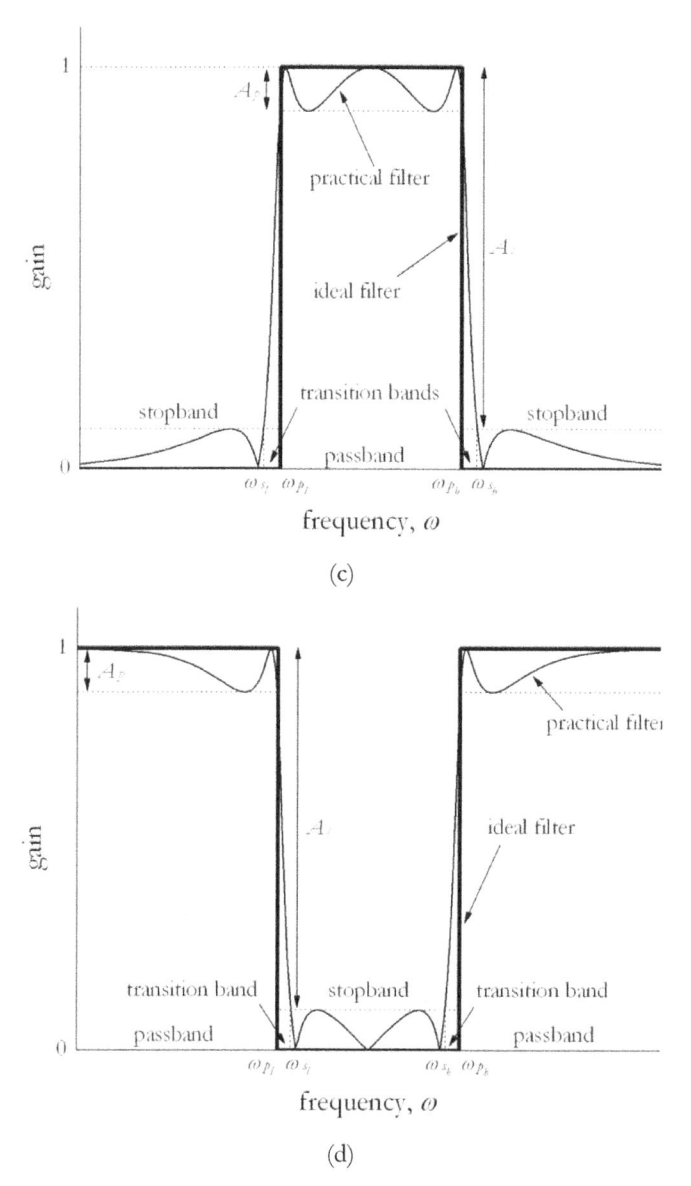

Figure 4.1. (Cont'd) Ideal (bold lines) and real or practical (solid lines) responses of (c) band-pass filter, (d) band-reject filter

Associated with the amplitude response of a filter is its *phase response*, which in some cases (e.g. transfer of rectangular pulses) has to be linear. *Linear phase filters* can be low-pass (Bessel-Thomson) or all-pass. In the latter the amplitude response is constant, is usually of unity amplitude and their phase linear up to a certain frequency. Linear phase filters, whether low-pass or all-pass, delay the signal, the frequency band of which falls within their linear phase range, by a certain amount of time according to the design specifications. Ideally, it would be desirable each filter to possess a linear phase characteristic within its passband, but in many cases (e.g. microphone signals) this is not necessary.

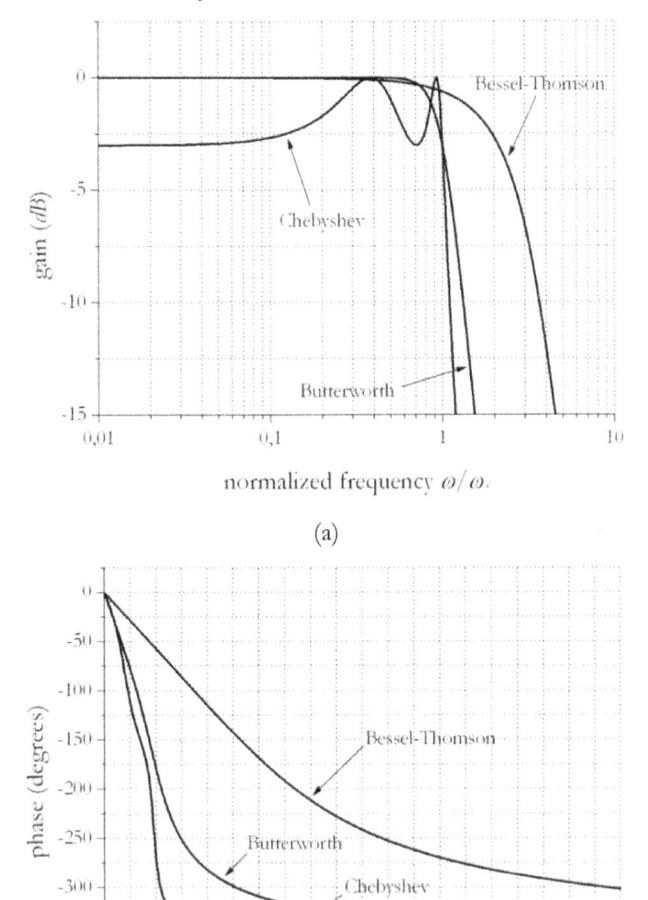

Figure 4.2. (a) Amplitude and (b) phase response of the fourth-order Butterworth, Chebyshev (3 dB ripple) and Bessel-Thomson low-pass filter

Mathematically, each filter is represented by its *transfer function*, which is a function of the complex frequency *s*. This is rational, i.e. the ratio of two real and finite polynomials. The roots of the numerator polynomial are the *zeros* and those of the denominator polynomial are the *poles* of the transfer function. For stability reasons, the degree of the numerator polynomial should not exceed the degree of the denominator. The highest power of s in the transfer function defines the *order* of the filter. Low-pass analog filter functions have been tabulated and are available for various orders and passband errors under names like *Butterworth*, *Chebyshev*, *Inverse Chebyshev*, *Elliptic* and *Bessel-Thomson*. In Fig. 4.2 the amplitude and phase responses of the fourth-order Butterworth, Chebyshev and Bessel-Thomson are shown for comparison. From these low-pass filter functions one can obtain the corresponding high-pass, band-pass or band-reject filter functions by applying certain transformations of the complex frequency variable s, as given in Table 4.1.

Table 4.1. Frequency transformations

LP to LP	LP to HP	LP to BP	LP to BR
$s_n \to \dfrac{s}{\omega_p}$	$s_n \to \dfrac{\omega_p}{s}$	$s_n \to \dfrac{\omega_o}{B}\left(\dfrac{s}{\omega_o} + \dfrac{\omega_o}{s}\right)$	$s_n \to \dfrac{1}{\dfrac{\omega_o}{B}\left(\dfrac{s}{\omega_o} + \dfrac{\omega_o}{s}\right)}$

In Table 4.1, s_n is the normalized frequency variable in the tabulated filter functions, in which the cut-off frequency is 1, ω_p the required cut-off frequency of the *denormalized* LP or HP filter, ω_o the center frequency of the denormalized BP or BR and B their bandwidth.

4.3 Passive and Active *RC* Filters

For high selectivity, passive filters must be built out of inductors and capacitors. However, at the low frequencies, which characterize for example the biomedical signals, the required inductors are bulky, heavy and non-ideal (nonlinear) and as such cannot be included in an integrated filter in microelectronic form. For this reason, filters for low frequency signals employ resistors, capacitors and active elements. Most suitable active element at these frequencies has been proved to be the *operational amplifier* (*opamp*), as described in Chapter 3. So, in the rest of this and subsequent chapters continuous analog filters will be built out of resistors, capacitors and opamps. Resulting filters will be termed *RC* active filters, and we will refer to them in the rest of this chapter as *active filters*.

4.4 First-Order Active *RC* Filters

The circuit shown in Fig. 4.3 has the following transfer function

$$F(s) = \frac{V_o}{V_i} = \frac{\dfrac{1}{RC}}{s + \dfrac{1}{RC}} = \frac{\omega_c}{s + \omega_c} \qquad (4.1)$$

where $\omega_c = 1/RC$.

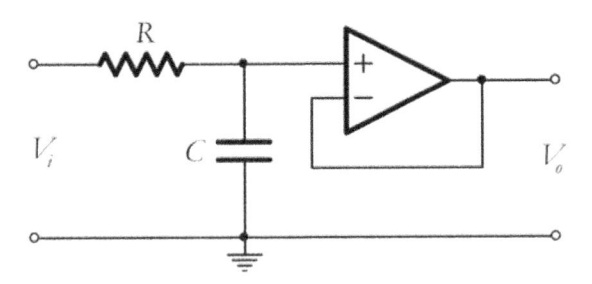

Figure 4.3. First-order low-pass active RC filter

Evidently, this is a first-order low-pass filter function with ω_c its 3dB cut-off frequency ω_{pl}. Here the use of the opamp is to provide the circuit with zero output impedance and does not affect the filter frequency response. Clearly, if the circuit following the filter has a very high input resistance (nominally infinite) the opamp can be eliminated.

It can be easily shown that, if the passive elements R and C exchange positions the resulting circuit will be high-pass having the following transfer function:

$$F(s) = \frac{V_o}{V_i} = \frac{s}{s + \frac{1}{RC}} = \frac{s}{s + \omega_c} \qquad (4.2)$$

Again $\omega_c = 1/RC$ is the 3dB cut-off frequency ω_{pl} of the high-pass filter.

As an example, consider the case of designing a low-pass filter of first-order with a cut-off frequency at 2Hz. Using Eq. (4.1) we will have

$$\omega_c = 2\pi f_c = \frac{1}{RC} \Rightarrow \frac{1}{RC} = 2\pi \cdot 2 = 4\pi$$

or

$$RC = \frac{1}{4\pi}$$

Since there is not another equation relating the components R and C, we may select a convenient value for one of these components and solve the above equation for the other. Such a value could be, for example, $C = 1\mu F$ when

$$R = \frac{1}{\omega_c C} \Rightarrow R = \frac{1}{4\pi \cdot 10^{-6}} = \frac{10^6}{4\pi} \Rightarrow R = 79.6 k\Omega$$

Notice that there are no first-order band-pass or band-reject filters.

Analog Filters 113

Example 4.M.1

Using MATLAB draw the amplitude and phase response of the first order low-pass filter (Eq. (4.1)) for $f_c = 2Hz$.

Code

$$F(s) = \frac{V_o}{V_i} = \frac{\frac{1}{RC}}{s + \frac{1}{RC}} = \frac{\omega_c}{s + \omega_c}$$

$$|F(j\omega)| = \frac{1}{\sqrt{\left(\frac{\omega}{\omega_c}\right)^2 + 1}}$$

$$\arg\{F(j\omega)\} = -\tan^{-1}\left(\frac{\omega}{\omega_c}\right)$$

```
>> w=0:0.01:20*pi;                  % define frequency range (in rad/sec)
```

% we present four methods to find the gain and phase (in degrees) of the transfer function

1st Method
```
>> mag=1./sqrt((w./(4*pi)).^2+1);
>> phase=-atan(w/(4*pi));
>> phase=phase*180/pi;
```

2nd Method
```
>> s=j*w;
>> h=4*pi./(s+4*pi);
>> mag=abs(h);
>> phase=angle(h);
>> phase=phase*180/pi;
```

3rd Method
```
>> num=4*pi;
>> den=[1 4*pi];
>> h=freqs(num,den,w);
>> mag=abs(h);
>> phase=angle(h);
>> phase=phase*180/pi;
```

4th Method
```
>> num=4*pi;
>> den=[1 4*pi];
>> [mag,phase]=bode(num,den,w);
```

```
>> magdb=20*log10(mag);             % convert gain to dBs
>> f=w/(2*pi);                      % convert frequency to Hz
```

% the plots below are derived with the following procedure (the reader is advised first to consult the MATLAB tutorial in Appendix D)

```
>> subplot(2,1,1)
>> semilogx(f,magdb)
>> axis([0.01 10 -10 1])
>> grid
>> xlabel('frequency (Hz)')
>> ylabel('gain (dB)')
>> title('frequency response of first order low-pass filter')
>> text(2,-3,'(2Hz,-3dB)')
```

```
>> subplot(2,1,2);semilogx(f,phase);axis([0 10 -80 10]);grid;xlabel('frequency (Hz)');
ylabel('phase (degrees)')
```

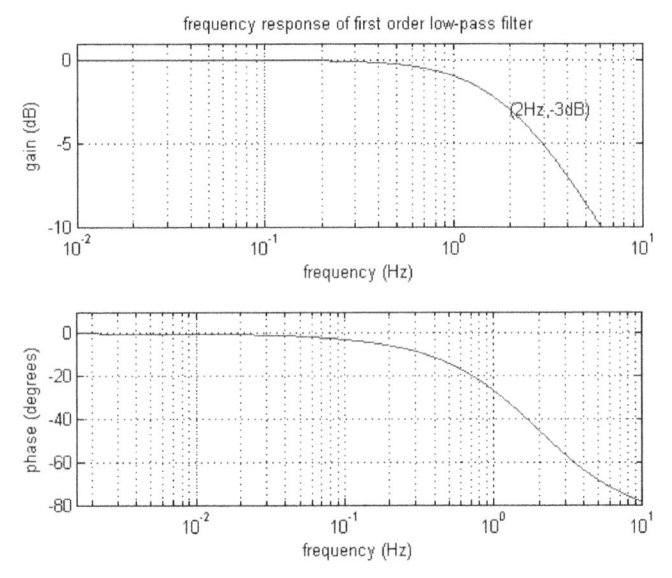

Figure 4.M.1. Frequency response of first order low-pass filter for $f_c=2Hz$.

Example 4.M.2

Repeat Example 4.M.1 for the first order high-pass filter (Eq. (4.2)) for $f_c=2Hz$.

Code

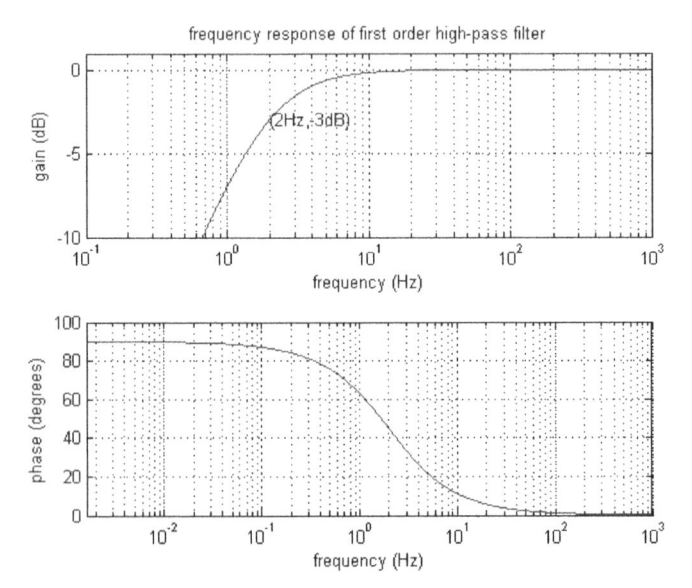

Figure 4.M.2. Frequency response of first order high-pass filter for $f_c=2Hz$.

```
>> w=0:0.01:2000*pi;
>> num=[1 0];
>> den=[1 4*pi];
>> [mag,phase]=bode(num,den,w);
>> magdb=20*log10(mag);
>> f=w/(2*pi);
>> subplot(2,1,1);semilogx(f,magdb);axis([0.1 1000 -10 1]);grid;xlabel('frequency (Hz)'); ylabel('gain (dB)');title('frequency response of first order high-pass filter');text(2,-3,'(2Hz,-3dB)')
>> subplot(2,1,2);semilogx(f,phase);axis([0 1000 0 100]);grid;xlabel('frequency (Hz)'); ylabel('phase (degrees)')
```

4.5 Second-Order Active *RC* Filters

There is an abundance of *RC* active circuits with second-order transfer functions. It has become customary to call such circuits *biquads*, since, in general, their transfer function is biquadratic. Apart from passive elements, these biquad circuits may employ one, two, three and even four opamps. In this Chapter we will give examples of biquad circuits employing one opamp, which subsequently will be called *SAB*s (from *Single Amplifier Biquads*), and a circuit employing three opamps, by which all useful second-order filter functions i.e. low-pass, high-pass, band-pass, band-reject (or *band-stop* or *notch*) and all-pass can be realized.

4.5.1 A Low-Pass *SAB*

Among the various *RC* active low-pass filter circuits, we select the Sallen and Key type, which happens to be the most popular one. This is shown in Fig. 4.4(a), where the voltage amplifier of gain G is realized using an opamp (see Chapter 3), as shown in Fig. 4.4(b).

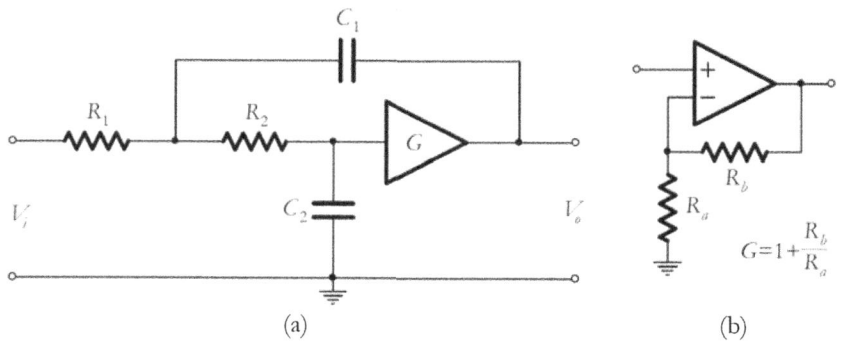

Figure 4.4. The Sallen-Key second-order low-pass filter

The transfer function of this circuit obtained using straightforward analysis is as follows:

$$F(s) = \frac{V_o}{V_i} = \frac{\dfrac{G}{R_1 R_2 C_1 C_2}}{s^2 + \left(\dfrac{1}{R_1 C_1} + \dfrac{1}{R_2 C_1} + \dfrac{1-G}{R_2 C_2}\right)s + \dfrac{1}{R_1 R_2 C_1 C_2}} \quad (4.3)$$

Supposing that the transfer function of the required filter is the following

$$F(s) = \frac{K}{s^2 + \beta s + \gamma} \quad (4.4)$$

where coefficients K, β and γ are known from the filter specifications, in order to determine the component values we match the coefficients of equal powers of s in Eqs. (4.3) and (4.4). Thus we will have

$$\frac{1}{R_1 C_1} + \frac{1}{R_2 C_1} + \frac{1-G}{R_2 C_2} = \beta \quad (4.5a)$$

$$\frac{1}{R_1 R_2 C_1 C_2} = \gamma \quad (4.5b)$$

$$\frac{G}{R_1 R_2 C_1 C_2} = K \quad (4.5c)$$

Clearly, the number of equations is smaller than the number of unknowns and some of the component values will have to be selected "arbitrarily". Also Eq. (4.5c) may not be satisfied exactly, since the value of K affects only the gain of the filter and not the shape of its response, which is of the main importance in any filter design. Usually, K is selected such that the gain of the filter at $\omega=0$ will be equal to unity. Following these, one may select

$$G = 1$$
$$R_1 = R_2 = R \quad (4.6)$$

Then, Eqs. (4.5) give

$$\frac{2}{RC_1} = \beta$$

$$\frac{1}{R^2 C_1 C_2} = \gamma$$

or

$$C_1 = \frac{2}{\beta R} \quad (4.7a)$$

$$C_2 = \frac{1}{\gamma R^2 C_1} = \frac{\beta}{2\gamma R} \tag{4.7b}$$

Next, we select any convenient value for R and from Eqs. (4.7) we calculate the values of C_1 and C_2. Finally, from Eq. (4.5c) the resulting value of K is equal to γ, as it is expected for unity gain of the filter at $\omega=0$.

Example 4.1

As an example consider the design of a low-pass filter with a cut-off frequency at $100 rad/s$, the frequency response of which is compatible with that of a second-order Butterworth filter.

Solution

The transfer function of the Butterworth filter is in normalized form the following:

$$F(s) = \frac{1}{s^2 + \sqrt{2}s + 1} \tag{4.8}$$

The cut-off frequency of this is $\omega_c = 1 rad/s$ and the gain at $\omega=0$ is also 1. In order to bring this F(s) in the form with the cut-off frequency at $100 rad/s$, substitute s by $s/100$ in Eq. (4.8) and get

$$F'(s) = \frac{100^2}{s^2 + 100\sqrt{2}s + 100^2} \tag{4.9}$$

Now referring to Fig. 4.4(a), we select $R_1 = R_2 = R = 10 k\Omega$ and then, using Eqs. (4.7), we get

$$C_1 = \frac{2}{100\sqrt{2} \cdot 10^4} = 1.414 \cdot 10^{-6} \Rightarrow C_1 = 1.414 \mu F$$

$$C_2 = \frac{100\sqrt{2}}{2 \cdot 10^4 \cdot 10^4} = 0.707 \cdot 10^{-6} \Rightarrow C_2 = 0.707 \mu F$$

For the amplifier gain to be 1, we remove R_a in Fig. 4.4(b) ($R_a = \infty$) and let $R_f \approx 20 k\Omega$ (for better operation of the opamp by reducing the effect of input off-set currents).

Another approach to the design of the circuit, which is the usual one in order to avoid large numbers for the coefficients as in Eq. (4.9), is to calculate the "normalized" component values using the normalized transfer function in Eq. (4.8), and then denormalize them so that the cut-off frequency would be at $100 rad/s$. Proceeding then this way, we select

$$R_1 = R_2 = R = 1 \Omega$$

and obtain using Eqs. (4.7)

$$C_1 = \frac{2}{\sqrt{2}} = \sqrt{2} F$$

$$C_2 = \frac{\sqrt{2}}{2} = \frac{1}{\sqrt{2}} F$$

which for the required cut-off frequency are denormalized by dividing the capacitor values by 100. This gives

$$R_1 = R_2 = 1\Omega$$
$$C_1 = 1.414 \cdot 10^{-2} F$$
$$C_2 = 0.707 \cdot 10^{-2} F$$

These values are impractical for use with the opamp, in which the currents are of the order of few milliamperes. Therefore, we raise the impedance level of the components up to $10k\Omega$ by multiplying the resistances by 10^4 and dividing the capacitances by 10^4. Thus we get

$$R_1 = R_2 = 10k\Omega$$
$$C_1 = 1.414 \mu F$$
$$C_2 = 0.707 \mu F$$

which are the same as those obtained by the previous approach.

Note that in applying the *"frequency denormalization"*, we could have divided the resistances by the required cut-off frequency and leave the capacitances intact, but this would have led to rather awkward component values. Usually, we divide the capacitance values to avoid having bulky capacitors in the actual circuit.

Example 4.M.3

Using MATLAB (a) find the transfer function of the second-order normalized Butterworth low-pass filter (Eq. (4.8)) in Example 4.1, (b) find the transfer function of the denormalized low-pass filter (Eq. (4.9)) for $\omega_c=100 rad/s$ and (c) draw the amplitude and phase response of the denormalized filter.

Code

(a)   `>> [z,p,k]=buttap(2);`   % create zeros, poles and gain of second order
                                 % normalized Butterworth low-pass filter
`>> [num,den]=zp2tf(z,p,k);`   % convert zero-pole to transfer function
`>> tf(num,den)`   % display the transfer function

```
Transfer function:
        1
  ---------------
  s^2 + 1.414 s + 1
```

(b)   `>> [num,den]=butter(2,100,'s');`   % second order Butterworth
                                           % low-pass filter with $\omega_c=100 rad/sec$
`>> tf(num,den)`   % display the transfer function

Analog Filters 119

```
Transfer function:
      1e004
-----------------------
s^2 + 141.4 s + 1e004
```

The same result can be derived as:

```
>> [z,p,k]=buttap(2);
>> [num,den]=zp2tf(z,p,k);          % normalized second order
                                    % Butterworth low-pass filter
>> [numt,dent]=lp2lp(num,den,100);  % transform the normalized low-pass
                                    % to a low-pass with ωc=100rad/sec
>> tf(numt,dent)                    % display the denormalized filter
```

(c)
```
>> w=0:0.01:1000;
>> num=10000;
>> den=[1 141.4 10000];
>> [mag,phase]=bode(num,den,w);
>> magdb=20*log10(mag);
>> subplot(2,1,1);semilogx(w,magdb);axis([0.1 1000 -20 2]);grid;xlabel('frequency (rad/sec)');ylabel('gain (dB)');title('frequency response of second order low-pass Butterworth filter');text(100,-3,'(100rad/sec,-3dB)')
>> subplot(2,1,2);semilogx(w,phase);axis([0 1000 -200 20]);grid;xlabel('frequency (rad/sec)'); ylabel('phase (degrees)')
```

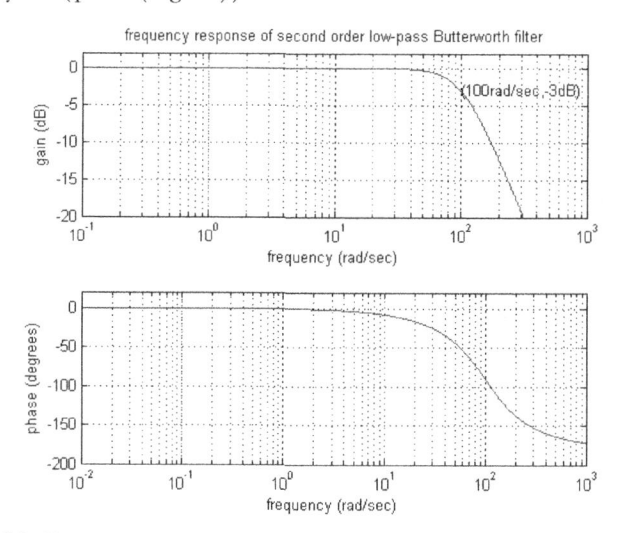

Figure 4.M.3. Frequency response of second order low-pass Butterworth filter for $\omega_c = 100 rad/s$

4.5.2 A High-Pass *SAB*

The low-pass circuit in Fig. 4.4(a) can be transformed to high-pass, if the positions of the passive components *R*'s and *C*'s are interchanged as shown in Fig. 4.5. By straightforward analysis, the transfer function of the circuit is found to be as follows:

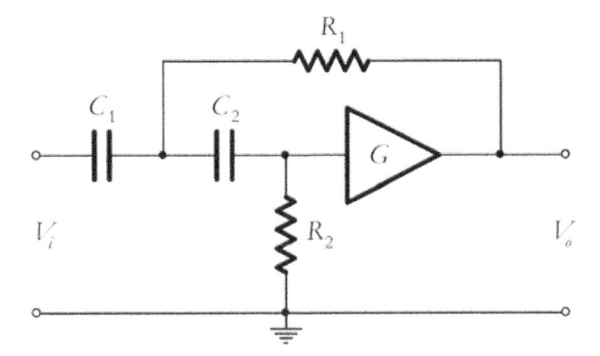

Figure 4.5. A high-pass circuit

$$F(s) = \frac{V_o}{V_i} = \frac{Gs^2}{s^2 + \left(\frac{1}{R_2C_1} + \frac{1}{R_2C_2} + \frac{1-G}{R_1C_1}\right)s + \frac{1}{R_1R_2C_1C_2}} \quad (4.10)$$

The required value of G is realized using an opamp as shown in Fig. 4.4(b).

The design of the circuit proceeds as in the case of the low-pass. For example in order to realize the second-order Butterworth high-pass filter function

$$F(s) = \frac{s^2}{s^2 + \sqrt{2}s + 1} \quad (4.11)$$

we could select $G=1$ and $C_1=C_2=C=1F$ when matching coefficients of equal powers of s in Eqs. (4.10) and (4.11) gives

$$\frac{2}{R_2} = \sqrt{2}$$

$$\frac{1}{R_1R_2} = 1$$

from where we find

$$R_1 = \frac{1}{\sqrt{2}}\Omega \qquad R_2 = \sqrt{2}\Omega$$

It can be observed that these values of the components could have been obtained from the corresponding ones for the low-pass circuit, if we had thought that the exchange of component positions actually means exchange of impedances. Thus, the value of R_1 in Fig. 4.5 should be equal to the value of the impedance of the capacitor C_1 at the cut-off frequency ω_s, which here happens to be 1 and so on. By dividing the capacitances by the actually required cut-off frequency and raising the impedance level to any convenient value, say, between $1k\Omega$ and $10k\Omega$ the circuit can be "denormalized" following the specifications.

Example 4.M.4

Using MATLAB (a) find the transfer function of the second-order normalized Butterworth high-pass filter (Eq. (4.11)) in Section 4.5.2, (b) find the transfer function of the denormalized high-pass filter for $\omega_c=100 rad/s$ and (c) draw the amplitude and phase response of the denormalized filter.

Code

```
(a)     >> [z,p,k]=buttap(2);
>> [num,den]=zp2tf(z,p,k);      % normalized second order Butterworth
                                % low-pass filter
>> [numt,dent]=lp2hp(num,den,1); % transform the normalized low-pass to a
                                 % normalized high-pass (ω_c=1 rad/sec)
>> tf(numt,dent)
```

```
            Transfer function:
                   s^2
            ------------------
            s^2 + 1.414 s + 1
```

The same result can be derived as:

```
>> [num,den]=butter(2,1,'high','s');  % second order Butterworth high-pass filter
                                      % with ω_c=1 rad/sec (normalized)
>> tf(num,den)
```

```
(b)     >> [num,den]=butter(2,100, 'high','s');
>> tf(num,den)
```

```
            Transfer function:
                      s^2
            ----------------------
            s^2 + 141.4 s + 1e004
```

The same result can be derived as:

```
>> [z,p,k]=buttap(2);
>> [num,den]=zp2tf(z,p,k);
>> [numt,dent]=lp2hp(num,den,100);
>> tf(numt,dent)
```

```
(c)     >> w=0:0.01:10000;
>> num=[1 0 0];
```

```
>> den=[1 141.4 10000];
>> [mag,phase]=bode(num,den,w);
>> magdb=20*log10(mag);
>> subplot(2,1,1);semilogx(w,magdb);axis([1 10000 -20 2]);grid;xlabel('frequency (rad/sec)');ylabel('gain (dB)');title('frequency response of second order high-pass Butterworth filter');text(100,-3,'(100rad/sec,-3dB)')
>> subplot(2,1,2);semilogx(w,phase);axis([0 10000 0 200]);grid;xlabel('frequency (rad/sec)'); ylabel('phase (degrees)')
```

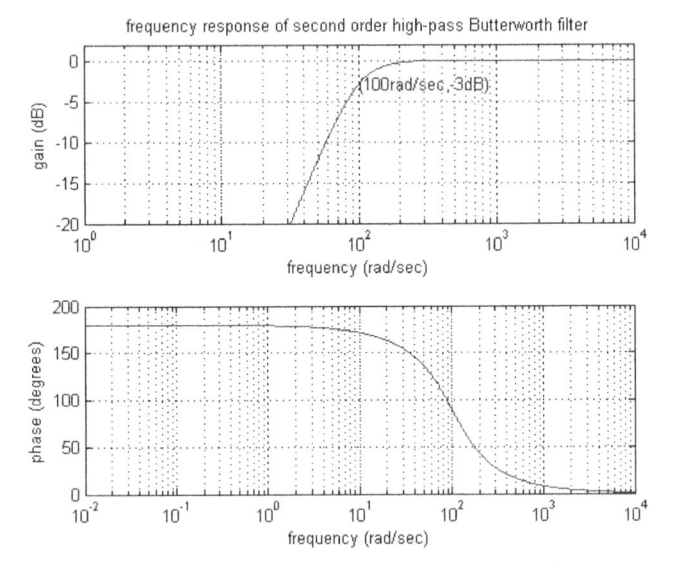

Figure 4.M.4. Frequency response of second order high-pass Butterworth filter for $\omega_r=100 rad/s$

4.5.3 A Band-Pass *SAB*

The second-order band-pass function is of the form

$$F(s) = \frac{\alpha s}{s^2 + \beta s + \gamma} \tag{4.12}$$

which can be written as follows:

$$F(s) = \frac{\alpha s}{s^2 + \frac{\omega_o}{Q}s + \omega_o^2} \tag{4.13}$$

Here, ω_o is the so called *center frequency*, while Q is related to the bandwidth of the frequency response i.e. how narrow or selective this is. By setting $s=j\omega_o$ in Eq. (4.13) the gain at the center frequency is easily found to be

$$K = \frac{\alpha Q}{\omega_o}$$

and usually is chosen to be unity by letting $\alpha = \omega_o/Q$.

Among possible SABs with band-pass frequency response, perhaps the most popular is that in Fig. 4.6. If properly designed its sensitivity to variations in element values is relatively low, while it is easy to tune and adjust the bandwidth of its frequency response. By straightforward analysis its transfer voltage ratio is found to be the following:

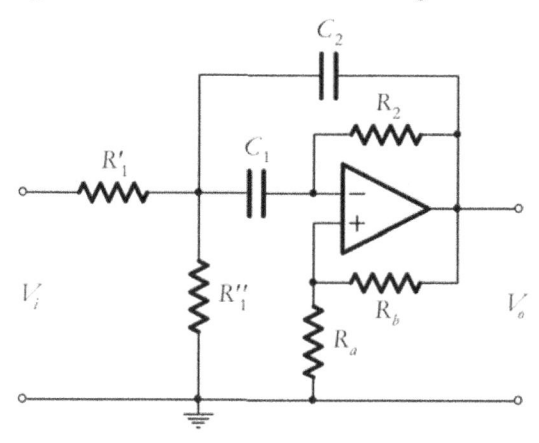

Figure 4.6. A band-pass SAB

$$F(s) = \frac{V_o}{V_i} = \frac{bs}{s^2 + \left(\frac{C_1 + C_2}{R_2 C_1 C_2} - k \frac{1}{R_1 C_2}\right)s + \frac{1}{R_1 R_2 C_1 C_2}} \quad (4.14)$$

where

$$k = \frac{R_a}{R_b} \qquad b = \frac{1+k}{R_1 C_2} \qquad R_1 = \frac{R_1' R_1''}{R_1' + R_1''}$$

A useful way to design this circuit is the following:

We select $C_1 = C_2 = C$ and $R_1 = r R_2$, where

$$r = 0.25 \frac{\omega_o}{\omega_T} \frac{\sigma_{\omega_T}}{\sigma_{R,C}}$$

with ω_T being the frequency at which the open-loop opamp gain is one and σ_{ω_T} and $\sigma_{R,C}$ the tolerances of ω_T and the components R_i, C_i (which will be used in the practical circuit) respectively.

Next we select

$$R_1C = \frac{\sqrt{r}}{\omega_o}$$

$$R_2C = \frac{1}{\omega_o\sqrt{r}}$$

$$k = \frac{R_a}{R_b} = \frac{2Q\sqrt{r}-1}{Q}\sqrt{r}$$

Clearly, for the value of k to be positive it is necessary that

$$\sqrt{r} > \frac{1}{2Q}$$

As an example consider the case when $\omega_o=1rad/s$ and $Q=10$. Selecting $r=0.01$ and $C=1F$ we obtain from the above equations

$$R_1 = 0.1\Omega \qquad R_2 = 10\Omega \qquad R_a = kR_b = 0.01R_b$$

and for $R_b=10\Omega$ then $R_a=0.1\Omega$

To complete the design, we have to calculate suitable values for R'_1, R''_1. The gain at the center frequency with the obtained values is hQ where

$$h = \frac{1+k}{R_1C} = \frac{1+0.01}{0.1} = 10.1$$

If we want to make this unity, we set

$$\frac{R''_1}{R'_1 + R''_1} = \frac{1}{hQ} = \frac{1}{101}$$

Also we have

$$\frac{R'_1 R''_1}{R'_1 + R''_1} = R_1 = 0.1\Omega$$

From these equations, we then obtain

$$R'_1 = 10.1\Omega \qquad\qquad R''_1 = 0.101\Omega$$

These normalized values can be denormalized to any center frequency ω_o and practical impedance level R_o by multiplying all resistances by R_o (e.g. $10k\Omega$) and dividing the capacitances by $R_o\omega_o$.

Example 4.M.5

Using MATLAB draw the amplitude and phase response of the band-pass filter that was designed in Section 4.5.3 (Eq. (4.13)) for $\omega_o=100rad/s$ and $Q=10$.

Code

```
>> w=0:0.01:1000;
>> num=[10 0];
>> den=[1 10 10000];
>> [mag,phase]=bode(num,den,w);
>> magdb=20*log10(mag);
>> subplot(2,1,1);semilogx(w,magdb);axis([10 1000 -40 5]);grid;xlabel('frequency (rad/sec)'); ylabel('gain (dB)');title('frequency response of second order band-pass filter')
>> subplot(2,1,2);semilogx(w,phase);axis([10 1000 -100 100]);grid;xlabel('frequency (rad/sec)') ;ylabel('phase (degrees)')
```

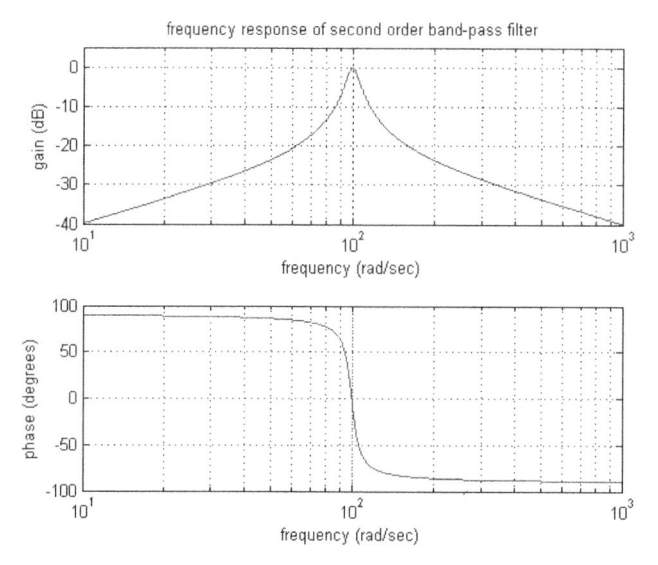

Figure 4.M.5. Frequency response of second order band-pass filter for $\omega_o=100 rad/s$ and $Q=10$

4.5.4 A Notch *SAB*

A second-order notch function is of the following form:

$$F(s) = k\frac{s^2 + \omega_o^2}{s^2 + \beta s + \gamma} \tag{4.15}$$

There are three possibilities:

a. $\omega_o^2 > \gamma$ low-pass notch

b. $\omega_o^2 < \gamma$ high-pass notch

c. $\omega_o^2 = \gamma$ symmetrical notch

Cases (a) and (b) are essential factors in the family of *elliptic filter functions* [1,3]. Case (c) is useful in measurements, when one needs to reject a certain frequency (e.g. 50 or 60Hz). In active *RC SABs* the notch characteristic is usually achieved by using a *Twin-Tee* (*TT*) two port, as shown in Fig. 4.7 in the case of the symmetrical notch circuit.

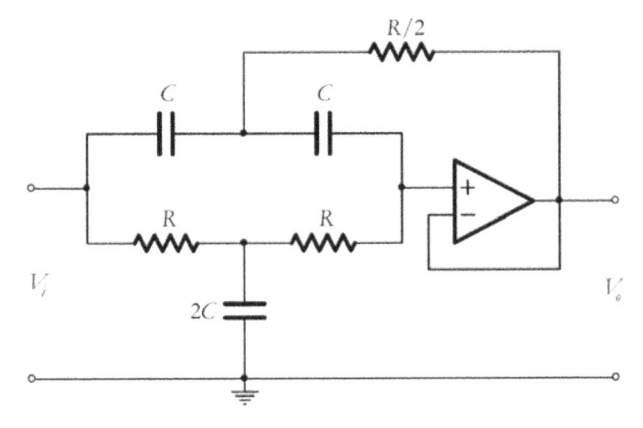

Figure 4.7. Symmetrical notch *SAB*

By straightforward analysis, the transfer function of this circuit is found to be the following:

$$F(s) = \frac{V_o}{V_i} = \frac{G(s^2 + \omega_o^2)}{s^2 + \frac{\omega_o}{Q}s + \omega_o^2} \tag{4.16}$$

where

$$G = 1 \qquad \omega_o^2 = \frac{1}{R^2C^2} \qquad \frac{\omega_o}{Q} = \frac{2}{RC}$$

Thus, the frequency of the notch may be altered by changing the time constant *RC*, but the value of *Q* is fixed.

More flexibility may be obtained by using other different values of *R*'s and *C*'s and even an additional resistance bridging the input and output terminals of the *TT* circuit (see [2]).

A different symmetrical notch *SAB*, without the use of a *TT*, can be obtained as a special case from the all-pass *SAB* described below in Fig. 4.8, but its selectivity will be low.

Example 4.M.6

Using MATLAB draw the amplitude and phase response of the notch filter that was designed in Section 4.5.4 (Eq. (4.16)) for $\omega_o=100\,rad/s$ and $Q=10$.

Analog Filters

Code

```
>> w=0:0.01:1000;
>> num=[1 0 10000];
>> den=[1 10 10000];
>> [mag,phase]=bode(num,den,w);
>> magdb=20*log10(mag);
>> subplot(2,1,1);semilogx(w,magdb);axis([10 1000 -40 5]);grid;xlabel('frequency (rad/sec)'); ylabel('gain (dB)');title('frequency response of second order notch filter')
>> subplot(2,1,2);semilogx(w,phase);axis([10 1000 -100 100]);grid;xlabel('frequency (rad/sec)') ;ylabel('phase (degrees)')
```

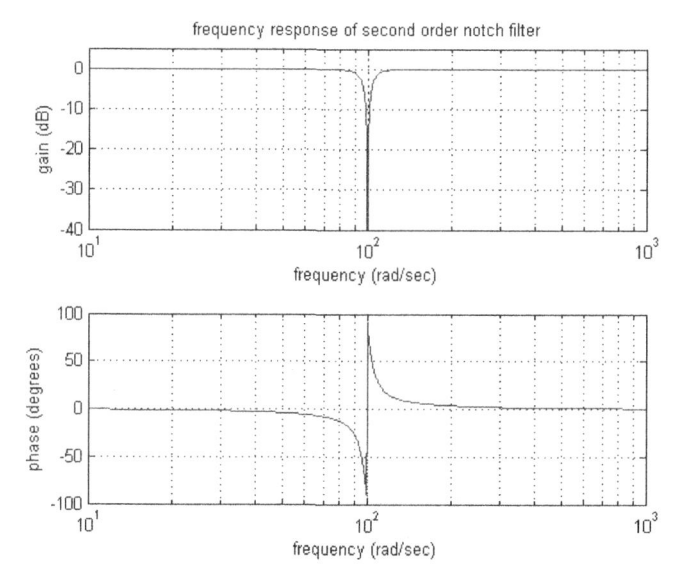

Figure 4.M.6. Frequency response of second order notch filter for $\omega_o=100 rad/s$ and $Q=10$

4.5.5 An All-Pass *SAB*

An all-pass biquadratic function is of the form

$$F(s) = \frac{s^2 - \beta s + \gamma}{s^2 + \beta s + \gamma} \quad (4.17)$$

Clearly, the zeros of this function are symmetrical with the poles across the imaginary axis and the magnitude part of the frequency response is unity at all frequencies. However, the phase is double the phase of the poles and can be made to be linear up to a certain frequency.

A simple active *RC SAB*, realizing this function, is that shown in Fig. 4.8. Its transfer voltage ratio is found to be

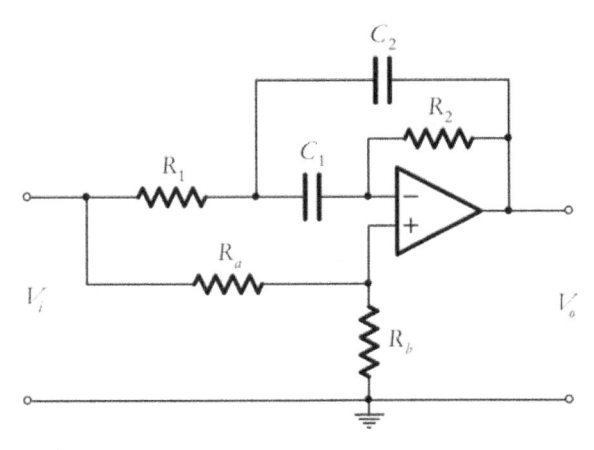

Figure 4.8. An all-pass SAB

$$F(s) = \frac{V_o}{V_i} = K \frac{s^2 - \left(\frac{R_a}{R_b}\frac{1}{R_1 C_2} - \frac{1}{R_2}\frac{C_1 + C_3}{C_1 C_2}\right)s + \frac{1}{R_1 R_2 C_1 C_2}}{s^2 + \frac{1}{R_2}\frac{C_1 + C_2}{C_1 C_2}s + \frac{1}{R_1 R_2 C_1 C_2}} \quad (4.18)$$

where

$$K = \frac{R_b}{R_a + R_b}$$

Since K is always less than one, we may compensate for the subsequent amplitude reduction of the output signal by feeding the right-hand terminals of R_2 and C_2 through the wiper (moving terminal) of a potentiometer, which is connected between the opamp output and the earth point.

In designing the circuit, we select $C_1 = C_2 = C$ and matching coefficients of equal powers of s in Eqs. (4.17) and (4.18) we get

$$R_1 = \frac{\beta}{2\gamma C} \qquad R_2 = \frac{2}{\beta C} \qquad \frac{R_a}{R_b} = \frac{\beta^2}{\gamma} = \frac{4R_1}{R_2} \quad (4.19)$$

Clearly, if we choose

$$\frac{R_a}{R_b}\frac{1}{R_1 C} = \frac{2}{R_2 C}$$

the circuit realizes a symmetrical notch function.

As an example, consider the realization of the following all-pass function:

Analog Filters

$$F(s) = \frac{s^2 - 6s + 12}{s^2 + 6s + 12}$$

Let $C_1=C_2=C=1F$. Substituting in the design Eqs. (4.19) we get

$$R_1 = \frac{6}{24} = 0.25\Omega \qquad R_2 = \frac{2}{6} = \frac{1}{3}\Omega$$

$$\frac{R_a}{R_b} = 3 \quad \text{or} \quad R_a = 3R_b$$

Then

$$K = \frac{R_b}{R_a + R_b} = 0.25$$

Another equation between R_a, R_b can be obtained by choosing $R_a R_b/(R_a+R_b)=R_2$ for better operation in practice of the opamp (minimize the effects of the off-set currents). Then we find

$$R_b = 0.4\Omega \qquad R_a = 1.2\Omega$$

If it is necessary, the amplitude of the output signal will have to be increased using a potentiometer as was explained above.

The circuit, as has been designed here, produces a time-delay of $1 sec$. It can be denormalized to any smaller time-delay (by scaling the time constants) and suitable impedance level as usual.

Example 4.M.7

Using MATLAB draw the amplitude and phase response of the all-pass function that was considered as an example in Section 4.5.5.

Code

```
>> w=0:0.01:100;
>> num=[1 -6 12];
>> den=[1 6 12];
>> [mag,phase]=bode(num,den,w);
>> magdb=20*log10(mag);
>> subplot(2,1,1);semilogx(w,magdb);axis([0 100 -20 10]);grid;xlabel('frequency (rad/sec)'); ylabel('gain (dB)');title('frequency response of second order all-pass function')
>> subplot(2,1,2);semilogx(w,phase);axis([0 100 -400 50]);grid;xlabel('frequency (rad/sec)'); ylabel('phase (degrees)')
```

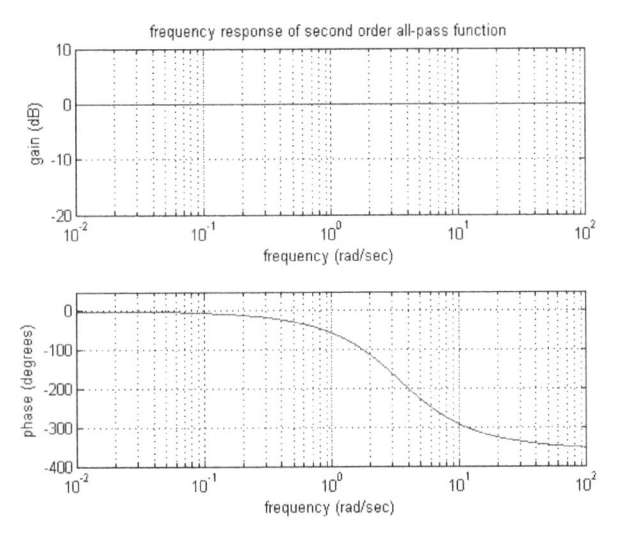

Figure 4.M.7. Frequency response of second order all-pass filter

4.5.6 A Three-Opamp Biquad

Using three opamps, one can produce at least three simultaneous filter responses. There have been suggested few such schemes in the past, which are modifications of a two-integrator-loop circuit. We will develop only this parental circuit here.

Consider the second-order low-pass filter function

$$F(s) = \frac{k}{s^2 + \beta s + \gamma} \qquad (4.20)$$

This is to be realized as the voltage ratio V_o/V_i, where V_o and V_i are defined in the complex frequency domain i.e. V_o and V_i are functions of the complex frequency variable. The above equation can then be written as follows:

$$\frac{V_o}{V_i} = \frac{k}{s^2 + \beta s + \gamma}$$

or

$$(s^2 + \beta s + \gamma)V_o = kV_i$$

which can also be written as

$$s^2 V_o = kV_i - s\beta V_o - \gamma V_o \qquad (4.21)$$

If we assume that the voltages $s\beta V_o$ and γV_o are available, and with V_i being the input voltage, we may add these three voltages according to

Eq. (4.21), using an opamp as in Fig. 4.9, to obtain the voltage s^2V_o. Next, having obtained s^2V_o we can integrate twice to obtain the voltages sV_o and V_o. These then can be added (properly signed and scaled) with kV_i to form s^2V_o according to Eq. (4.21). The whole configuration is shown in Fig. 4.9, where the sign reversing of the adder and the integrator, produced by the opamp, has been taken into consideration, while the time constant of the integrators has been taken to be 1 sec.

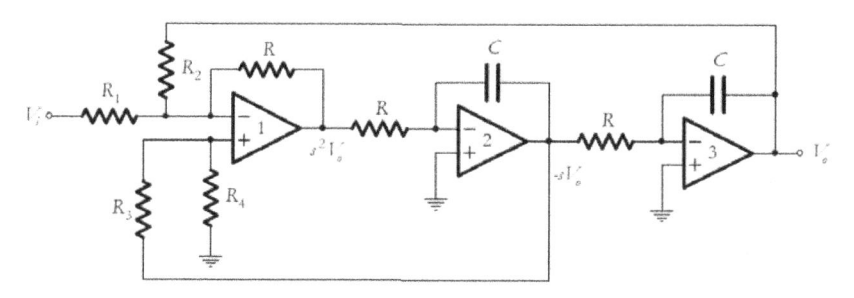

Figure 4.9. A three-opamp biquad

Straightforward analysis of the circuit will show that V_i and V_o have opposite signs, but this is of no concern in filter design, as the sign of V_o can always be reversed using an extra opamp, if this is required. On the other hand, the sign of V_o and V_i will be the same, if V_i is applied through R_1 to the non-inverting input of opamp No 1. Also, the analysis will give the equations required to determine the values of the components i.e.

$$\frac{R}{R_1} = k \qquad \frac{R}{R_2} = \gamma \qquad \frac{R_4}{R_3 + R_4}(1 + k + \gamma) = \beta \qquad (4.22)$$

The importance of this circuit is due to the fact that, from the output of each opamp we get a different filter function i.e.

From amplifier No 3 $\qquad \dfrac{V_o}{V_i} = -\dfrac{k}{s^2 + \beta s + \gamma} \qquad$ low-pass

From amplifier No 2 $\qquad \dfrac{V'_o}{V_i} = \dfrac{s}{s^2 + \beta s + \gamma} \qquad$ band-pass

From amplifier No 1 $\qquad \dfrac{V''_o}{V_i} = -\dfrac{s^2}{s^2 + \beta s + \gamma} \qquad$ high-pass

This flexibility of the circuit has been used advantageously to introduce various modifications to it to obtain also notch and all-pass responses.

One final point is that denormalization of the filter to any cut-off frequency will be achieved by choosing the proper value of the integrators time constants.

Example 4.2

Consider again the realization of the low-pass Butterworth filter with a cut-off frequency $\omega_c=100\,rad/s$ and gain 1.

Solution

For $\omega_c=100\,rad/s$ the RC time constant should be

$$\frac{1}{RC} = \omega_c = 100\,\text{sec}^{-1}$$

Selecting $R=10k\Omega$ we get

$$C = \frac{1}{10^2 \cdot 10^4} = 10^{-6} F = 1\mu F$$

From the normalized Butterworth low-pass function

$$F(s) = \frac{1}{s^2 + \sqrt{2}s + 1}$$

and Eqs. (4.22) we have

$$R_1 = \frac{R}{k} = \frac{10^4}{1} = 10k\Omega$$

$$R_2 = \frac{R}{\gamma} = 10k\Omega$$

$$(R_3 + R_4)\beta = R_4(1 + k + \gamma)$$

or

$$R_3 + R_4 = \frac{3R_4}{\sqrt{2}}$$

or

$$\sqrt{2}R_3 = (3 - \sqrt{2})R_4$$

or

$$\frac{R_3}{R_4} = \frac{3 - \sqrt{2}}{\sqrt{2}} = 1.121$$

If we select $R_4=10k\Omega$ then

$$R_3 = 1.121 R_4 = 11.21 k\Omega$$

4.6 Higher-Order Active *RC* Filters

In most cases, e.g. telecommunications, the selectivity that can be achieved by the biquads is not satisfactory. It is then imperative to realize higher-order filter functions.

Analog Filters

Three general methods have been proved most suitable and useful in practice and these are the following:

a. Cascade connection of lower-order sections (first- and second-order)

b. Multiple-loop feedback in a cascade connection of lower-order sections

c. Simulation of doubly and equally resistive terminated LC ladders

In the first method, the high-order filter function is written as the product of first-and second-order functions. Actually, if the filter is of even order, only second-order functions will result while if it is of odd-order, there will be one first-order factor and the rest of second order. Then, each low-order function is realized by the methods explained earlier and the resulting subnetworks are connected in cascade. Two examples are given later in this section for demonstrating this method.

The application of multiple feedback in the cascade of low-order sections, method b, has some advantages as well as disadvantages when compared with method a, which make it more involved and we will not explain it, but the interested reader can consult references [1,4,5] for details.

Finally, the third method makes use of existing filter designs of passive LC doubly resistively terminated ladders, in which either the inductors are simulated by active subnetworks namely, gyrators, *PICs*, *GICs* or, the whole ladder is functionally simulated. The main characteristic of the simulated ladder filters is their low sensitivity with respect to changes in component values, however, there are disadvantages too. Since we have not introduced the inductance simulation, we will not be concerned with this method either, but the interested reader can be referred to specialized texts for example [1,4].

Here we will demonstrate the use of the first method for the design of the third- and fourth-order Butterworth low-pass functions.

Example 4.3

The third-order filter function is the following:

$$F(s) = \frac{1}{s^3 + 2s^2 + 2s + 1}$$

which is written as

$$F(s) = \frac{1}{s+1} \cdot \frac{1}{s^2 + s + 1} \qquad (4.23)$$

The first-order term is realized by the circuit in Fig. E.4.1(a) and the second-order term by the circuit in Fig. 4.4, which are repeated here.

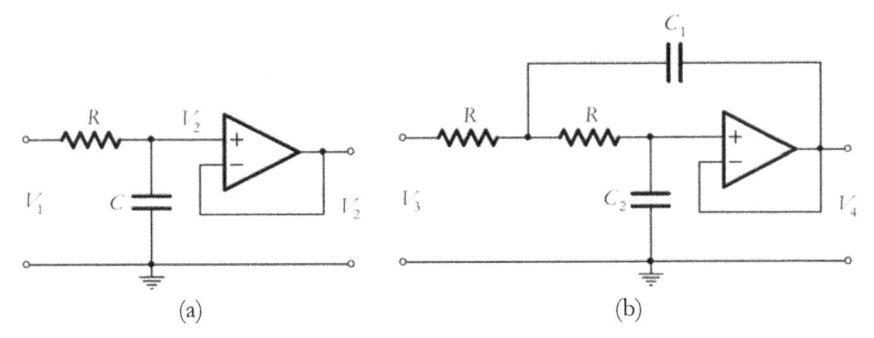

Figure E.4.1. (a) Realization of the first-order term and (b) realization of the second-order term in Eq. (4.23)

The transfer voltage ratio V_2/V_1 in circuit (a) is

$$\frac{V_2}{V_1} = \frac{\frac{1}{RC}}{s + \frac{1}{RC}}$$

which when compared with the first-order term in Eq. (4.23) gives $RC=1s$. Choose for example $R=1\Omega$, $C=1F$.

Following the procedure that was explained in Section 4.5.1, we can easily find for the 2nd-order section in Fig. 4.10(b) that for $R=1\Omega$ and $G=1$

$$C_1 = 2F \qquad C_2 = 0.5F$$

Then the two sections in Fig. E.4.1 are connected in cascade and the realization is completed. The obtained circuit will have a cut-off frequency at $1 rad/s$. Denormalization to any cut-off frequency and suitable impedance level will follow the procedure outlined previously.

Example 4.M.8

Using MATLAB (a) find the transfer function of the third-order normalized Butterworth low-pass filter in Example 4.3, (b) find the transfer function of the denormalized low-pass filter for $\omega_c=100 rad/s$ and (c) draw the amplitude and phase response of the denormalized filter.

Code

(a)   `>> [z,p,k]=buttap(3);`
`>> [num,den]=zp2tf(z,p,k);`
`>> tf(num,den)`

Analog Filters 135

```
Transfer function:
            1
---------------------
   s^3 + 2 s^2 + 2 s + 1
```

The same result can be derived as:

\>> [num,den]=butter(3,1,'s');
\>> tf(num,den)

(b)   \>> [num,den]=butter(3,100,'s');
\>> tf(num,den)

```
Transfer function:
            1e006
---------------------------------
   s^3 + 200 s^2 + 2e004 s + 1e006
```

(c)   \>> w=0:0.01:1000;
\>> num=1000000;
\>> den=[1 200 20000 1000000];
\>> [mag,phase]=bode(num,den,w);
\>> magdb=20*log10(mag);
\>> subplot(2,1,1);semilogx(w,magdb);axis([0.1 1000 -20 2]);grid;xlabel('frequency (rad/sec)');ylabel('gain (dB)');title('frequency response of third order low-pass Butterworth filter');text(100,-3,'(100rad/sec,-3dB)')
\>> subplot(2,1,2);semilogx(w,phase);axis([0 1000 -300 20]);grid;xlabel('frequency (rad/sec)'); ylabel('phase (degrees)')

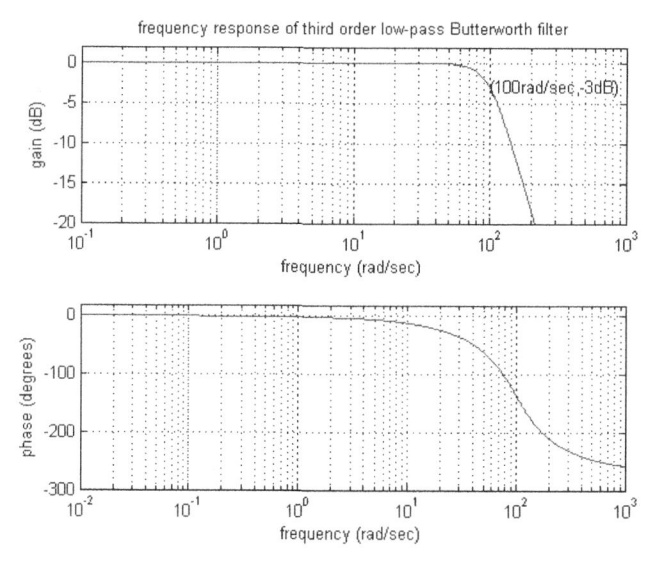

Figure 4.M.8. Frequency response of third order low-pass Butterworth filter for $\omega_c=100 rad/s$

Example 4.4

The fourth-order filter function is as follows:

$$F(s) = \frac{1}{s^4 + 2.613s^3 + 3.414s^2 + 2.613s + 1}$$

This is written as

$$F(s) = \frac{1}{s^2 + 0.765s + 1} \cdot \frac{1}{s^2 + 1.848s + 1} \qquad (4.24)$$

Each of these two second-order functions will be realized by the SAB in Fig. E.4.1(b). The component values will be calculated as before and are tabulated in Table 4.2.

Table 4.2. Component values

Component	First term	Second term
R	1Ω	1Ω
C_1	2.614F	1.082F
C_2	0.383F	0.924F
C	1F	1F

In connecting the two second-order sections in cascade, it is preferable in practice, to place the second section first. This is due to the fact that the Q factor of the second section is lower than that of the first section and thus, the overall circuit would handle higher input signals at all frequencies. For more details on this and other practical points the interested reader is again referred to [1].

References

[1]. Deliyannis T., Sun Y. & Fidler J. K., *Continuous-Time Active Filter Design*, CRC Press, 1999.

[2]. Moschytz G. S., *Linear Integrated Circuits: Design*, Van Nostrand Reinhold, New York, 1974.

[3]. Daryanani G., *Principles of Active Network Synthesis and Design*, Wiley, New York, 1976.

[4]. Sedra A. S. & Brackett P. O., *Filter Theory and Design: Active and Passive*, Pitman, London, 1978.

[5]. Schaumann R., Ghausi M. S. & Laker K. R., *Design of Analog Filters: Passive, Active RC and Switched Capacitor*, Prentice Hall Intl. Ed., 1990.

[6]. Su K. L., *Analog Filters*, Chapman Hall, 1996.

[7]. Mitra S. K. & Kurth C. F. Eds, *Miniaturized and Integrated Filters*, Wiley, 1989.

MATLAB Problems

❖ *The reader is advised first to consult the MATLAB tutorial in Appendix D*

❖ *Useful MATLAB functions: freqs, abs, angle, bode, buttap, butter, cheb1ap, cheby1, cheb2ap, cheby2, ellipap, ellip, lp2lp, lp2hp, lp2bp, lp2bs*

4.M.1. Using MATLAB (a) find the transfer function of the fourth-order normalized Butterworth low-pass filter in Example 4.4 and (b) draw the amplitude responses of the corresponding denormalized low-pass, high-pass, band-pass and notch filters for $\omega_c=100rad/s$, $\omega_o=100rad/s$, $Q=10$ all in the same plot for comparison.

4.M.2. Repeat Problem 4.M.1 for a fourth-order normalized Chebyshev low-pass filter with $3dB$ ripple in the passband.

4.M.3. Repeat Problem 4.M.1 for a fourth-order normalized Inverse Chebyshev low-pass filter with $20dB$ attenuation in the stopband.

4.M.4. Repeat Problem 4.M.1 for a fourth-order normalized Elliptic low-pass filter with $3dB$ ripple in the passband and $20dB$ attenuation in the stopband.

4.M.5. Using MATLAB draw the amplitude and phase responses of the normalized low-pass filters of Problems 4.M.1, 4.M.2 and the following fourth-order Bessel-Thomson in the same plots for comparison. The resulting plots should be the same with those in Fig. 4.2(a) and (b).

$$H(s) = \frac{105}{s^4 + 10s^3 + 45s^2 + 105s + 105}$$

4.M.6. Repeat Problem 4.M.5 for the normalized low-pass filters of Problems 4.M.2 and 4.M.3.

Problems

4.1. Derive the transfer function of the following circuit and determine its passive components for the $3dB$ cut-off frequency to be at $10Hz$.

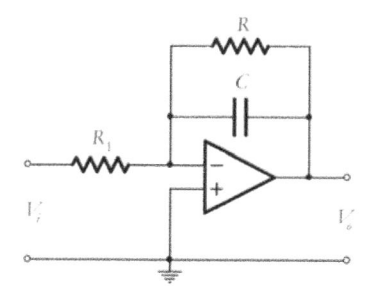

Figure P.4.1.

4.2. Derive the transfer function of the following circuit and sketch its amplitude and phase responses.

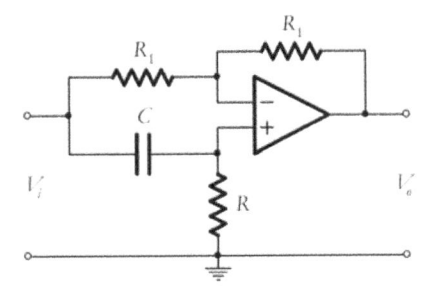

Figure P.4.2.

4.3. Derive the transfer voltage ratio V_o/V_i of the circuit in Fig. 4.4(a).
4.4. Repeat Problem 4.3 for the circuit in Fig. 4.5.
4.5. Repeat Problem 4.3 for the circuit in Fig. 4.6.
4.6. Repeat Problem 4.3 for the circuit in Fig. 4.7.
4.7. Repeat Problem 4.3 for the circuit in Fig. 4.8.
4.8. Repeat Problem 4.3 for the circuit in Fig. 4.9 and prove Eqs. (4.22).

Chapter 5

Digital Combinational Electronics

5.1 Introduction

The electronic circuits we have presented so far are *analog*, in that the voltage or current at the output of a circuit is continuously proportional to the corresponding quantity at the input. This happens at any time instant thus explaining the meaning of "*continuously*". On the other hand, the function of *digital electronics* is based on the fact that a large category of electronic circuits operates in one of two discrete stable states. They change their state responding to a suitable pulse at their input. This switching of states is actually achieved very fast.

Digital electronic circuits can be distinguished in *Combinational* (or *Combinatorial*) and in *Sequential*. The former include all digital circuits, the output signal of which depends only on their structure and the input signals. In the sequential digital circuits the output signal is dependent, apart from their structure, on past and present values of the input signal as well as on past values of the output signal. The latter requires the presence of memory elements in their structure.

In this book all digital circuits are presented in the form of functional blocks, i.e. without giving any details on their internal structure. Since these circuits operate in two discrete stable states, it is most suitable for their study to use a different arithmetic system, namely, the *Binary Arithmetic System*, and, accordingly, a suitable algebra, the *Algebra Boole*. It is most appropriate to treat the combinational electronics first and this is followed here.

A comparison of analog to digital systems is discussed first followed by an introduction to the binary system of numbers. Next the representation of the voltages in the two states is explained. Then follows the introduction of *Boolean Algebra* and the logic gates which implement the operations of this algebra and logic functions. Also, using these gates, various digital functional blocks like comparators, adders, encoders, decoders, multiplexers, demultiplexers and error detectors are implemented.

5.2 Digital Systems Against Analog Systems

The electronic circuits that have been introduced so far in this book, are suitable for processing continuous in time signals, although in some cases these signals are in pulse form like those coming from the heart, muscles etc. These signals have a value at any instant that is proportional to the physical quantity, which has caused it.

Digital signals on the other hand are series of equal height pulses representing numbers, which correspond to the height of pulses obtained from the analog signal at discrete time instances. This means that the analog signal has been sampled at discrete time instances first. In each of the obtained pulses a certain number has been assigned according to its height and then this number has been represented by a series of pulses of equal height. Clearly, this type of translation entails the use of a conversion device called *Analog to Digital Converter (ADC)*. On the other hand if one wants to convert a digital signal to analog, one requires the use of a *Digital to Analog Converter (DAC)*.

Accordingly, the processing of a digital signal requires the employment of a digital system. In general digital systems have important advantages over analog systems, like better noise immunity, reliability, accuracy etc, and for this reason they are preferred in signal processing nowadays. A good example of the superiority of a digital system performance over that of an analog is the reproduction by a *CD* player of sound stored in a compact disc instead of been stored on a magnetic tape or a vinyl record which is played by analog sound reproduction players.

If the input and output signals are in analog form, it is quite common to use digital techniques for processing them, with the disadvantage that data converters must also be used to translate signals between analog and digital forms. However in some cases, the advantages of digital processing make this worthwhile.

5.3 Binary System of Numbers

The digits in the usual decimal system of numbers are 0, 1, 2, ..., 9. The base of this system is 10. Powers of 10 determine the weights of the digits in a multidigital number depending on their position in it. Thus, the digits in the decimal number 295 have the following meaning:

$$(295)_{10} = 2\times 10^2 + 9\times 10^1 + 5\times 10^0$$

Therefore digits 2, 9, 5 are the coefficients in writing the decimal number in powers of 10. In the binary system of numbers the digits are two, namely 0 and 1, while the base is 2. Accordingly, in a multidigital binary number the position of each digit in it determines the weight of the digit expressed as some power of 2. For example in the binary number 01 the digit 0 has a weight 2^1, while the weight of 1 is 2^0. This leads to the conversion of a binary number to the corresponding decimal, for example

$$(11101)_2 = 1\times 2^0 + 0\times 2^1 + 1\times 2^2 + 1\times 2^3 + 1\times 2^4$$

or

$$(11101)_2 = (29)_{10}$$

The first ten numbers of the binary system are as follows:

Binary	Decimal
0000	0
0001	1
0010	2
0011	3
0100	4
0101	5
0110	6
0111	7
1000	8
1001	9

The conversion of a decimal number, e.g. 29, to the corresponding binary number can proceed according to the scheme

Quotient of division by 2	0	1	3	7	14	29	Decimal number
	1	1	1	0	1	←	Binary number

That is, divide the decimal number (29) by 2 and write the quotient (14) to the left and the remainder (1) below the quotient. The resulting quotient (14) is divided by 2 and the new quotient (7) is written to the left of the previous and the remainder (0) below the new quotient. The division by 2 continues each time writing the new quotient to the left of the previous one and the remainder below the new quotient until the quotient becomes 0. The binary number is that formed by the successive remainders.

In accordance with decimal numbers, the binary point is used to represent a non-integer binary number. For example

01101.101

The conversion to the corresponding decimal number follows the previous argument i.e.

$$(01101.101)_2 =$$
$$= 0 \times 2^4 + 1 \times 2^3 + 1 \times 2^2 + 0 \times 2^1 + 1 \times 2^0 + 1 \times 2^{-1} + 0 \times 2^{-2} + 1 \times 2^{-3} =$$
$$= (13.625)_{10}$$

On the other hand, for the conversion of the fractional part of the decimal number to the corresponding part of the binary number the procedure is the following: The fractional part is multiplied by 2 and, if the result is greater than 1, the digit 1 is kept as the first digit after the binary point, and the new fractional part is multiplied again by 2 and so on. If the product is smaller than 1 the corresponding digit is 0. For example the conversion of $(0.625)_{10}$ to binary follows the following steps

```
      0.625              0.250              0.500
      × 2                × 2                × 2
      -----              -----              -----
      1.250              0.500              1.000
       ↑                  ↑                  ↑
```

and the binary number is 0.101.

For the representation of a signed binary number the digit 0 is placed as the first from the left digit if the sign is plus (+) and 1 if the sign is minus (-). Thus the binary number 1010.11, if positive, is written as 01010.11 and as 11010.11 if it is negative.

The binary digit 0 or 1 is called *bit* from the words **binary digit**. A string of bits when they represent a number, an instruction etc, is a *word*, while a word of 8 bits is called a *byte*.

The arithmetic operations addition, subtraction, multiplication and division can be executed in a manner similar to that with decimal numbers. As an example consider the addition of the numbers 11101 and 10101 which is executed as follows:

```
        Binary numbers              Corresponding decimal numbers
          1 1 1 0 1                           2 9
          1 0 1 0 1                           2 1
carry     1 1 1 0 1          carry            1
        -----------                         -----
        1 1 0 0 1 0                           5 0
```

$$(110010)_2 = 0 \times 2^0 + 1 \times 2^1 + 0 \times 2^2 + 0 \times 2^3 + 1 \times 2^4 + 1 \times 2^5 = (50)_{10}$$

In Digital Electronics the circuits for performing the addition are relatively simple. However those performing subtraction are rather complex and difficult to design. For this reason the operation of subtraction of two binary numbers is executed by a method which in effect converts the subtraction to addition. This is usually achieved by adding to the minuend the 2's complement of the subtrahend. The 2's complement of a natural binary number is found by adding 1 to the Least Significant Bit (*LSB*) of the 1's complement of the number. So

$$2\text{'s complement} = 1\text{'s complement} + 1$$

The 1's complement of the natural binary number is obtained if all 0 in the number become 1 and all 1 become 0. Thus the 1's complement of 1010 is 0101. Then the 2's complement of 1010 will be

$$2\text{'s complement of } 1010 = 0101 + 1 = 0110$$

The 1's and 2's complements of the first 16 numbers are given in Table 5.1.

Table 5.1.

Decimal	Binary	1's Complement	2's Complement
0	0000	1111	10000
1	0001	1110	1111
2	0010	1101	1110
3	0011	1100	1101
4	0100	1011	1100
5	0101	1010	1011
6	0110	1001	1010
7	0111	1000	1001
8	1000	0111	1000
9	1001	0110	0111
10	1010	0101	0110
11	1011	0100	0101
12	1100	0011	0100
13	1101	0010	0011
14	1110	0001	0010
15	1111	0000	0001

As an example of converting the subtraction to addition using the 2's complement consider the subtraction of decimal 12 from decimal 21 using 8 bit numbers.

$$(12)_{10} = (00001100)_2$$

```
1's complement      1 1 1 1 0 0 1 1
2's complement      1 1 1 1 0 1 0 0
   plus (21)₁₀     0 0 0 1 0 1 0 1
         sum    1 0 0 0 0 1 0 0 1
```

The last carry in the addition is dropped and the final result is $(9)_{10}=(00001001)_2$.

With regard to multiplication, it is observed that as with decimal numbers, multiplication by 0 gives 0 as result, while multiplication by 1 leaves the binary number unaltered. Multiplication by $(10)_2$ is equivalent to moving all bits of the number one position to the left completing the position of the least significant bit by a 0. This implies that the multiplication of two binary numbers entails consecutive shifting of digits to the left and additions. Thus multiplication can be executed as consecutive additions. Note that multiplication by $(10)_2$ of a binary number is similar to the multiplication of a decimal number by $(10)_{10}$.

On the other hand, division can be achieved by consecutive shifts of digits to the right and subtractions. Since subtraction can be turned to addition, it follows, that division is also turned to addition. Note again the similarity of the procedure when dividing a binary number by $(10)_2$ with the procedure when dividing a decimal number by $(10)_{10}$.

Thus the relatively simple digital circuit for the addition can be used to execute all four arithmetic operations, and this is, in fact, what happens in digital practice.

Example 5.M.1

Using MATLAB perform the following operations (unsigned numbers):
a) $(10010011)_2 + (10101000)_2$
b) $(11100011)_2 - (10000001)_2$

Code

a)
```
>> x=base2dec('10010011',2);      % convert to decimal number
>> y=base2dec('10101000',2);      % convert to decimal number
>> z=x+y;
>> w=dec2base(z,2)                % convert to binary number

w =

100111011
```

The same result can be derived as follows:
```
>> x=bin2dec('10010011');         % convert to decimal number
>> y=bin2dec('10101000');         % convert to decimal number
>> z=x+y;
>> w=dec2bin(z)                   % convert to binary number
```
b)
```
>> x=base2dec('11100011',2);      % convert to decimal number
>> y=base2dec('10000001',2);      % convert to decimal number
```

```
>> z=x-y;
>> w=dec2base(z,2)                    % convert to binary number

w =

1100010
```

5.3.1 Octal and Hexadecimal Number Systems

The octal number system is a base 8 number system and has 8 digits the following: 0, 1, 2, 3, 4, 5, 6, 7. Digits 8 and 9 do not appear in an octal number as shown in Table 5.2 which gives its correspondence to the decimal and binary number systems. The conversion from octal to decimal follows the rule set for the conversion of a binary to decimal number. For example the octal number $(53.4)_8$ is converted to decimal as follows:

$$(53.4)_8 = 5 \times 8^1 + 3 \times 8^0 + 4 \times 8^{-1} = (43.5)_{10}$$

Table 5.2. Correspondence of number systems

Decimal	Octal	Hexadecimal	Binary
0	00	0	0000
1	01	1	0001
2	02	2	0010
3	03	3	0011
4	04	4	0100
5	05	5	0101
6	06	6	0110
7	07	7	0111
8	10	8	1000
9	11	9	1001
10	12	A	1010
11	13	B	1011
12	14	C	1100
13	15	D	1101
14	16	E	1110
15	17	F	1111

The hexadecimal is a base 16 number system and its correspondence to the other systems is given in Table 5.2. Accordingly its conversion to decimal follows the same rule set above. For example

$$(A7)_{16} = 10 \times 16^1 + 7 \times 16^0 = (167)_{10}$$

The octal and hexadecimal number systems are used in some computers.

Example 5.M.2

Using MATLAB perform the following operations:
a) $(3355)_8 + (1432)_8$
b) $(6771)_8 - (1567)_8$
c) $(BC83)_{16} + (1A38)_{16}$
d) $(EE34)_{16} - (AF56)_{16}$

Code

a)   >> x=base2dec('3355',8);   % convert to decimal number
>> y=base2dec('1432',8);   % convert to decimal number
>> z=x+y;
>> w=dec2base(z,8)   % convert to octal number
w =

5007

b)   >> x=base2dec('6771',8);   % convert to decimal number
>> y=base2dec('1567',8);   % convert to decimal number
>> z=x-y;
>> w=dec2base(z,8)   % convert to octal number
w =

5202

c)   >> x=base2dec('BC83',16);   % convert to decimal number
>> y=base2dec('1A38',16);   % convert to decimal number
>> z=x+y;
>> w=dec2base(z,16)   % convert to hexadecimal number
w =

D6BB

The same result can be derived as:

>> x=hex2dec('BC83');   % convert to decimal number
>> y=hex2dec('1A38');   % convert to decimal number
>> z=x+y;
>> w=dec2hex(z)   % convert to hexadecimal number

d)   >> x=base2dec('EE34',16);   % convert to decimal number
>> y=base2dec('AF56',16);   % convert to decimal number
>> z=x-y;
>> w=dec2base(z,16)   % convert to hexadecimal number
w =

3EDE

5.4 Voltage Representation of Binary Numbers

In digital circuits bits 0 and 1 are represented by voltages, which are at discrete states, i.e. one high and the other low. In *positive logic* the high state represents bit 1 and the low state bit 0. Thus, in Fig. 5.1(a) the bits representation corresponds to positive logic. On the other hand one refers to *negative logic*, if one represents bit 1 by the lower voltage and bit 0 by the higher voltage. Positive logic is adopted in this book in accordance with the predominant situation in international literature. An alternative characterization of states 1 and 0 are state H (high) and state L (low) respectively.

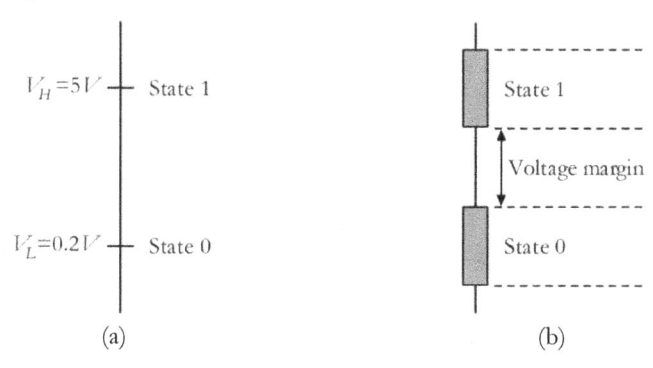

Figure 5.1. (a) Voltage representation for positive logic and (b) voltage margin between the two states

In practice, states 1 and 0 refer to two regions of voltages rather than to two clear voltages, as shown in Fig. 5.1(b). These two voltage regions should be separated by a *voltage margin*, which must be wide enough for the two states to be discrete. This is so because for reliability, the input voltages in digital circuits should lead to correct states at their outputs even in the presence of noise or changes in component values, bias voltage, etc.

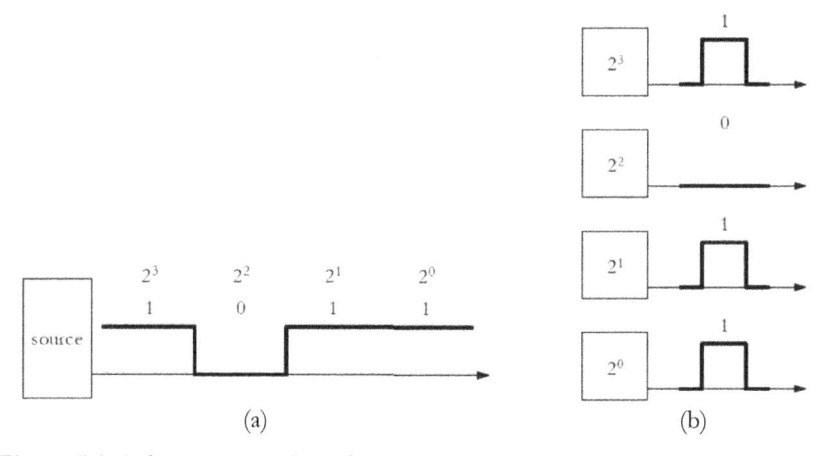

Figure 5.2. Pulse representation of number 1011 in (a) serial and (b) parallel form

The representation of a binary number by voltage pulses can be achieved by a sequence of pulses in time, which correspond to increasing powers of 2 in the number. This representation is called *serial* and is shown in Fig. 5.2(a). However the representation can be *parallel*, as shown in Fig. 5.2(b), where all pulses representing powers of 2 in the number appear simultaneously.

In Fig. 5.3 the pulse representation of numbers 1010 and 0011 is given together with the corresponding representation of their sum 1101.

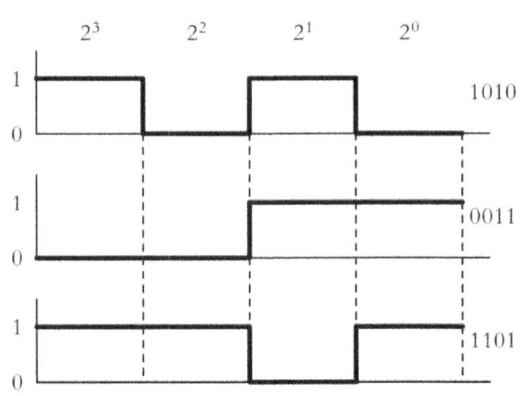

Figure 5.3. Pulse representation of numbers 1010, 0011 and their sum 1101

Both serial and parallel representation can be used, even in the same digital system. Comparing these two forms of representation, it is noted that the serial is more economical, since for the transmission of the signal only one path is required whereas in the parallel, the required paths are as many as the bits of the number. However the parallel transmission is advantageous as far as speed is concerned, because for the transmission of the whole number, the required time is that for the transmission of one bit, while for the transmission in serial form the required time is much longer, depending on the multitude of bits that represent the number. In practice, the selected form, serial or parallel, depends on which characteristic, economy or speed, is most important for the application at hand each time.

5.5 Boolean Algebra

The *Boolean Algebra* is most suitable for studying logic circuits, i.e. circuits that are characterized by two discrete states. These two states are identified as the two values of the variables, which are logic **1** and logic **0** and, in general, true and false, yes and no, etc. Sufficient operations in Boolean Algebra are the *complementation*, the *logical addition* and the *logical multiplication*.

Complementation is executed by the negation of a variable i.e.

Digital Combinational Electronics

NOT A

This is represented by $\overline{A}$. Since variable **A** is **1** or **0** it follows that

$$\overline{1} = 0 \quad \text{and} \quad \overline{0} = 1$$

It is also valid that

$$\overline{\overline{A}} = A \quad \text{Theorem of } \textit{Double Negation}$$

The logic addition of variables **A** and **B** is represented as **A+B** and read as **A** OR **B**. On the other hand the logic multiplication of variables **A** and **B** is represented as **A·B** and read as **A** AND **B**. The results of OR and AND operations on the two variables **A** and **B** are the following:

```
A + B = C        A · B = C
0 + 0 = 0        0 · 0 = 0
0 + 1 = 1        0 · 1 = 0
1 + 0 = 1        1 · 0 = 0
1 + 1 = 1        1 · 1 = 1
```

These results can be presented collectively in a table, which is called *truth table*, and given in Table 5.3.

Table 5.3. Truth table

A	B	A+B	A·B
0	0	0	0
0	1	1	0
1	0	1	0
1	1	1	1

Following the above presentation one can easily prove the following simple theorems:

$$A + 0 = A \qquad A \cdot 0 = 0$$
$$A + 1 = 1 \qquad A \cdot 1 = A$$
$$A + A = A \qquad A \cdot A = A$$
$$A + \overline{A} = 1 \qquad A \cdot \overline{A} = 0$$

5.5.1 Theorems of Boolean Algebra

For the two-value variables the following theorems are valid:

Theorem of *Commutation*
$$A + B = B + A$$
$$A \cdot B = B \cdot A$$

Theorem of *Association*
$$A + (B + C) = (A + B) + C$$
$$A \cdot (B \cdot C) = (A \cdot B) \cdot C$$

Theorem of *Distribution*
$$(A + B)(A + C) = A + BC$$
$$AB + AC = A(B + C)$$

Theorem of *Absorption*
$$A(A + B) = A$$
$$A + AB = A$$

De Morgan Theorems
$$\overline{A \cdot B} = \overline{A} + \overline{B}$$
$$\overline{A + B} = \overline{A} \cdot \overline{B}$$

The proof of the above theorems can be easily achieved using truth tables. As an example the proof of De Morgan Theorems is given in Table 5.4.

Table 5.4. Proof of De Morgan Theorems

A	B	A+B	A·B	$\overline{A}$	$\overline{B}$	$\overline{A \cdot B}$	$\overline{A} \cdot \overline{B}$	$\overline{A + B}$	$\overline{A} + \overline{B}$
0	0	0	0	1	1	1	1	1	1
0	1	1	0	1	0	0	1	1	0
1	0	1	0	0	1	0	1	1	0
1	1	1	1	0	0	0	0	0	0

                                        ↑ same ↑

                          ↑   same   ↑

The above theorems are useful for the simplification of Boolean expressions or functions in order to reduce the complexity of their implementation by digital circuits. As an example consider the function

$$F = A(AB + C)$$

Using the theorem of distribution gives

$$F = AAB + AC$$

or

$$F = AB + AC$$

Then using again the distribution theorem **F** can also be written as

$$F = A(B + C)$$

There are a number of more systematic and efficient simplification methods of a logic function. For example one such method for functions of up to 6 variables is that of *Karnaugh Maps*. Details of using these methods are

Digital Combinational Electronics

beyond the scope of this book though and the interested reader is advised to refer to specializing books like [1,2] for example.

5.6 Logic Gates

In logic circuits the operations of the Boolean Algebra are implemented by circuits called *gates*. Each gate takes the name of the operation it implements. So the basic gates are the following:

- NOT gate or *Inverter*
- OR gate
- AND gate

In logic circuits these gates are represented by their symbols shown in Fig. 5.4. NOT gate has one input and one output, whereas the OR and AND gates have two or more inputs and one output. The input(s) of more than one other gate can be connected to the output of each gate, this being considered, theoretically, an ideal voltage source. Of course in practice there are some limitations. Equivalently, the input impedance of each gate is considered, theoretically, infinite in order not to load the output of the gate it is connected to.

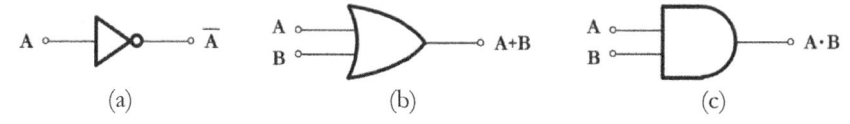

Figure 5.4. Symbols of (a) NOT gate, (b) OR gate and (c) AND gate

The above basic gates can be combined to give other gates, which have been proved very useful in practice. Two such simple gates are the NAND gate and the NOR gate. The NAND gate is formed by an AND gate followed by a NOT gate, while the NOR gate can be obtained by an OR gate followed by a NOT gate. These combinations are shown in Fig. 5.5 together with the symbols of the NAND and NOR gates. The truth tables of these two gates are as given in Table 5.5, which is in fact part of Table 5.4.

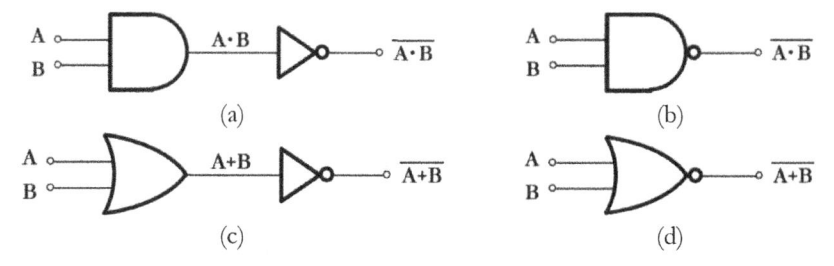

Figure 5.5. Implementation and symbols of NAND and NOR gates

Table 5.5. The truth tables of gates NOR and NAND

A	B	A+B	$\overline{A+B}$	A·B	$\overline{A \cdot B}$
0	0	0	1	0	1
0	1	1	0	0	1
1	0	1	0	0	1
1	1	1	0	1	0

It should be mentioned that as NAND and NOR gates are obtained from the three basic gates NOT, AND and OR, so can the latter be obtained from the former, and, in fact, from only NAND gates or only from NOR gates. This is shown in Fig. 5.6 and Fig. 5.7 for the case of NAND and NOR gates respectively.

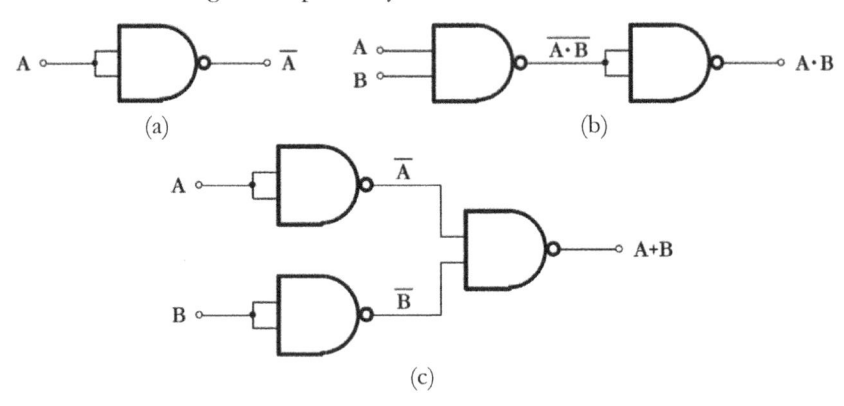

Figure 5.6. Implementation of gates (a) NOT, (b) AND and (c) OR using NAND gates only

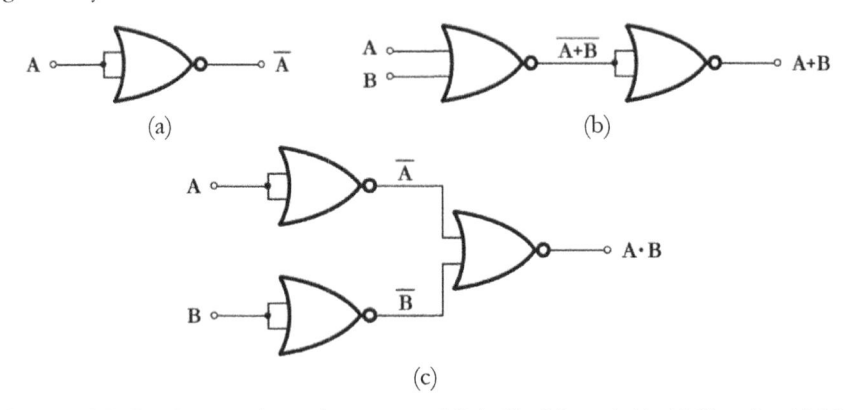

Figure 5.7. Implementation of gates (a) NOT, (b) OR and (c) AND using NOR gates only

Clearly, since NAND or NOR gates can implement the three basic gates NOT, OR and AND, it follows, that any logic function, which is

Digital Combinational Electronics 153

implemented by the three basic gates, can also be implemented by NAND or NOR gates only. This makes NAND and NOR gates to be universal. However, the same functions can be implemented by fewer NAND or NOR gates only, if this substitution is not followed as is shown below.

5.7 Implementation of Logic Functions

Consider a logic function which, after simplification can be written in the following form:

$$F = AB + AC + BC \qquad (5.1)$$

Using the basic gates AND and OR, the implementation of **F** would require three AND gates and one OR gate, as shown in Fig. 5.8(a). The same function can be implemented using NAND gates only, as shown in Fig. 5.8(b). The justification of this is as follows. The output **Y** in Fig. 5.8(b) is

$$Y = \overline{\overline{AB} \cdot \overline{AC} \cdot \overline{BC}}$$

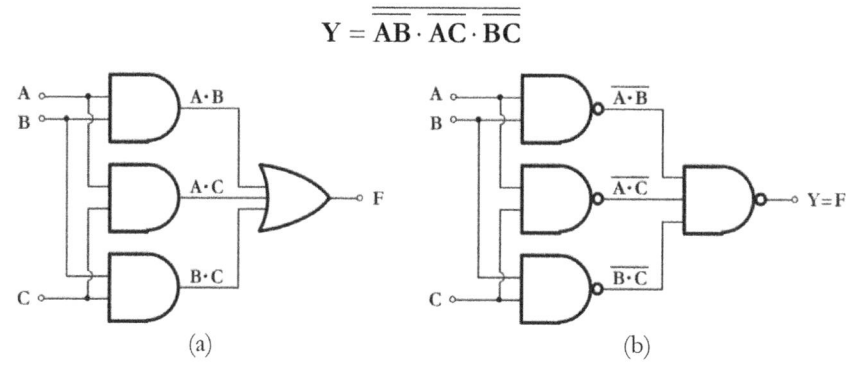

Figure 5.8. Implementation of function **F** using (a) AND and OR gates and (b) NAND gates only

Applying De Morgan Theorem, **Y** becomes

$$Y = \overline{\overline{AB}} + \overline{\overline{AC}} + \overline{\overline{BC}} = AB + AC + BC = F$$

The number of NAND gates here is equal to the total number of AND and OR gates required in the implementation in Fig. 5.8(a), and much smaller than the number of NAND gates that would have been required, if the substitution of AND and OR gates by NAND gates had been utilized.

The form of **F** in Eq. (5.1) is called sum (*S*) of products (*P*), *SOP*, and this is most suitable for implementing a logic function with minimum number of NAND gates, provided of course, that the function cannot be simplified any further.

In accordance with the above discussion, consider a function **F** that, after simplification, can be written in the form

$$F = (A + B) \cdot (A + C) \cdot (B + C) \qquad (5.2)$$

This can be implemented using three OR gates and one AND with three inputs. However, it can be also implemented using NOR gates only, as shown in Fig. 5.9.

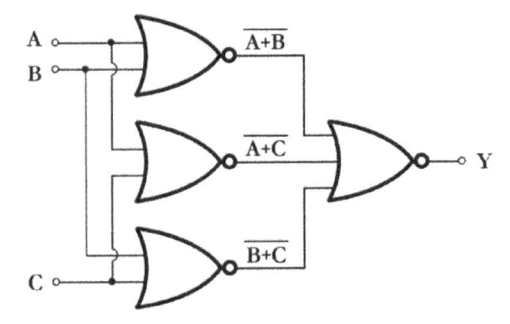

Figure 5.9. Implementation of function **F** in Eq. (5.2) using NOR gates only

To show that this is so, write

$$F = \overline{\overline{(A+B)} + \overline{(A+C)} + \overline{(B+C)}}$$

which, when applying De Morgan Theorem, becomes

$$F = \overline{\overline{(A+B)}} \cdot \overline{\overline{(A+C)}} \cdot \overline{\overline{(B+C)}} = (A+B) \cdot (A+C) \cdot (B+C)$$

It can be seen that writing **F** after simplification in the form of product (*P*) of sums (*S*), *POS*, leads to the implementation of the function with minimum number of NOR gates only.

5.8 Exclusive-OR and Exclusive-NOR Gates

The *Exclusive-OR* (EX-OR) and *Exclusive-NOR* (EX-NOR) gates are two composite gates, but their function is simple, which makes them very useful in practice.

The function of the EX-OR gate on variables **A** and **B**, symbolized as **A⊕B**, is defined as

$$A \oplus B = A\overline{B} + \overline{A}B$$

Similarly the function of the EX-NOR gate, symbolized as **A⊙B**, is defined as

$$A \odot B = AB + \overline{AB}$$

Their truth table is given in Table 5.6.

From the truth table, it follows, that the output of the EX-OR gate is **1** only when the two variables are not the same, whereas the output of the EX-NOR gate is **1** only when the two variables are equal. For this reason,

Digital Combinational Electronics

the EX-NOR is a coincidence gate, whereas the EX-OR is an anticoincidence gate. It is also seen from the truth table that

$$\overline{A \oplus B} = A \odot B \qquad \text{or} \qquad A \oplus B = \overline{A \odot B}$$

Table 5.6.

A	B	A⊕B	A⊙B
0	0	0	1
0	1	1	0
1	0	1	0
1	1	0	1

The symbols of these gates are given in Fig. 5.10(a), while their implementation using NAND gates only is shown in Fig. 5.10(b). The proof that, Fig. 5.10(b) indeed implements these gates, is left for the reader as an exercise.

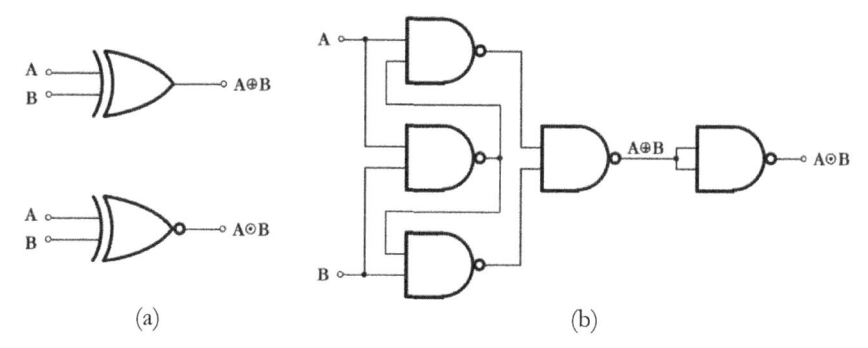

Figure 5.10. (a) Symbols of EX-OR and EX-NOR and (b) their implementation using NAND gates only

The EX-NOR gate is used in practice as an equality detector.

Example 5.M.3

Using Simulink confirm that the circuit of Fig. 5.10(b) indeed implements gates EX-OR and EX-NOR verifying their truth tables in Table 5.6.

Model

% The reader is advised first to consult the Simulink tutorial in Appendix D

In Fig. 5.M.1 the circuit of Fig. 5.10(b) has been implemented as a Simulink model using five *logic operator* blocks operating as NAND gates, two *constant* blocks for the inputs **A** and **B** and two *display* blocks for displaying the outputs of gates EX-OR and EX-NOR. The truth tables of Table 5.6 can be easily derived defining the inputs **A** and **B** in MATLAB command prompt. The outputs in Fig. 5.M.1 have been derived for **A=B=1**.

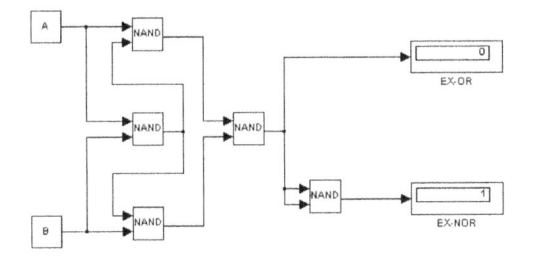

Figure 5.M.1. Simulink model for the circuit of Fig. 5.10(b)

Another approach could be seen in the Simulink model of Fig. 5.M.2. A *signal builder* block together with two *data type conversion* blocks is used instead of the two *constant* blocks in order to create the input waveforms of Fig. 5.M.3. At the output a *scope* block has taken the place of the two *display* blocks in order to observe the output waveforms together with the input ones.

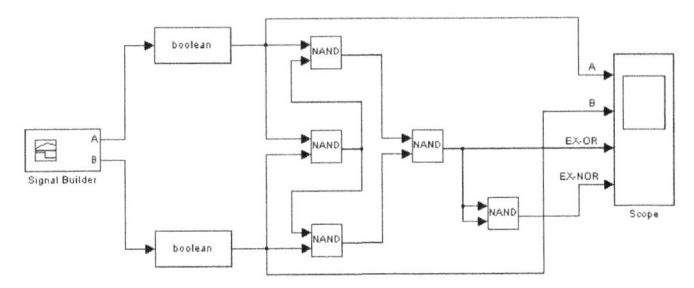

Figure 5.M.2. Alternative Simulink model for the circuit of Fig. 5.10(b)

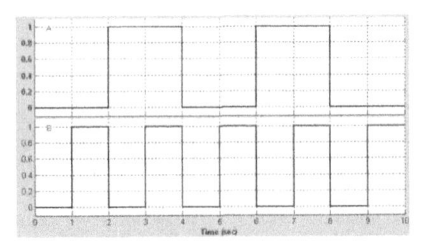

Figure 5.M.3. Input waveforms generated by the *signal builder* block of Fig. 5.M.2

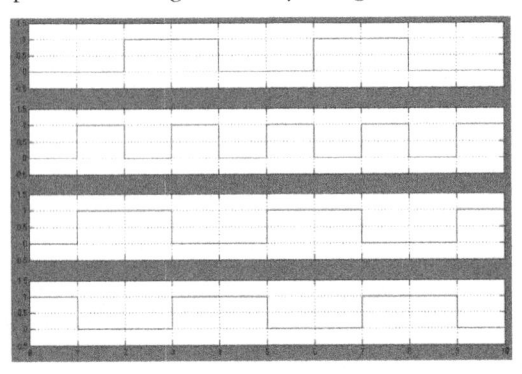

Figure 5.M.4. Output waveforms taken by the block *scope* of Fig. 5.M.2

5.9 A Digital Comparator

On some occasions in practice, it is required to compare two binary numbers. This can be achieved by comparing their corresponding bits. Consider for example bit A_i in number A and the corresponding bit B_i in number B. Clearly if $A_i > B_i$ then $A_i = 1$ and $B_i = 0$, and

$$A_i \cdot \overline{B_i} = 1$$

This can be detected as shown in Fig. 5.11(a).

Similarly if $A_i < B_i$ then $A_i = 0$ and $B_i = 1$ and

$$\overline{A_i} \cdot B_i = 1$$

This can be detected as shown in Fig. 5.11(b). Finally if $A_i = B_i$, they can both be either **0**, or **1**, and this can be detected by an EX-NOR gate. Putting these subcircuits together, the required comparator is obtained, as shown in Fig. 5.11(c).

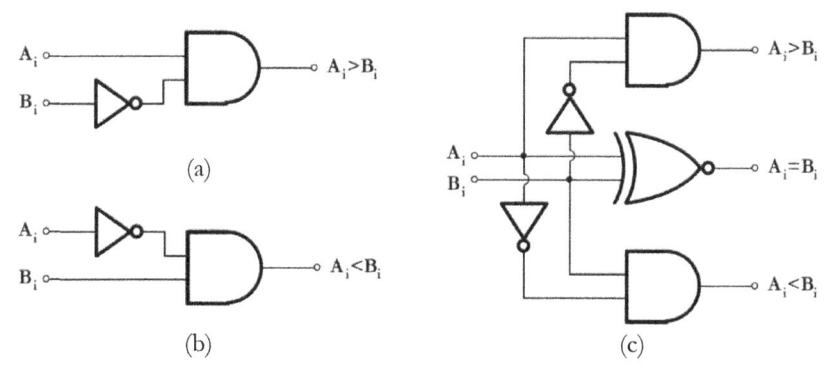

Figure 5.11. Developing the digital comparator circuit

Example 5.M.4

Using Simulink confirm that the circuit of Fig. 5.11(c) indeed implements a digital comparator extracting its truth table.

Model

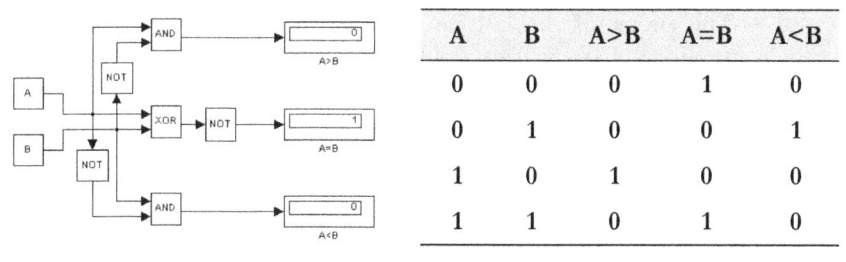

A	B	A>B	A=B	A<B
0	0	0	1	0
0	1	0	0	1
1	0	1	0	0
1	1	0	1	0

Figure 5.M.5. Simulink model for the comparator of Fig. 5.11(c) and its truth table

In Fig. 5.M.5 the circuit of Fig. 5.11(c) has been implemented as a Simulink model using six *logic operator* blocks, two *constant* blocks for the inputs **A** and **B** and three *display* blocks for displaying the outputs. The truth table of the comparator can be easily derived defining the inputs **A** and **B** in MATLAB command prompt. The outputs in Fig. 5.M.5 have been derived for **A=B=1**.

5.10 Binary Adder

Consider the addition of bits **A** and **B**. Their sum **S** will be **0** if **A=B**, **1** if **A** or **B**, but not both, is **1**, and carry **C=1** if **A=B=1**. This can be summarized in the truth table of Fig. 5.12(a). This truth table can be implemented as shown in Fig. 5.12(b), which is called a *Half-Adder (HA)*.

Truth table

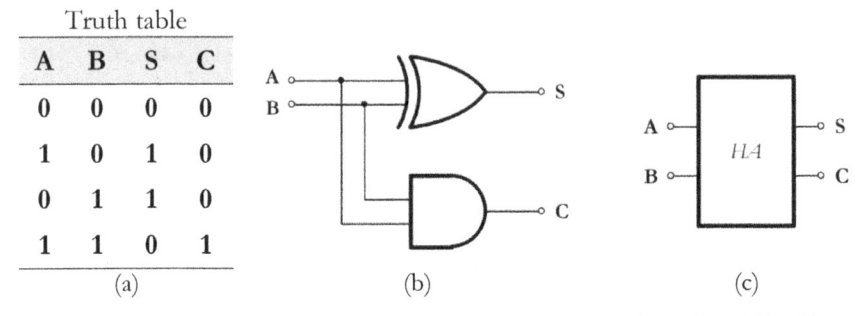

A	B	S	C
0	0	0	0
1	0	1	0
0	1	1	0
1	1	0	1

(a)   (b)   (c)

Figure 5.12. (a) Truth table, (b) implementation and (c) symbol of a Half-Adder

However in the addition of multibit numbers there may be the need to add a carry to **S**, and this can be achieved by using a second *HA*, as shown in Fig. 5.13(a), which is now called a *Full-Adder (FA)*.

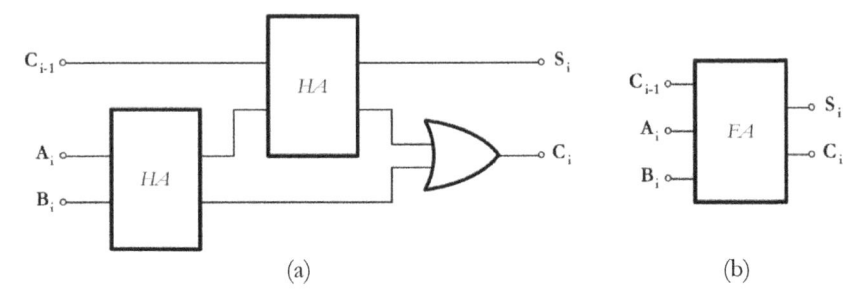

(a)   (b)

Figure 5.13. (a) Implementation of a Full-Adder and (b) its symbol

The circuit of the *FA* can be used to add two multibit numbers serially provided that the carry output C_i is connected to the C_{i-1} input through a delay of one bit. On the other hand, a parallel multibit adder employs as many *FA*s as the number of bits in the numbers to be added. However, provision should be taken concerning the timing of the addition of the carry each time two bits are added. In *IC* parallel adders this is achieved by the method of Look-Ahead-Carry, which requires additional suitable circuitry. An available 4-bit *IC* parallel adder is the 7483, while more

Digital Combinational Electronics 159

such IC's can be combined for the addition of two binary numbers with 16 or 32 bits each.

As was said in Section 5.3, the subtraction of two numbers can be obtained by adding the 2's complement of the subtrahend to the minuend. A 4-bit parallel adder can be used as an adder/Subtractor with additional logic, as shown in Fig. 5.14. When the circuit operates as a Subtractor, the finally dropped bit **1** is suitably fed to input **C₀** and is called *end-around carry*.

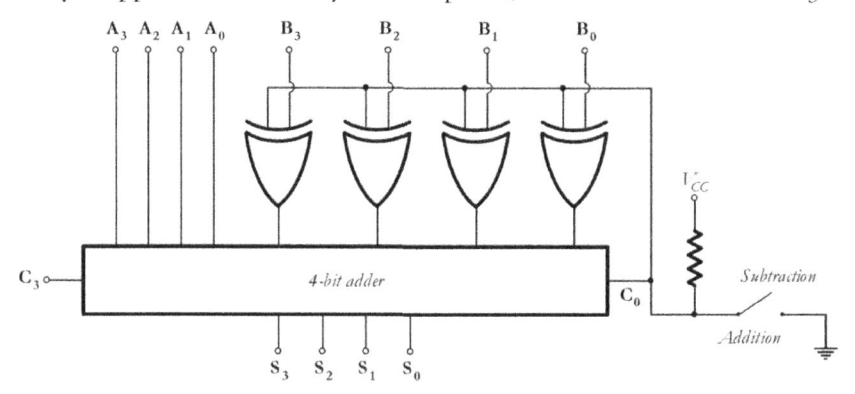

Figure 5.14. A 4-bit adder/subtractor

Example 5.M.5

Using Simulink build a 4-bit adder using the full adder of Fig. 5.13. Confirm its operation.

Model

First a *subsystem* block is built that implements the full adder of Fig. 5.13. The implementation of the subsystem as well as its symbol, named 1-bit full adder, are shown in Fig. 5.M.6. Then, the 4-bit adder is implemented by connecting four such subsystems in cascade, as shown in Fig. 5.M.7. The outputs in Fig. 5.M.7 have been derived by adding the numbers $A_3A_2A_1A_0$=**1001** and $B_3B_2B_1B_0$=**1110** with **C=1**.

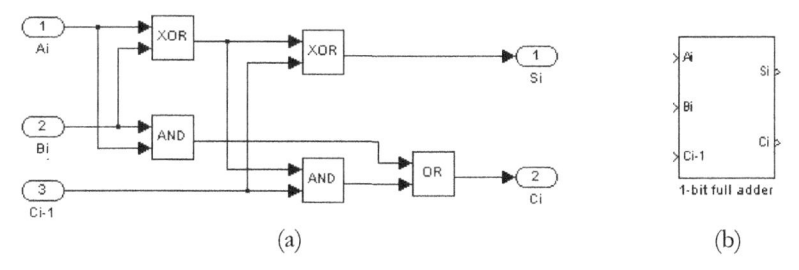

Figure 5.M.6. (a) Simulink subsystem for a full adder and (b) its symbol

It is useful to mention here that the full adder could be modeled without designing the circuit itself but taking into account its truth table only. In Fig. 5.M.8

the full adder has been implemented using a *combinatorial logic* block, where all possible outputs of the truth table of Fig. 5.M.8 have been specified as [0 0;1 0;1 0;0 1;1 0;0 1;0 1;1 1] (see Appendix *D*). The sum is referred first and the carry second. The reader is encouraged to build the 4-bit adder using the subsystem of Fig. 5.M.8 instead of that in Fig. 5.M.6. Attention must be paid to the fact that the inputs in the *constant* blocks must be of Boolean type.

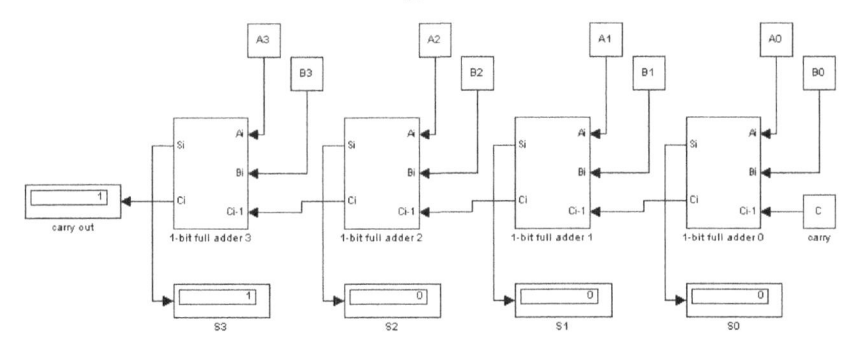

Figure 5.M.7. Simulink model for a 4-bit adder

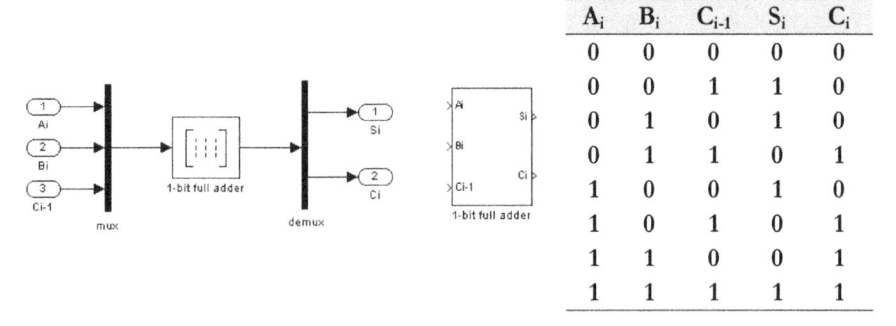

A_i	B_i	C_{i-1}	S_i	C_i
0	0	0	0	0
0	0	1	1	0
0	1	0	1	0
0	1	1	0	1
1	0	0	1	0
1	0	1	0	1
1	1	0	0	1
1	1	1	1	1

Figure 5.M.8. Simulink subsystem for a full adder, its symbol and its truth table

5.11 Binary Codes

Very often electronic circuits process, store and transmit information in the form of binary numbers. For someone to comprehend it, this has to be represented by a decade number, a letter or a special symbol.

A binary code is a unique representation of a set of symbols by a set of binary symbols. With n bits, 2^n different symbols can be represented. There are various binary codes each of which has advantages and disadvantages and is suitable for special applications. In Section 5.3 the "physical" binary code was introduced as well as the 1's complement and the 2's complement codes.

A very useful binary code is the *BCD* (Binary Coded-Decimal) 8421 code, that relates decimal with binary numbers. In this code each decade digit is represented by a group of four bits. The conversion of the decimal digits in *BCD* 8421 code is given in Table 5.7. The 8421 code gets its name from the fact that its four bits $A_3A_2A_1A_0$ represent the digit n on the following basis:

$$n = 8 \cdot A_3 + 4 \cdot A_2 + 2 \cdot A_1 + 1 \cdot A_0$$

Table 5.7. Some binary codes

Physical Binary	8421 BCD	Gray Code	Equivalent Decimal
0000	0000	0000	0
0001	0001	0001	1
0010	0010	0011	2
0011	0011	0010	3
0100	0100	0110	4
0101	0101	0111	5
0110	0110	0101	6
0111	0111	0100	7
1000	1000	1100	8
1001	1001	1101	9
1010	0001 0000	1111	10
1011	0001 0001	1110	11
1100	0001 0010	1010	12
1101	0001 0011	1011	13
1110	0001 0100	1001	14
1111	0001 0101	1000	15

It is understood that the sum takes a value between 0 and 9. A two digit decimal number is represented by two groups of four bits etc. For example the decimal number 35 is represented in the *BCD* code by 0011 0101.

The 8421 code is one of many *BCD* codes, others being the 3311, the 5211 etc. In what follows, when we refer to *BCD* code we will mean the 8421 code. This code is easy in its use by humans, but is not very useful for binary arithmetic operations in digital circuits.

The *Gray* code given on Table 5.7 is characterized by the fact that any two consecutive numbers of the code differ by one bit only.

Other codes use more bits e.g. 7 or 8 for the representation of letters and other symbols like the *ASCII* code (from the initials of American Standard Code for Information Interchange), which is used in the encoding of the computer keyboard. With the 7 bits of this code up to $2^7 = 128$ different symbols can be represented, which include 26 capital and 26 small letters of the alphabet, 10 decimal digits, special characters like >, ., % etc as well as abbreviations of control words.

5.11.1 Encoders

An encoder is a circuit that converts signals corresponding to symbols like A, <, 5, ... to groups of simpler symbols e.g. binary numbers 1010, 0110, i.e. in groups of symbols containing lower information. For example, use of encoders is made in the conversion of signals from the keyboard of a computer to digital signals for subsequent processing by the computer.

The encoder has N input lines, of which only one is at state 1 each time, and M output lines for the representation of a unique word of M bits for each state of the input lines. The operation principle of an encoder is shown in Fig. 5.15. The input lines are connected to the output lines in a unique way according to the decimal code. This determines a set of equations connecting each output to the corresponding state of the inputs, and these equations are then implemented using a diode matrix or otherwise.

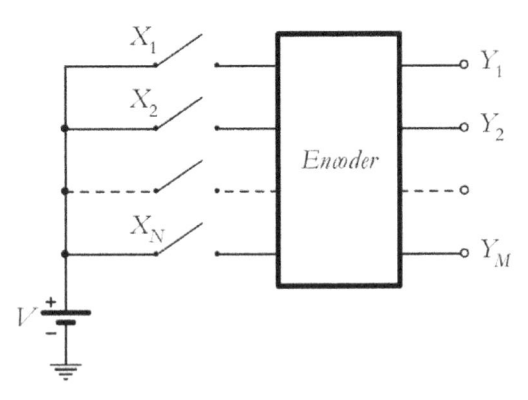

Figure 5.15. Basic operation diagram of an encoder

As an example of determining the required set of equations consider that the truth table of the encoder is the following:

Truth Table

Inputs				Outputs		
X_1	X_2	X_3	X_4	Y_1	Y_2	Y_3
1	0	0	0	1	1	0
0	1	0	0	0	1	1
0	0	1	0	0	0	1
0	0	0	1	1	0	1

It can be seen, that the outputs are given by the following relationships:

$$Y_1 = X_1 + X_4 \qquad Y_2 = X_1 + X_2 \qquad Y_3 = X_2 + X_3 + X_4$$

Digital Combinational Electronics

> **Example 5.M.6**
>
> Using Simulink build an encoder for the truth table of Section 5.11.1 and confirm its operation.

Model

The Simulink model is shown in Fig. 5.M.9. The outputs have been derived for $X_1=0$, $X_2=0$, $X_3=0$ and $X_4=1$.

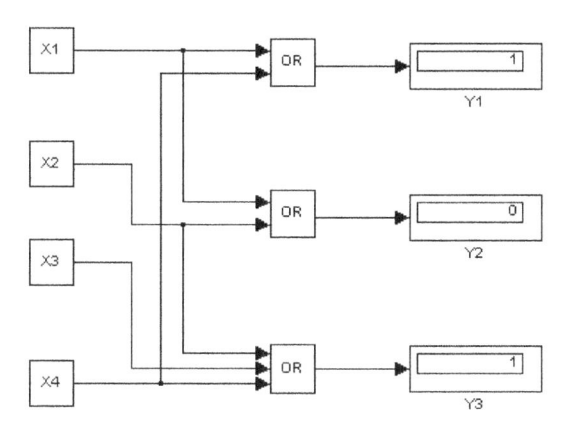

Figure 5.M.9. Simulink model for the encoder of Section 5.11.1

5.12 Decoders

A decoder is a circuit that converts binary symbols (0 and 1) to symbols containing more information e.g. to A, >, 6 etc. A binary decoder converts binary numbers to characters. In the general case they have n input lines and 2^n output lines. The decoders are available as *IC MSI* circuits.

Other decoders can drive displays and special devices apart from decoding. These are called decoders/drivers e.g. those converting *BCD* signals to decimal digits and drive *LEDs*, as explained below. Decoders are also used in the implementation of Boolean functions and combined with memories.

The display of states in digital circuits can be achieved using *LEDs*, liquid crystals (*LCD*) etc. In Fig. 5.16 an *LED* is used to detect a 0 or 1 at the output of an inverter and, consequently, at its input. To emit light each *LED* draws a current of about 15-20mA, when the voltage drop across it is about 1.6V. If the voltage at the output of the inverter at zero state is 0.2V, say, then the value of resistance R, which is used for the protection of the *LED*, should be

$$R = \frac{5 - 0.2 - 1.6}{18 \times 10^{-3}} \approx 180\Omega$$

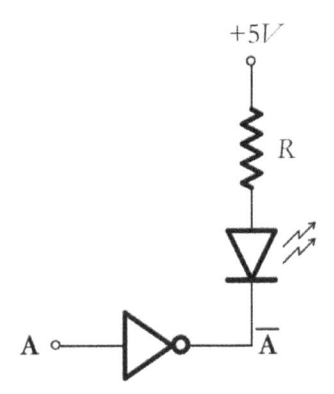

Figure 5.16. Detection of the state of A

A decoder/driver is also used to activate a seven-segment display for the display of any decimal digit from 0 to 9. Seven line segments are arranged to form number 8, as shown in Fig. 5.17(a), while each segment can be lit separately to display any decimal digit. In place of each segment can be an *LED*, but the same effect can be achieved by an *LCD* element.

If the seven-segment element is common-anode, all anodes of the segments are connected together and to the high voltage of the circuit, while each cathode is connected through a protection resistor to a corresponding output line of the decoder/driver. The opposite happens, if the element is of common-cathode. In Fig. 5.17(b) the situation is shown for the display of the decimal digit 2 using a common-anode seven-segment display.

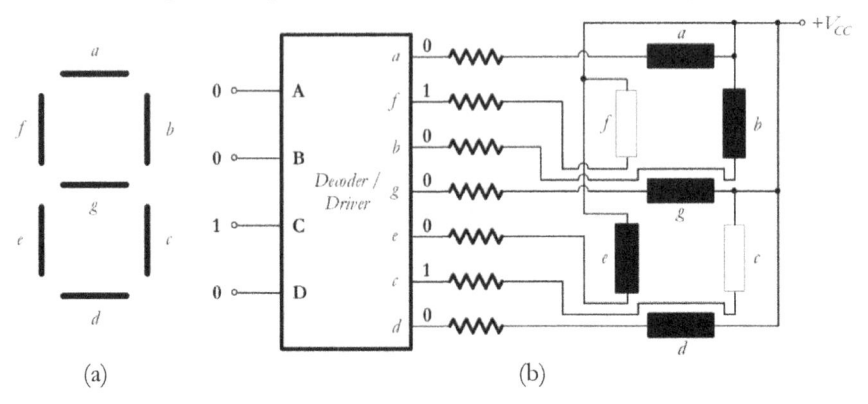

Figure 5.17. (a) A seven-segment display element, (b) the display of number 2 using a common-anode element

5.13 Multiplexers and Demultiplexers

On many occasions there is the need for transferring data serially from different sources through a single channel only. The method of sending data from different lines through one is called *multiplexing* and the device achieving this is called a *multiplexer*. The basic principle of the operation of a multiplexer is shown in Fig. 5.18(a). Only one switch is closed each time, so that only the corresponding input signal should appear at the output. In the digital multiplexer apart from the input lines and the output line, there are control lines, which are called *select* or *address lines*. The latter determine which input line is connected to the output each time. If the number of address lines is n the number of input lines is 2^n.

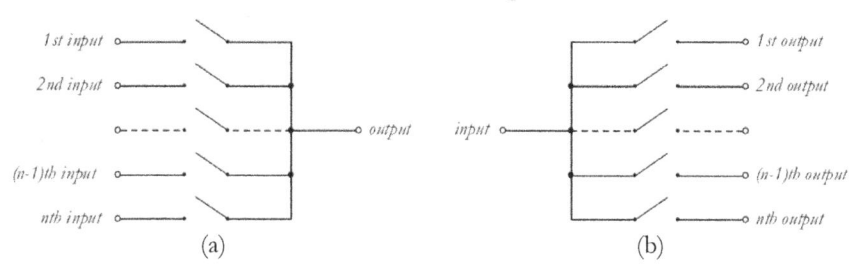

Figure 5.18. (a) Basic multiplexer, (b) basic demultiplexer

It is possible to multiplex the outputs of two (or more) multiplexers to obtain multiplexer with double the number of input lines. Integrated circuit multiplexers are available with additional control lines (strobe or enable) for further control of their operation.

Apart from the application referred to above, a multiplexer can be used for the conversion of parallel data to serial.

The operation of separation the different data from the output of a multiplexer is called *demultiplexing* and the corresponding device *demultiplexer*. The basic demultiplexer is shown in Fig. 5.18(b). Only one switch is closed each time for transferring the data to the corresponding output. The demultiplexer has one input and n output lines. Additionally it has address or select lines for determining the output line that should be connected to the input each time.

Note that a decoder may be used as a demultiplexer, if one of its inputs is used as data line and the rest input lines as address lines. The latter are used for selecting the output, which each time is connected to the input data.

For better understanding of the operation of the multiplexer and the demultiplexer refer to Fig. 5.19. In this figure the communication system needed for transferring the data AB, CD, EF, GH is shown. The address lines A_0, A_1 control the switches S_1, S_2, S_3, S_4 so that they will open and close simultaneously and successively in the multiplexer and the demultiplexer. Thus, the data will be sent serially from the multiplexer through the communication line as $ACEGBDFH$, and the demultiplexer will separate the four different data AB, CD, EF, GH.

166 Chapter 5

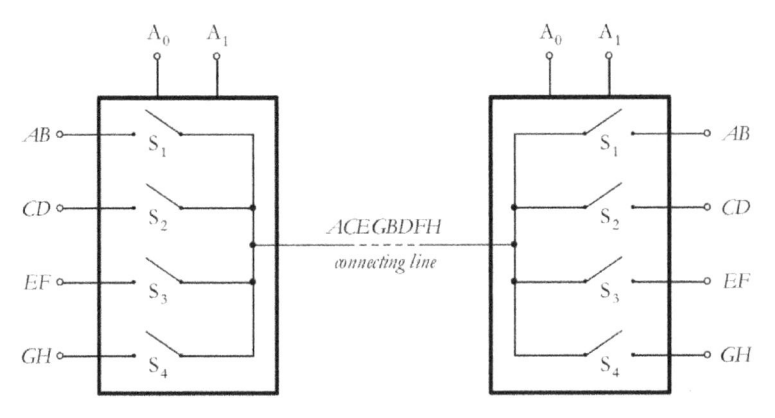

Figure 5.19. Communication system with a multiplexer-demultiplexer

Example 5.M.7

Using Simulink build the four input multiplexer of Fig. 5.19 and confirm its operation.

Model

The Simulink model is shown in Fig. 5.M.10. It has been derived in accordance with the following truth table and equation. The output has been derived for $A_1=0$, $A_0=1$ and $S_2=1$.

Select		Output Y			
A_1	A_0	S_1	S_2	S_3	S_4
0	0	1	0	0	0
0	1	0	1	0	0
1	0	0	0	1	0
1	1	0	0	0	1

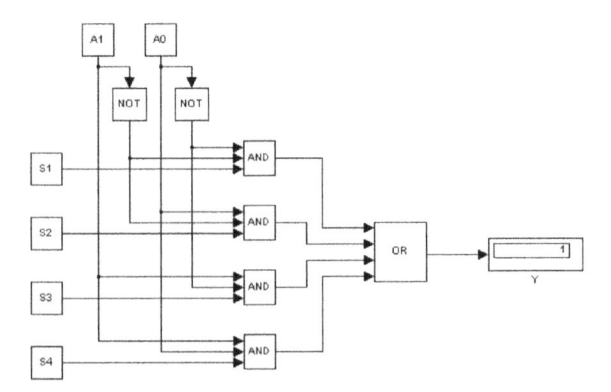

$$Y = S_1 \overline{A_1 A_0} + S_2 \overline{A_1} A_0 + S_3 A_1 \overline{A_0} + S_4 A_1 A_0$$

Figure 5.M.10. Simulink model for the multiplexer of Fig. 5.19

Digital Combinational Electronics

5.14 Error Detectors – Parity Checkers

When transmitting data in digital form, there is always a probability for the introduction of errors in the corresponding signal. Errors can be caused by the incorrect operation of a device, noise etc. Therefore, the need arises for checking the signals at the receiver for assuring that the data are correct.

This check is economically done by means of *parity*. In the binary word to be transmitted, a bit is added, so that the word will have an odd number or an even number of 1's. In the first case we talk about an odd parity and in the second about an even parity. If, for example, the correct word to be transmitted is 10100 in the case of odd parity a bit is added so that the under transmission word will be 110100. In case the under transmission word has an odd number of 1's, for example 01011, the additional bit will be a 0, so that the transmitted word will be 001011 in this example. In the case of even parity, the extra bit will be a 1 or a 0 to make the total number of 1's even.

In the receiver the number of 1's in the received word is checked and, if it does not have the correct odd or even parity, the word is considered erroneous and is rejected.

The circuit for the parity check in the signal source should be capable of adding the correct extra bit so that the under transmission word will have the correct parity. This circuit is then called parity generator and checker. Such circuits are made of Exclusive-OR gates, but are available also as integrated (e.g. the 74180).

References

[1]. Mano M. M. & Ciletti M. D., *Digital Design*, 4th Edition, Pearson Prentice Hall, 2007.

[2]. Greenfield J. D., *Practical Digital Design Using IC's*, Wiley, New York, 1977.

[3]. Millman J. & Tamb H., *Pulse Digital and Switching Waveforms*, McGraw-Hill, New York, 1973.

[4]. Millman J. & Grabel A., *Microelectronics*, McGraw-Hill, 1987.

[5]. Pascoe R. D., *Solid-State Switching: Discrete and Integrated*, Wiley, New York, 1973.

[6]. Malmstadt H. V. & Enke C. G., *Digital Electronics for Scientists*, W. A. Benjamin, New York, 1969.

[7]. Storey N., *Electronics – A Systems Approach*, Addison-Wesley, 1992 (reprinted 1995).

MATLAB Problems

❖ *The reader is advised first to consult the MATLAB tutorial in Appendix D*

❖ *Useful MATLAB functions: base2dec, dec2base, bin2dec, dec2bin, hex2dec, dec2hex*

5.M.1. Using MATLAB perform the following operations:

    a. $(11111011)_2 + (10001100)_2$
    b. $(11111000)_2 - (10000000)_2$

5.M.2. Using MATLAB perform the following operations:

    a. $(7653)_8 + (3611)_8$
    b. $(7156)_8 - (2377)_8$
    c. $(AF13)_{16} + (5EED)_{16}$
    d. $(FFAB)_{16} - (9ABC)_{16}$

5.M.3. Using MATLAB perform the following operations and find the results as decimal numbers:

    a. $(10011001)_2 + (5732)_8$
    b. $(4531)_8 - (3AB)_{16}$
    c. $(99)_{16} + (11001010)_2$
    d. $(456)_{16} - (456)_8$

5.M.4. Using Simulink confirm that the circuits of Fig. 5.6(a)-(c) indeed implement gates NOT, AND and OR verifying their truth tables.

5.M.5. Repeat problem 5.M.4 for the circuits of Fig. 5.7(a)-(c).

5.M.6. Using Simulink confirm that the circuit of Fig. 5.8(b) indeed implements the function of Eq. (5.1) verifying its truth table.

5.M.7. Using Simulink confirm that the circuit of Fig. 5.9 indeed implements the function of Eq. (5.2) verifying its truth table.

5.M.8. Using Simulink plot the output waveforms of the circuits in Fig. 5.8(b) and Fig. 5.9 for the input waveforms of Fig. P.5.M.1.

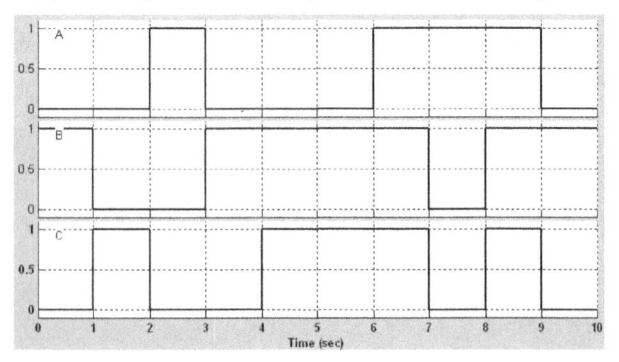

Figure P.5.M.1. Input waveforms

5.M.9. Using Simulink build the 4-bit adder/subtractor of Fig. 5.14 and confirm its operation.

5.M.10. Using Simulink build the four input demultiplexer of Fig. 5.19 and confirm its operation.

Problems

5.1. Convert the following decimal numbers to binary.

   a. $(85)_{10}$
   b. $(-110)_{10}$

5.2. Convert the following octal and hexadecimal numbers to decimal and then to binary.

   a. $(63)_8$
   b. $(1A4)_{16}$

5.3. Convert the following non-signed binary numbers to decimal and give their equivalents in *BCD* code.

   a. 101011
   b. 110111001

5.4. Using Boolean Algebra theorems or otherwise show the validity of the following identities:

$$AB + \overline{A}C + BC = AB + \overline{A}C$$
$$(A+B)(\overline{A}+C)(B+C) = (A+B)(\overline{A}+C)$$
$$A + \overline{A}B = A + B$$
$$A(\overline{A}+B) = AB$$

5.5. Implement the EX-OR gate using a) NAND gates only and b) NOR gates only.

5.6. Draw a logic diagram with three inputs **A, B, C** and an output **Y** using NAND gates only, that will give **Y=1** only when at least two of the inputs are in state **1**. Use as few gates as possible.

5.7. Simplify the following Boolean expressions and implement the results using NAND gates only.

$$F_1 = (\overline{A}+B)(A+B)$$
$$F_2 = (\overline{A}+B+C)(A+B+\overline{C})$$

5.8. Given two binary numbers A_1A_0 and B_1B_0 derive a logic diagram that will detect their equality. Use any suitable gates.

5.9. A group of signals for the decimal numbers 0 to 9 is sent in parallel in the binary system. Derive a logic diagram using the minimum number of NAND gates only that will recognize when the numbers 5 to 9 are sent.

5.10. Using any available gates design a logic diagram, that will implement the following code conversion.

X_1	X_0	Y_3	Y_2	Y_1	Y_0
0	0	1	0	1	0
0	1	1	0	0	1
1	0	0	1	1	0
1	1	1	1	0	0

Chapter 6

Digital Sequential Electronics

6.1 Introduction

In the combinatorial circuits we have seen so far, the output states of the basic gates depend on the signals at their inputs. Based on this, the combinatorial circuits "take decisions". However, in digital electronics, circuits are required to store previous values of input and output signals that, together with the present input signals, will determine the new value of the output signal. Such circuits are called *sequential*. Basic sequential circuits are the various types of *latches* and *flip-flops*, which can be built using gates, and in particular NAND gates. The flip-flops are useful in building more complicated sequential circuits like counters, shift-registers, memories, etc. All these circuits are available in integrated form and are examined to some extent here below.

In this Chapter the latch and flip-flops are introduced first. The important race problem in flip-flops is discussed together with its remedy, the use of master-slave flip-flops. Also the characteristics of the integrated circuit flip-flops are explained. Included in this Chapter are three circuits related to the latch, namely, the *monostable* or *one-shot*, the *astable multivibrator*, useful as clock in nearly all sequential circuits and the *Schmitt Trigger*. The use of flip-flops in building counters and registers is introduced next. Then the memories are discussed and microprocessors and microcomputers are briefly explained. Modern design of digital circuits using a *Hardware Description Language* (*HDL*) is also introduced. Finally, the state of the art and future projection of digital integrated circuits are briefly discussed.

6.2 Latches

The latch, realized using two NAND gates is shown in Fig. 6.1 together with its truth table. It must be noticed, that it has two complementary outputs **Q** and $\overline{\mathbf{Q}}$, a facility offered by all types of flip-flop. Clearly, with a **0** at the **S** input the latch output **Q** is SET to **1**, independently of the state at the input **R**, while when **S=1** and **R=0** the latch is RESET, i.e. **Q** comes to **0** independently of its previous state. Finally, when **S=R=1** the latch does not change, i.e. it remains at its previous state, unless it follows the input states **S=R=0**. In fact, if **S=R=0**, the output states will be **Q**=$\overline{\mathbf{Q}}$=**1**, and there will be an indetermination as to which values **Q** and $\overline{\mathbf{Q}}$ will take. The circuit operation can be explained by means of the NAND gate truth table and is left to the reader as an exercise.

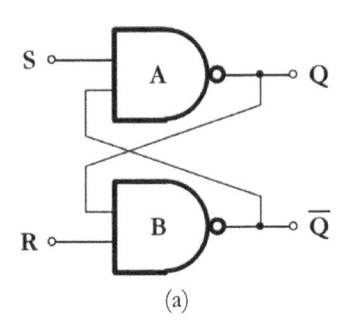

Truth table

S	R	Q	$\overline{Q}$
0	0	1	1
1	0	0	1
0	1	1	0
1	1	Q	$\overline{Q}$

(a) (b)

Figure 6.1. (a) Latch and (b) its truth table

Similar results are obtained, if NOR gates are used instead of NAND gates. However, the truth tables will be different in the two cases **S=R=0** and **S=R=1**.

It is clear from the above discussion, that the latch changes states, only when there is a change in the states at its inputs, which means, that it "remembers", which had been the last states at its inputs. Thus, the latch is a memory circuit. This simple bistable circuit is usually referred to as *Latch* to be distinguished from the other more complicated flip-flops that are introduced next.

Example 6.M.1

Create a Simulink model for the latch of Fig. 6.1(a) and confirm its operation verifying the truth table of Fig. 6.1(b).

Model

% The reader is advised first to consult the Simulink tutorial in Appendix *D*

In Fig. 6.M.1 the latch of Fig. 6.1(a) is shown implemented as a Simulink model using two *logic operator* blocks operating as NAND gates, one *memory* block for defining the previous output state **Q**, two *constant* blocks for the inputs **S** and **R** and two *display* blocks for displaying the outputs **Q** and $\overline{\mathbf{Q}}$. The truth table of Fig. 6.1(b) can be easily derived defining the inputs **S** and **R** in MATLAB command prompt.

Digital Sequential Electronics 173

The outputs in Fig. 6.M.1 have been obtained for **S=0** and **R=1**. For the case **S=R=1** the initial condition of the *memory* block has to be defined.

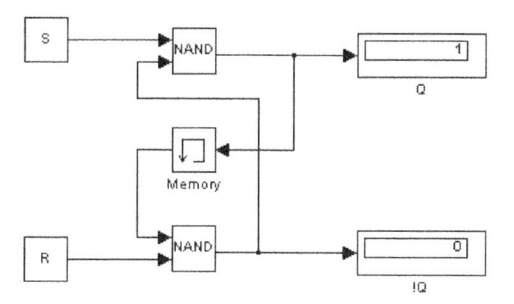

Figure 6.M.1. Simulink model for the latch of Fig. 6.1(a)

Another approach can be seen in Fig. 6.M.2, where the latch is modeled without designing the circuit itself, but taking into account its truth table only. The latch has been implemented using a *combinatorial logic* block, where all possible outputs of the truth table of Fig. 6.M.2 have been specified as [1 1;1 1;1 0;1 0;0 1;0 1;0 1;1 0] (see Appendix D). **Q** is stated first and $\overline{\mathbf{Q}}$ second. Attention must be paid to the fact that the inputs in the *constant* blocks must be of Boolean type.

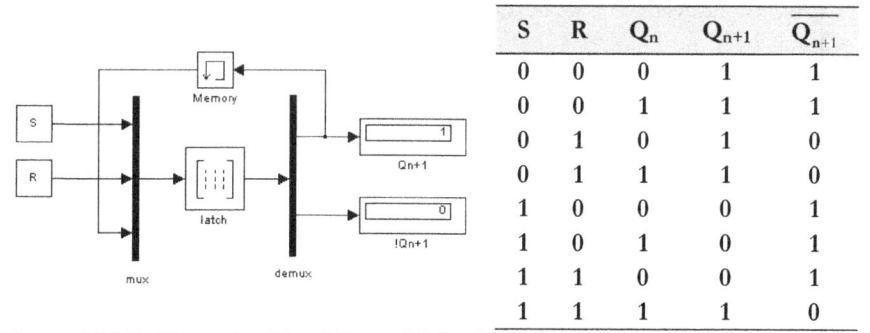

S	R	Q_n	Q_{n+1}	$\overline{Q}_{n+1}$
0	0	0	1	1
0	0	1	1	1
0	1	0	1	0
0	1	1	1	0
1	0	0	0	1
1	0	1	0	1
1	1	0	0	1
1	1	1	1	0

Figure 6.M.2. Alternative Simulink model for the latch and its truth table

6.3 Flip-Flops

The various flip-flops constitute a family of circuits, which are used in all processing systems of digital data. They all are variations of a bistable circuit that has two stable (permanent) states. It remains in one state until it gets triggered to change state, an operation that is executed very fast. The basic flip-flop can be designed using transistors, resistors and capacitors, but their construction using NAND gates is considered here. The simplest R-S flip-flop is introduced first.

6.3.1 The *R-S* Flip-Flop

The realization of the R-S flip-flop, its symbol and its truth table are shown in Fig. 6.2. Two gates have been added to the latch, while a third input **Ck** is available. When input **Ck** is at zero state, the outputs of the new gates are at level **1** and no change can happen to the output of the latch part according to its truth table. However, when the **Ck** input is at level **1** the R-S flip-flop can change according to the states of **S** and **R** inputs. As in the case of the latch, though there will be indetermination in the states of **Q** and $\overline{Q}$ when the states at the inputs **R=S=0** follow the states **R=S=1**.

Truth table (**Ck=1**)

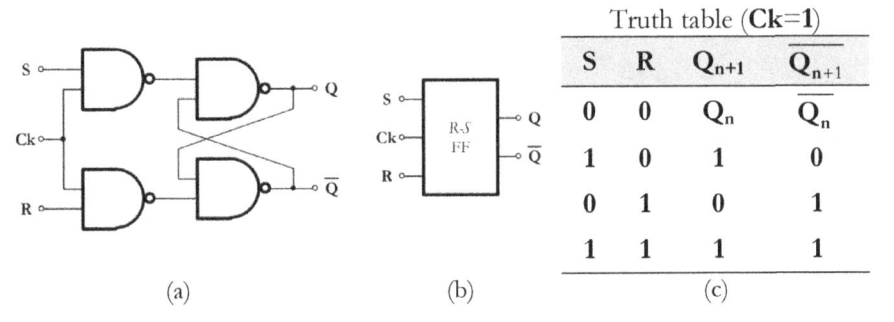

S	R	Q_{n+1}	$\overline{Q}_{n+1}$
0	0	Q_n	$\overline{Q}_n$
1	0	1	0
0	1	0	1
1	1	1	1

(a)   (b)   (c)

Figure 6.2. R-S flip-flop (a) block diagram, (b) its symbol and (c) its truth table

In the truth table Q_n represents the state of the output **Q** before the **Ck** input changes to **1**, while Q_{n+1} represents the state at the **Q** output after **Ck** has become **1**.

Clearly, the change of the R-S flip-flop is controlled by the input **Ck**, since, only when **Ck=1**, it is possible for the flip-flop to respond to signals at its inputs **R** and **S**. Usually, at the input **Ck** of the R-S flip-flop and of the other flip-flops that follow next, a series of square pulses is applied. These pulses are called *clock pulses* and are required for the synchronization of the various flip-flops in the same sequential system. They are produced by a non-harmonic oscillator, which is called *astable multivibrator* and, when used in the sequential circuits, it is called the *clock*.

6.3.2 The *T* Flip-Flop

Apart from the property to "remember", which makes the flip-flop useful as a memory device, it can be used to count pulses. In such a case, the flip-flop must change state each time a pulse is applied to its input. Then, for every second pulse the flip-flop will return to the initial state. This will be exploited later in the design of counters.

Suitable flip-flop for counting is the *T* flip-flop (toggle), which can be obtained from the R-S flip-flop by connecting the input **S** to the $\overline{Q}$ output and the **R** input to the output **Q**. Its symbol and truth table are shown in Fig. 6.3.

To avoid having the flip-flop changing states continuously as long as the clock is kept in the **1** state, the clock pulse is differentiated to appear

Digital Sequential Electronics

as a spike. Commercial more sophisticated T flip-flops are triggered either by the positive going edge of the clock pulse, *positive edge triggered*, or by its falling edge, *negative edge triggered* T flip-flop. In this case the **1**s in the clock column in the truth table are replaced by arrows pointing up or pointing down, depending on whether the T flip-flop is positive or negative edge-triggered respectively. One more point, the pulses at the input **Ck** do not necessarily have to be clock pulses i.e. applied regularly.

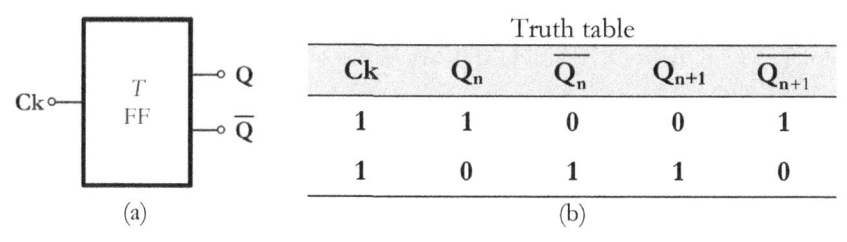

Ck	Q_n	$\overline{Q_n}$	Q_{n+1}	$\overline{Q_{n+1}}$
1	1	0	0	1
1	0	1	1	0

(a)   (b)

Figure 6.3. T flip-flop (a) symbol and (b) truth table

6.3.3 The *D* Flip-Flop

The D flip-flops have one input **D** for data and one control input **Ck** (enable). The most common D flip-flop is introduced here only.

Its symbol is as shown in Fig. 6.4(a), and it is triggered by the positive edge of the clock pulse (positive-edge triggered). Thus during the clock transition **0→1** the state of the input **D** is transferred to the output **Q**. The two extra inputs **Pr** (preset) and **Cr** (clear) are synchronized with the **Ck** pulses and during the operation of the D flip-flop should be both **1**. The truth table is given in Fig. 6.4(b) for the synchronous operation and in Fig. 6.4(c) for the asynchronous operation of the D flip-flop.

Pr=1, Cr=1

Ck	D_n	Q_{n+1}	$\overline{Q_{n+1}}$
↑	0	0	1
↑	1	1	0

Ck	Pr	Cr	Q_{n+1}	$\overline{Q_{n+1}}$
×	0	0	1	1
×	0	1	1	0
×	1	0	0	1
×	1	1	Q_n	$\overline{Q_n}$

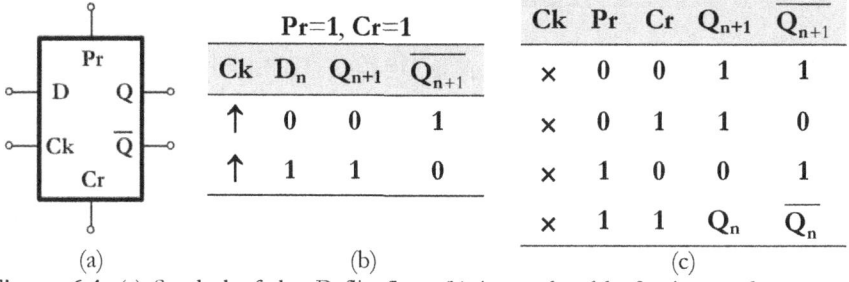

(a)   (b)   (c)

Figure 6.4. (a) Symbol of the D flip-flop, (b) its truth table for its synchronous operation and (c) truth table for tits asynchronous operation. The symbol ↑ has the meaning that the flip-flop is triggered by the positive-going edge of the clock pulse and the symbol ×, don't care whether the clock pulse is **0** or **1**.

The D flip-flop is used to delay pulses by 1 bit.

Example 6.M.2

Using Simulink confirm the truth table of a D flip-flop.

Model

In Fig. 6.M.3 a *D flip-flop* block together with a *clock* block have been used. The input **D** is of Boolean type and is defined in MATLAB command prompt. The state of input **CLR** is defined via a *manual switch* block. The flip-flop is triggered by the positive edge of the clock pulse and the state of the input **D** is transferred to the output **Q** if **CLR=1** (Fig. 6.M.4(a)) while the output is cleared if **CLR=0** (Fig. 6.M.4(b)).

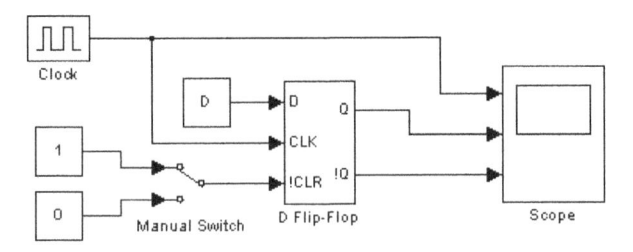

Figure 6.M.3. D flip-flop in Simulink

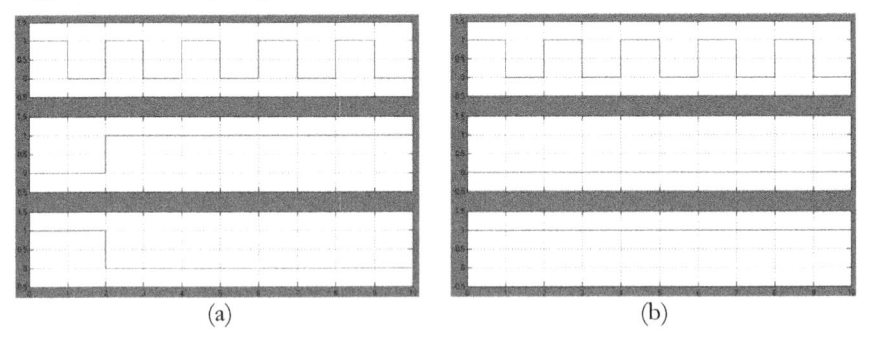

Figure 6.M.4. Output of D flip-flop for **D=1** and (a) **CLR=1** and (b) **CLR=0**

6.3.4 The *JK* Flip-Flop

The *JK* flip-flop and, in particular, its master-slave version is the most useful flip-flop. It can be used to obtain the *R-S* flip-flop, the *T* flip-flop and the *D* flip-flop. Its symbol and truth table for the negative-edge trigger type are given in Fig. 6.5.

All types of *JK* flip-flop also have two asynchronous preset (**Pr**) and clear (**Cr**) inputs. Also, some integrated circuit *JK* flip-flops have AND gates at the inputs **J** and **K** (gated *JK* flip-flops) with their inputs characterized as $J_1, J_2, \ldots$ and $K_1, K_2, \ldots$ respectively. These multiple inputs can be used to perform extra logic, as this happens, for example, in designing synchronous counters.

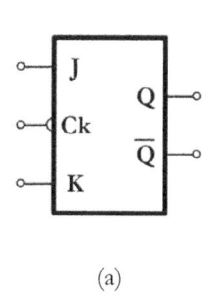

Digital Sequential Electronics 177

J_n	K_n	Q_{n+1}	$\overline{Q_{n+1}}$
0	0	Q_n	$\overline{Q_n}$
1	0	1	0
0	1	0	1
1	1	$\overline{Q_n}$	Q_n

(a)   (b)

Figure 6.5. *JK* flip-flop (a) symbol and (b) truth table

Example 6.M.3
Using Simulink make a *JK* flip-flop operate as *T* flip-flop and confirm its operation.

Model

The *JK* flip-flop in Fig. 6.M.5 operates as *T* flip-flop when **J=K=1** according to the truth table of Fig. 6.5(b). The waveforms of Fig. 6.M.6 verify its operation.

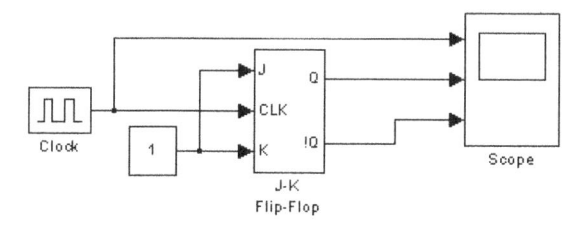

Figure 6.M.5. *JK* flip-flop operating as *T* flip-flop

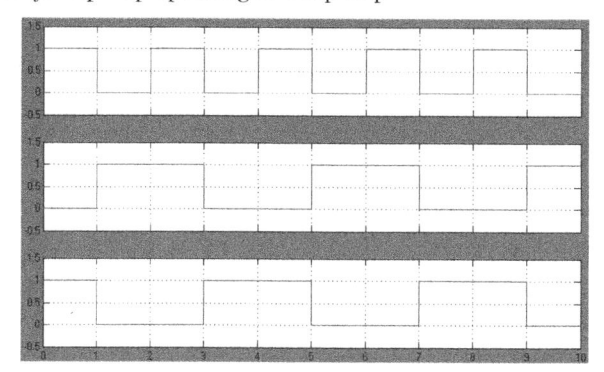

Figure 6.M.6. Clock pulses and outputs of *JK* flip-flop for **J=K=1**

6.3.5 The Race Problem in Flip-Flops. Master-Slave Method

Consider the case of two flip-flops connected in cascade, as shown in Fig. 6.6. Let the conditions in the flip-flop *A* be **S=1, R=0, Q₁=0, Q₂=0** before the active transition of the clock. The active transition of the clock pulse will be applied simultaneously to both clock inputs of the flip-flops. At that moment flip-flop B has $S_2=Q_1=0$, since Q_1 does not change instantaneously with the active transition of the clock. It changes after a short delay. If flip-flop B reacts quickly to the clock pulse, it may respond before Q_1 changes to state **1** and thus Q_2 will remain at **0**. But if flip-flop B reacts more slowly, it may have the time to see Q_1 change to state **1** and thus, Q_2 will change to state **1**. There is therefore an uncertainty in the operation of flip-flop B and this produces what is called the *race problem*.

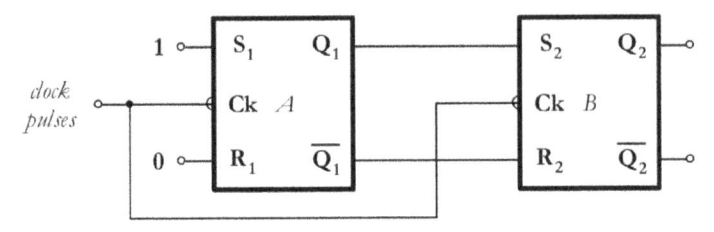

Figure 6.6.

Race problems could occur in chains of flip-flops such as in counters and shift-registers. However, there are methods that prevent these uncertainty problems from occurring by paying attention to the delay and response times of the device, as it is clarified by the characteristics of the flip-flops.

A method that overcomes the problems of race is the use of *master-slave* flip-flops. Each master-slave flip-flop consists of two individual flip-flops, such as *RS*, *D* or *JK*, connected in cascade, but with the clock pulse being applied to the second flip-flop, the slave, through an inverting gate. Thus the first flip-flop, the master, sets to the proper state when the clock is active, while during that period the slave remains unchanged. When the clock becomes inactive for the master flip-flop, it becomes active for the slave, because of its inversion, which then takes the state of the master. This way, there is no uncertainty regarding the state at the input of the slave; therefore, no race problems will occur.

IC master-slave *RS*, *D* and *JK* are also available together with the other types of *IC* flip-flops.

6.3.6 Flip-Flop Characteristics

The various types of flip-flop are produced as integrated circuits in TTL or MOS technologies (see Section 6.11), which of course determine some of their characteristics. Most of these characteristics are given in the manufacturers' data and they are the following:

a. Highest clock frequency f_m that can be applied for reliable operation of the flip-flop.

Digital Sequential Electronics 179

b. Set up time t_s. For time t_s the signal has to be applied to an input terminal before the appearance of the triggering signal at another terminal. Thus for a D flip-flop $t_s=20ns$ has the meaning that the voltage at **1** or **0** state should be kept constant for $20ns$ at the input D before the appearance of the active edge ↑ or ↓, depending on the D flip-flop, of the clock pulse.

c. Hold time t_{hold}. The signal at one input terminal should remain constant for time t_{hold} after the active edge of the clock pulse has been appeared. For most flip-flops this time is zero.

d. Clock high pulse width is the minimum time that the clock pulse should remain at the high state for reliable operation.

e. Clock low pulse width is the minimum time that the clock pulse should remain at low state for reliable operation.

f. Preset or clear low is the minimum time the preset or clear input should remain in low state for reliable presetting or clearing the flip-flop.

Also, the flip-flop manufacturers give the delay times after which clock, preset and clear pulses affect the output, so that the designer will be able to estimate the total delay time of signal transmission through a series of gates and flip-flops.

6.4 Clocks

As was mentioned earlier in this chapter, the sequential circuits require a train of regular pulses for their operation and synchronization. The clock is an astable circuit that produces this type of pulses. Together with the bistable and the monostable, which are introduced in the next section, they belong to a family of electronic circuits called *multivibrators*.

An astable multivibrator can be produced using two gates NOT, NAND or NOR, two resistors and a capacitor, as shown in Fig. 6.7 for example. A brief explanation of the operation of the circuit is as follows: Let the output of gate A at a certain moment be at high state (when its input will be at the low state). Then, the output of gate B will be at the low state and the capacitor will start charging through the resistance R. When the voltage of the capacitor reaches a value that corresponds to state **1** for the input of gate A, this gate will change state with its output coming to low state, which will bring the output of gate B to state **1**. Then, the capacitor will start discharging through R. But when the voltage at the left plate of the capacitor reaches the low state, which through R_s is the input to gate A, this will cause this gate to change to high state and the process will be repeated. The period of pulses generated this way is proportional to the product RC.

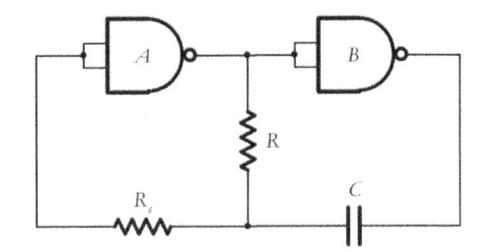

Figure 6.7. Astable multivibrator (clock) using NAND gates

Astable multivibrators are available in all integrated circuit technologies. A popular circuit is under the code number 555 and can give frequencies ranging from $0.1Hz$ up to $100kHz$. It requires the external connection of two resistors and a capacitor, which determine the frequency, and the duty cycle of each pulse. Another useful circuit is the 74S14 of *TTL* technology giving a frequency range from $0.12Hz$ up to $85MHz$, and requires the external connection of only one capacitor to determine the frequency.

This and some other astable integrated circuits may have their frequency controlled by an externally applied voltage. Such circuits are called *Voltage-Controlled Oscillators*, or *VCO*.

A high frequency clock can be turned to a polyphase one using flip-flops or even a counter. Such polyphase clocks are useful in many digital systems.

Finally, to ensure high frequency stability of the clock pulses to variations of applied voltage or changes in temperature, piezoelectric crystals are used instead of *RC* circuits.

6.5   Monostables or One-Shots

The monostable multivibrator or *One Shot*, *OS*, is very useful in pulse circuitry. It can be built using two transistors, two NAND or NOR gates and *RC* components. A version using *IC* NOR gates is shown in Fig. 6.8.

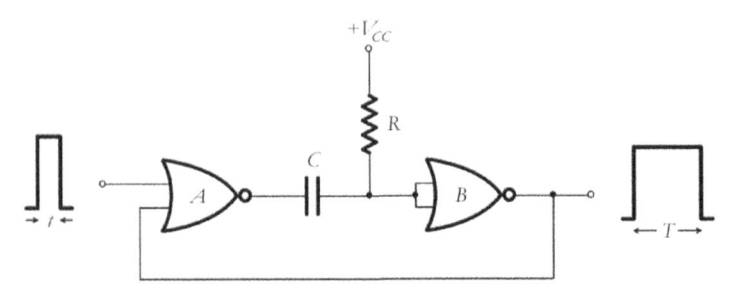

Figure 6.8. An *OS* using NOR gates and corresponding pulses at the input and output

Its operation is briefly as follows: With no input pulse (input low) the output of gate A is high and the capacitor is uncharged, since both its plates are high. Therefore, the output of gate B is low, since both its inputs are high. When an input pulse is applied, as shown in Fig. 6.8, the *OS* is triggered, meaning that with one of its inputs high, gate A is changed, its output becoming low. Consequently, the left plate of the capacitor is at low (**0**) voltage and, since the voltage in a capacitor cannot change suddenly, the right plate of the capacitor comes to the same low voltage. This causes the output of gate B to go high, while the capacitor starts charging, with time constant RC. However, when the voltage across it reaches a value that the input of gate B considers as high, this gate changes to low, which causes the output of gate A to go high. Thus depending on the time constant RC, the width of the output pulse can be much wider than that of the input pulse, as shown in the figure. In fact, the output pulse width is found to be

$$T = 0.69RC \qquad (6.1)$$

In the circuit in Fig. 6.8 the resistor and the capacitor are added externally to the integrated circuit of NOR gates package.

Clearly, the monostable as presented above can be turned to an astable circuit (see Fig. 6.7), while it provides two outputs **Q** and $\overline{\mathbf{Q}}$, and can be controlled by a gate. It is also used to restore the square form of pulses with different widths and can be useful as a pulse amplifier.

IC monostables (not including the R and C components, which are connected externally to determine the required pulse width) are produced in different technologies and can be triggered with the leading (up-going) edge of the input pulse, *leading edge-triggered monostables*, or with the falling edge of the input pulse, *trailing edge-triggered monostables*.

Also, *IC* monostables can be distinguished as non-retriggerable and retriggerable. The latter can be retriggered, when their output is still high, which may produce pulses of long duration, something that is not possible with the non-retriggerable *OS*.

The *IC* 555, which was referred to in the previous section, apart from being used as an astable or clock circuit, can also be used as a monostable with externally added passive components.

6.6 Schmitt Trigger

The *Schmitt Trigger* is a useful circuit in any pulse circuitry. This circuit will not turn on until the input voltage is greater than a certain value, called positive-going threshold voltage or upper threshold voltage, and will not turn off unless the input voltage is less than another voltage, called the negative-going threshold voltage or lower threshold voltage. This characteristic operation can be presented more clearly by means of Fig. 6.9.

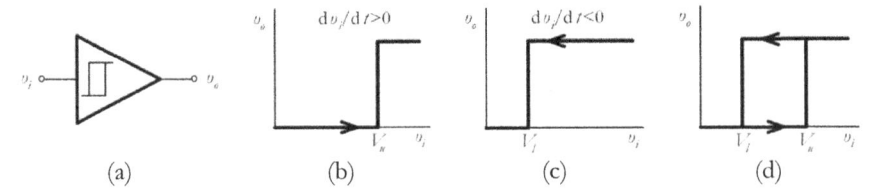

(a) (b) (c) (d)

Figure 6.9. (a) Symbol, (b), (c) operation of the Schmitt Trigger and (d) formation of a hysteresis loop

The usual symbol is a triangle with a hysteresis symbol placed inside. As the input voltage v_i increases from a low value, ($dv_i/dt>0$), the output v_o remains low until v_i passes the upper-threshold voltage V_u when v_o becomes high, Fig. 6.9(b). When v_i decreases ($dv_i/dt<0$) from values higher than V_ℓ and v_o is high, the circuit output does not go low unless v_i becomes lower than V_ℓ, Fig. 6.9(c). Combining figures 6.9(b) and 6.9(c) a hysteresis loop is formed, as shown in Fig. 6.9(d), which is the input-output characteristic of the Schmitt Trigger.

The result of the characteristic behaviour of the Schmitt Trigger can be demonstrated by means of Fig. 6.10. Consider that the input waveform to the circuit is as shown in Fig. 6.10(a), where the threshold voltages V_u and V_ℓ are marked. The output waveform of the circuit is shown in Fig. 6.10(b). Note how the noisy waveform in Fig. 6.10(a) has been changed to a clear waveform in Fig. 6.10(b).

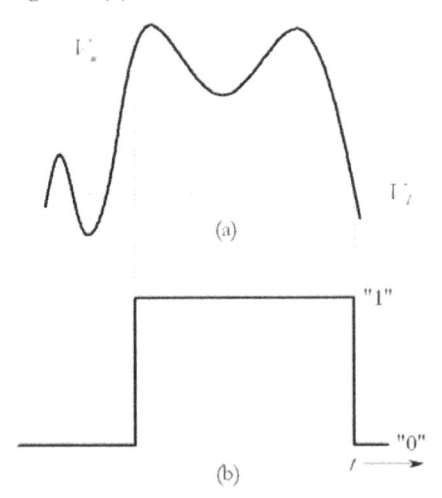

Figure 6.10. (a) A noisy waveform and (b) the Schmitt Trigger output waveform

Apart from providing noise immunity to waveforms, Schmitt Trigger can be used as a pulse-height discriminator. Each time the pulse-height of the input voltage becomes higher than the upper-threshold

Digital Sequential Electronics

voltage, a pulse is produced at the output of the circuit. Further, on connecting two such circuits with different V_u levels in parallel, as shown in Fig. 6.11, the output of the EX-OR gate will produce **1** for the pulses whose heights are between V_{u1} and V_{u2} and **0** for all the other pulses. Thus the circuit behaves like a "window", and if one can move both V_u levels up or down keeping the difference $\Delta V_u = V_{u1} - V_{u2}$ constant, one can produce a record of the number of pulses with heights between $V_u + \Delta V_u$, like a spectrum, useful in the case for example of studying the energy distribution of particles from a radioactive material.

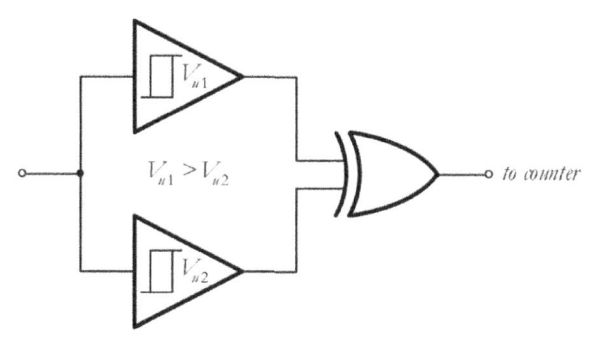

Figure 6.11. A simple arrangement to count the pulses with heights between V_{u1} and V_{u2}

An obvious application of a Schmitt Trigger is to obtain square waves out of a sine wave. *IC* Schmitt Triggers are available mainly in *TTL* and *CMOS* (see Section 6.11) technologies. Their characteristic is the hysteresis voltage V_{11}, with $V_{11} = V_u - V_\ell$. However, it should be noted that generally, the voltages V_u and V_ℓ of the *IC* Schmitt Trigger cannot be changed. A popular Schmitt Trigger *IC* is the 7413.

6.7 Counters

A counter is a digital circuit that can count pulses, which regularly or irregularly appear at its input. These pulses may represent discrete events, as for example radiation from a radioactive material, the number of vehicles crossing a certain line in a road, etc.

Digital counters are built as a chain of flip-flops and are based on the property of the flip-flops to return to their initial state for every second pulse applied to their input. Counters can be binary i.e. counting in the binary system, or decade, counting in the decimal system of numbers. However, using suitable logic, they can count up to a certain number, say up to seven, and return to their initial state. There are counters that can count-up or count-down depending on the application. Counters are also distinguished as ripple or asynchronous and as synchronous.

6.7.1 Ripple Counters

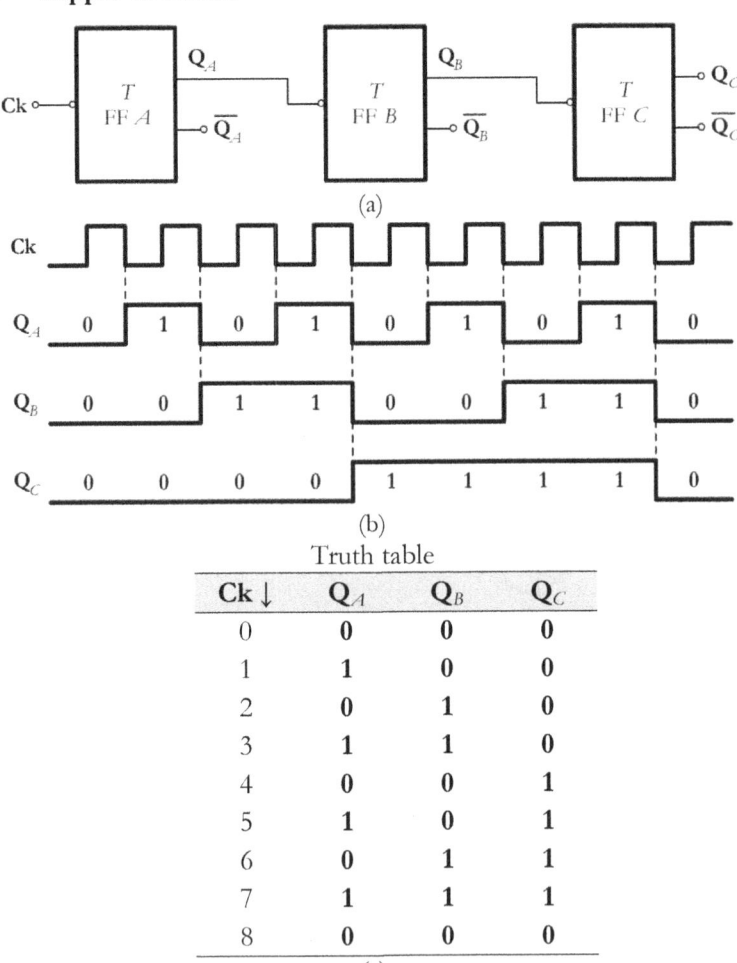

Figure 6.12. (a) A simple binary counter (modulo-8), (b) waveforms and (c) truth table

Ck↓	Q_A	Q_B	Q_C
0	0	0	0
1	1	0	0
2	0	1	0
3	1	1	0
4	0	0	1
5	1	0	1
6	0	1	1
7	1	1	1
8	0	0	0

As an example and demonstration of the operation of a simple binary ripple counter consider the chain of three T flip-flops shown in Fig. 6.12(a). The flip-flop here is triggered during the negative pulse transition, which is indicated by the little circle (bubble) at the clock input of each flip-flop. The output **Q** of each flip-flop is connected to the clock input of the following one. Let all three flip-flops be in the zero state initially, i.e. Q_A, Q_B, Q_C=**0**. With the fall of the first clock pulse, (see the waveforms in Fig. 6.12(b)), flip-flop *A* changes state and will have Q_A=**1**, while Q_B and Q_C will remain at **0** state. The fall of the second clock pulse will cause Q_A to change to **0**, but this will trigger flip-flop *B* to change and have Q_B=**1**, while Q_C will remain at **0**, since flip-flop *C* cannot change with positive transition of the

Digital Sequential Electronics

pulse at its clock input. Next, at the end of the third clock pulse flip-flop A will change to $Q_A=1$, flip-flop B will remain unchanged, i.e. $Q_B=1$, while Q_C will also remain unchanged, i.e. $Q_C=0$. Now writing the states of the three flip-flops in the sequence $Q_C Q_B Q_A$ i.e. **011**, gives the total number of the clock pulses in the flip-flop A. Following this procedure for each new clock pulse at the input of flip-flop A, the waveforms at the outputs of the three flip-flops will be as shown in Fig. 6.12(b). It should be noted that with the fall of the eighth clock pulse all flip-flops return to the zero state. Thus, the set-up can count up to 8 pulses, i.e. the counter has a cycle of 8 pulses (modulo-8). The operation described here can also be followed using the truth table in Fig. 6.12(c). Because each flip-flop changes state after the preceding one has been changed, this counter is called a *ripple-counter*. It is also an asynchronous counter since all flip-flops do not change at the same time.

Clearly, if a fourth flip-flop were connected after flip-flop C, the counter would count up to sixteen. So with n flip-flops the counter would count up to 2^n pulses. For this reason this is also called a scale of 2- counter.

Asynchronous decade counters can be obtained using a chain of four flip-flops and suitable logic. Such a counter will return to the reset state (0000) after each tenth pulse. For numbers of pulses above 10 the first decade counter can feed a second similar one and so on as required.

Example 6.M.4

Build a Simulink model for the binary ripple counter of Fig. 6.12 using *JK* flip-flops and verify that it is indeed a modulo-8 counter.

Model

In Fig. 6.M.7 the *JK* flip-flops operate as *T* flip-flops (see Example 6.M.3). The waveforms of Fig. 6.M.8 verify that the counter counts up to 8 pulses.

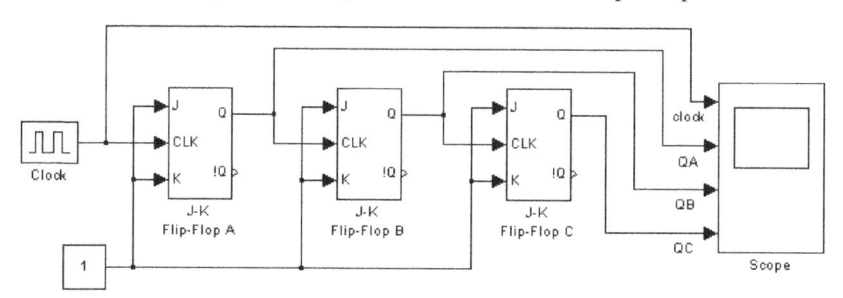

Figure 6.M.7. Modulo-8 binary up counter

It is interesting to observe what will happen, if the outputs are taken from the complementary outputs $\overline{Q}$ instead of Q (Fig. 6.M.9). It is clear from the waveforms of Fig. 6.M.10 that the counter counts down to 8 pulses that is, the counter is a modulo-8 down counter.

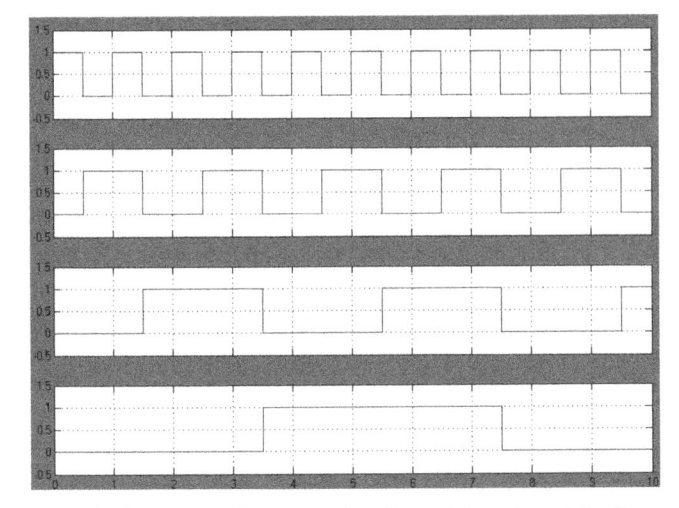

Figure 6.M.8. Clock pulses and outputs Q_A, Q_B and Q_C of modulo-8 up counter

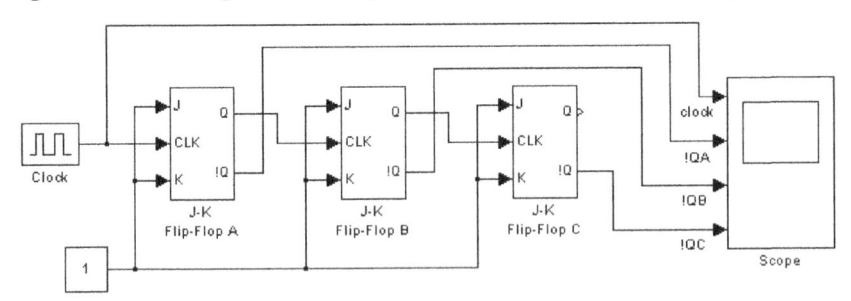

Figure 6.M.9. Modulo-8 binary down ripple counter

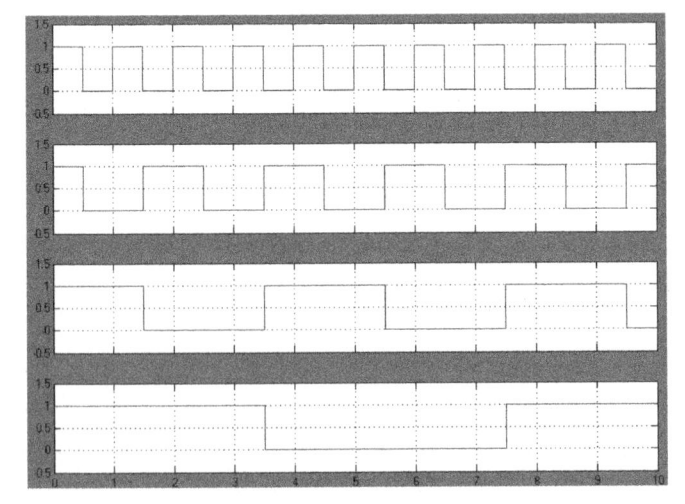

Figure 6.M.10. Waveforms of modulo-8 down counter

6.7.2 Synchronous Counters

As was mentioned above, in the ripple counter, each flip-flop in the chain changes state only after the preceding one has changed from state **1** to state **0**. Thus the flip-flops do not change their state all at the same time, i.e. they operate asynchronously. This may result in erroneous counting at high pulse rates, because the first flip-flop may be triggered by the second pulse before the last flip-flop has obtained its final state having responded to the previous pulse. Thus there is going to be an uncertainty as to what the correct count will be at any instant. In synchronous counters, whether binary or decade, this defect does not exist, since all flip-flops change their state at the same time, i.e. synchronously. This makes synchronous counters faster than corresponding ripple counters and, therefore, more suitable for counting higher pulse rates.

A ripple counter can be changed to the corresponding synchronous one by using flip-flops with Preset (**Pr**) and Clear (**Cr**) inputs and additional logic gates (e.g. NAND gates). A simpler design of a synchronous binary counter using T flip-flops and AND gates is shown in Fig. 6.13. The explanation of its operation is left to the reader as an exercise.

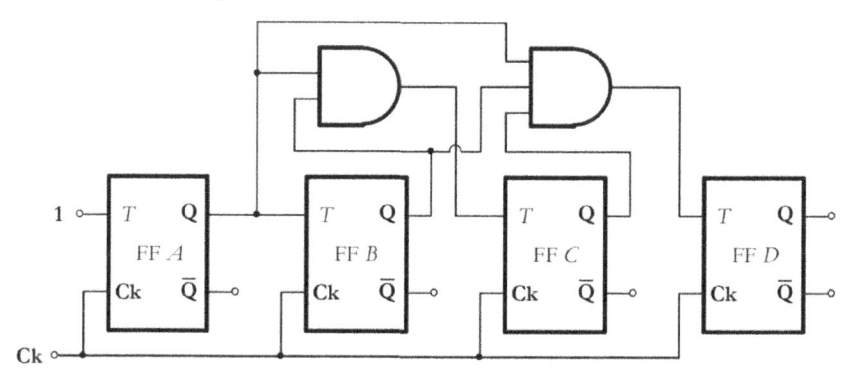

Figure 6.13. A synchronous binary counter with parallel carry

All counters are available in integrated form in *TTL* and *MOS* technologies and the designer's choice depends on the application as far as the pulse rate to be counted is concerned.

Some popular *IC* counters are as follows:

	7490 decade counter
Ripple Counters	*7492 divide-by-twelve counter*
	7493 4-bit binary counter
Synchronous Counters	*74192 synchronous up/down BCD decade (8421) counter*
	74193 synchronous up/down 4-bit binary counter

6.8 Registers

In digital electronics, on many occasions, for example, when binary arithmetic is carried out in the computer, it is necessary to store binary numbers as words in circuits called *registers*. A register is a combination of flip-flops suitable for the storage of digital signals. Each bit in the word requires a flip-flop for its storage and thus the storage of an N-bit word requires N flip-flops, usually D-latches.

6.8.1 Shift-Registers

Very often though, it is useful to shift a word in the register along the chain of flip-flops without, of course, any change in the relative waveform of the signal. Such a register is then a *shift register* (SR).

In Fig. 6.14(a) a four-bit shift-register is shown. All flip-flops are of the R-S type, but flip-flop FFD has been changed to D flip-flop. Note that with this change, the unwanted condition $S=R$ is avoided and, thus, there will not be any ambiguity in the states of the flip-flops. Of course, all flip-flops could be of D-type, as was said above.

Ck	Q_D	Q_C	Q_B	Q_A
0	0	0	0	0
1	1	0	0	0
2	0	1	0	0
3	1	0	1	0
4	0	1	0	1

(b)

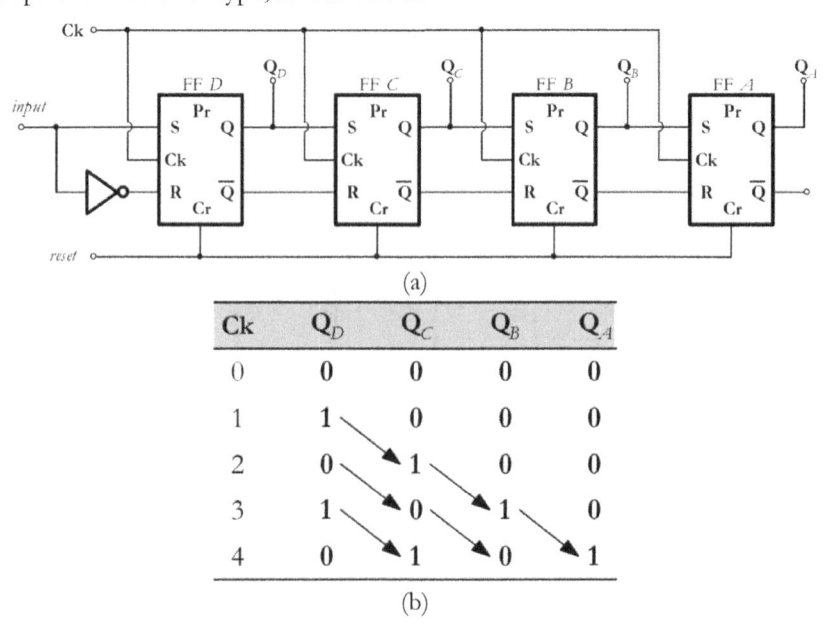

Figure 6.14. (a) Four-bit shift-register and (b) table of consecutive data shifts along the register

The operation of the circuit as a shift-register is as follows: Suppose that the word to be registered is **0101**, coming to the input of the register serially with the *LSB* first, and that all flip-flops have been initially cleared. With the first clock pulse the *LSB*, bit **1**, causes the output of flip-flop D to change to $Q_D=1$. There will be no change in the output of the other flip-flops. The second clock pulse will cause the next data bit, **0**, to enter the circuit and this will result in $Q_D=0$. However, now Q_C will become **1** with Q_B and Q_A remaining unchanged i.e. **0**. With the third clock pulse the third data bit will appear at the output Q_D, while Q_C will become **0**, Q_B will become **1** with Q_A remaining unchanged. Finally, with the fourth clock pulse the *MSB* of the data will come to Q_D, while all the other data bits, that had already entered the circuit, will move along the register one place to the right i.e. $Q_C=1$, $Q_B=0$ and $Q_A=1$. The consecutive data shifts along the register are given in detail as a Table in Fig. 6.14(b). Thus, with the fourth clock pulse the data word **0101** has been stored in the register and is available as $Q_D Q_C Q_B Q_A$. Since all bits can be obtained at the same time from the outputs of the flip-flops, the data word has been changed from serial to parallel. Thus, the shift-register can be used to convert data from serial to parallel.

If the clock pulses continue to be applied, the stored bits will continue to shift to the right and will be obtained from Q_A serially. Thus, with the fifth clock pulse the *LSB* of the data will be shifted out of Q_A, while with the eight clock pulse the *MSB* will be shifted out of Q_A. What has been achieved this way actually, apart from converting serial data back to serial, is to cause a four-clock pulse delay in the serial transmission of the data, which is also an important application of shift-registers.

Yet another application of a shift-register is to convert parallel data to serial. This is because data can enter the *SR* in parallel form using the preset (**Pr**) inputs of the flip-flops and then be obtained from the output Q_A serially by the application of clock pulses. Thus, the *SR* can be used for converting parallel data to serial. Of course data, which have been stored in parallel in the *SR*, can be again obtained in parallel form from the outputs Q_D, Q_C, Q_B and Q_A after a certain desired time.

6.8.2 Additional Applications of Shift-Registers

a.   Division by 2

In the *SR* of Fig. 6.14(a), shifting the data one place to the right with **0** at its input, changes the stored data to **0010**. This data number is half the initially stored one but with an error of ½ bit. However, if the *LSB* were **0**, the error of the division by 2 would have been 0. Thus, the *SR* can be used as a divider of a binary number by 2.

b.   Multiplication by 2

Clearly, if the stored binary number **0101** in the *SR* could be shifted one place to the left, the new stored number would be **1010**, which is double the initially stored number.

Thus, multiplication by 2 and division by 2 of binary numbers can be achieved by the SR very simply, provided that there exists the possibility of left and right shifts in the SR. This can be achieved using suitable logic and control signals. Such SRs are available in integrated form.

c.   Delay

As was said before, the LSB of the word **0101** at the output Q_4, when this enters the SR serially, appears with the fourth clock pulse. Thus, the data can be obtained from Q_4 serially but delayed in time. Therefore, the SR can be used for a digital signal as a delay circuit, which depends on the number of its stages, i.e. the number of its flip-flops and the rate of the clock pulses.

Other applications of the SR include the generation of random sequences, as memories and as special counters (Ring-Counters and Twisted-Ring Counters) etc.

6.8.3   Integrated Circuit Shift-Registers

Shift-Registers are available as integrated circuits in bipolar (TTL) and MOS technologies. The bipolar SRs are static, i.e. use latch flip-flops and can store data indefinitely under the condition that the biasing power supply is not removed. On the other hand the unipolar (MOS) SR can be static or dynamic. In the dynamic SR the data is not stored in flip-flops, but in the capacitance existing between the source and the gate of the MOS transistor. Due to the charge leakage appearing in such a capacitor, it is required to recharge (refresh) this at regular time intervals. Without this recharging or refreshing, the stored information in the dynamic register will be lost after sometime. However, the dynamic registers are advantageous over the static MOS, since they can store a larger number of bits in a given chip area.

In general, the unipolar SR compared to the bipolar have the advantages of lower power consumption, smaller size for certain number of bits and lower cost for a large number of bits.

On the other hand, the bipolar technology SRs have the advantage over the unipolar and, particularly over the dynamic ones, that they can be used at higher clock frequencies for operating reliably. As for the number of stored bits there are available SRs of TTL (see Section 6.11) technology of eight bits, while dynamic SR of MOS technology can be over 2048 bits.

6.9   Memories

In many electronic circuits and systems the need arises for storing certain information for short or long periods of time. The means in which information can be stored and from which it can be obtained are called memories. Computers of any type, certain types of oscilloscopes, digital multimeters and industrial control systems need memories for their operation.

Information is stored in the memory of a system in the form of binary digits and is used for the control of the system, when it is required.

There are various types of memory, which are characterized by the method and the material they are made of. Most important nowadays are those made of magnetic material (e.g. magnetic discs), optical recordings (e.g. compact discs *CD* or *DVD*), charge-coupled devices and semiconductors. In computer systems, memory stores programs that will never be changed, while in other cases, memory is used for data that is constantly modified during the system's operation. For bulk storage of data and programs, a secondary memory is used that, in most, cases, is in the form of a disc drive. Here, we will concentrate on semiconductor memories. These can be split into two categories: *RAM* or *ROM*, which are available nowadays in integrated form using bipolar or *MOS* transistors.

The size of a memory is defined by the number of its locations, which is always a power of 2. This is so, because its content is addressed by n lines, which specify 2^n memory addresses (locations). Normally, the memory size is expressed in Kilobytes, Kbytes, where 1 Kbyte (*KB*) is 1024 bytes and the memory will be addressed by 10 lines since 2^{10} is 1024. Computer memories can now be of sizes of many Megabytes, Mbytes (*MB*) and even many Gigabytes, Gbytes (*GB*).

Independently of the type of their formation a memory is divided into groups of bits, the *words*. The number of bits in each word varies from system to system, e.g. in computers it may be 8, 16, 32 bits depending on the computer. Each word is stored in a selected address in the memory. Thus, there are two parameters that characterize each word in the memory: its location and its content, i.e. the data that it includes.

6.9.1 Random Access Memory, *RAM*

Random Access Memory, *RAM*, is a term that has, rather incorrectly, prevailed to call the read/write memories, i.e. memories the content of which can be read unlimited times and can also be changed by external control unlimited times.

The read/write memories are characterized by the read or access time and the write time. Read or access time is the time required for the memory to present correct data after the address and the select signals have become constant. Write time is the duration of time for which the address and the data must remain constant in order to be registered in the memory.

The name RAM comes from the fact that for these memories the access time is independent of the address of the stored data. However, there are other types of memory e.g. the ROM described below, they can also be accessed randomly.

A RAM can be either word organized or bit organized. The former have m data inputs and m data outputs with additional inputs for n address bits (one address decoder), Read/Write input and a Chip Select or Enable input.

Bit organized memories, which are used in larger memories, have one input and one output for the data and two address decoders. The memory matrix is usually square e.g. 16×16.

Larger size memories than $1MB$ can be obtained from smaller sizes IC memories that are properly connected. For this purpose, their outputs are three-state or open collector.

Most semiconductor Read/Write memories are made in bipolar or MOS technology.

In bipolar technology memories, the elementary cell is a bistable latch. These memories are called *static*, as they keep the stored information as long as the power supplies are applied.

The MOS memories can be static or dynamic. The elementary cell in the static MOS memories is a bistable latch using $NMOS$ or $PMOS$ or Complementary MOS ($CMOS$) transistors. The main advantage of the static $CMOS$ (see Section 6.11) memories over the bipolar and the $NMOS$ or $PMOS$ ones is the smaller standby consumed power.

In *dynamic* memories the information is stored as charge (or no charge) in an MOS capacitor, which is usually the gate-source capacitance in the MOS transistor. Since this charge leaks with time, periodic refreshing is required. On the other hand, reading the content of the cell leads to its destruction, therefore, after each reading, it is necessary to rewrite the information in the cell. This is achieved automatically by the proper logic and amplification circuitry, which is built on the same memory chip.

Main disadvantage of static MOS memories compared to corresponding bipolar is their lower speed (longer access times). On the other hand, their advantages are lower power consumption, higher package density (components/cm^2) and lower cost. Static memories are non-destructive, i.e. the stored information is not lost when they are read, which is the case with the dynamic memories. However, dynamic memories can store more bits per package than static memories and in particular than the bipolar ones. In general, a RAM is volatile, that is it loses its contents when the power is removed.

6.9.2 Read Only Memory, *ROM*

Read Only Memory is a memory with predetermined content, that is, the stored information is determined by the physical arrangement of its circuits in the matrix. Thus, a ROM does not lose its content even if power is removed, that is, it is non-volatile.

The organization of a ROM is either in words or in bits, each word can be accessed randomly and its content read by selecting a unique address. Basically, one ROM consists of an n-input decoder followed by an n-output encoder. The decoder has 2^n output lines and consequently the encoder has an equal number of input lines. A 4×4 bit ROM is shown in Fig. 6.15 having its encoder been built using a diode matrix.

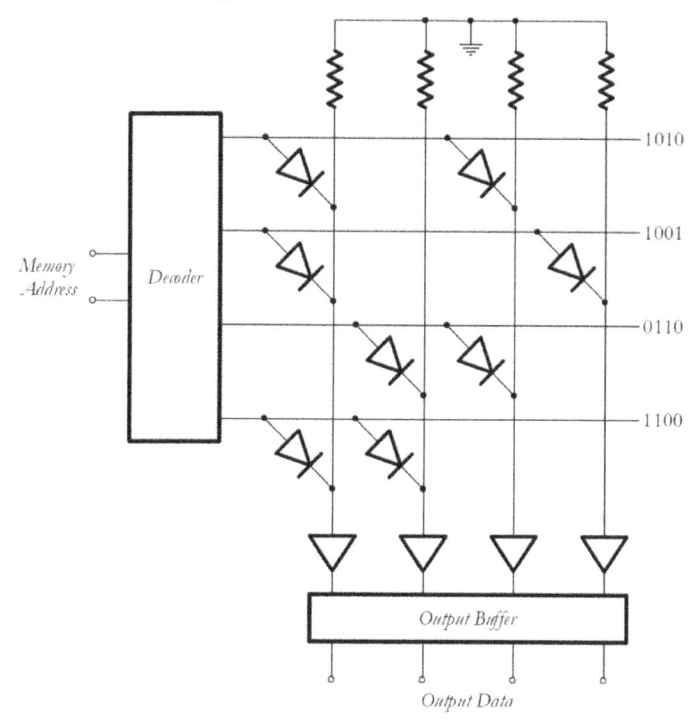

Figure 6.15. 4×4 bit *ROM* using a diode matrix

The ROM programming, that is the introduction of its content for certain applications can be done by its manufacturer. However, it is possible for someone to program himself a ROM for a special application after its construction, by applying suitable electric signals. The programming of these *Programmable* ROMs (PROMs) is achieved by special programming equipment, in which, depending on the method of construction of the PROM, sending suitable electric signals destroy (break) certain connections by burning the corresponding connecting conductors and thus, changing the state of corresponding cells.

In certain *PROMs* their content may be altered, after their previous content had been erased by having exposed them to ultraviolet radiation. Such a *ROM* takes the name *EPROM* (*Erasable PROM*) and can be reprogrammed many times. These constitute a type of the so-called Read-mostly Memories. On the other hand *MOS* technology *ROMs* are available under the name *EAROM* (*Electrically Alterable ROM*), in which by applying suitable positive or negative voltage to the gates of some *MOS* transistors, charge is stored in (logic **1**) or removed from (logic **0**) the cells, thus, achieving or altering their programming.

In certain applications of a *ROM* with n inputs, not all 2^n output lines of the decoder and not all 2^n input lines of the encoder are necessary. In such cases, the matrix size of the memory can be substantially reduced and this can be achieved by making the decoder programmable also. Such a *ROM* is called *Programmable Logic Array*.

ROMs in general are used for code conversion, as look-up-tables for the special control programs in the computers, as character generators, etc.

6.10 Digital Computers

Digital Computers are electronic systems suitable for data processing. An electronic system is a set of electronic circuits, properly interconnected, for the total execution of a job. On the other hand, data processing is referred to the way of manipulation of data, such as numbers, letters or symbols, i.e. information elements or characters.

During the processing of data and, in general, during its operation the computer takes decisions automatically operating. It may be also used for taking measurements automatically and making adjustments.

The digital computer is generally composed of certain units, as shown in Fig. 6.16.

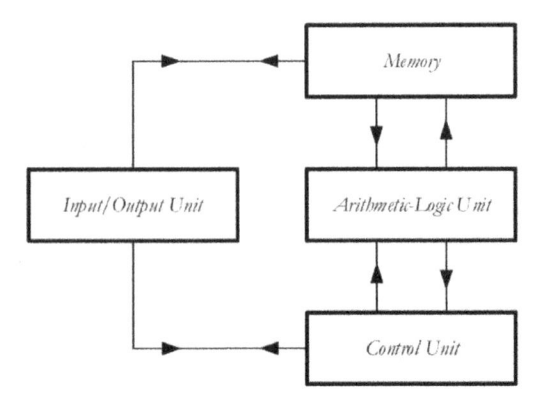

Figure 6.16. General block diagram of a digital computer

Digital Sequential Electronics 195

The arithmetic-logic and control units are usually referred to as the *Central Processing Unit, CPU*.

In what follows, the role of each of these units during the computer operation is examined separately.

Memory

In the computer memory information is stored such as numbers, words or any other data. In general the information stored in the memory is referred either to data or to instructions. The instructions determine the way the computer processes the data. A list of successive instructions determines the program on the basis of which the computer will process the data by a unique manner, that is, the computer will solve a problem step by step. During the operation of the computer data and instructions are stored in the memory, are recorded in it and are recalled, i.e. read. The memory may be magnetic, like disks, or semiconductor as was explained earlier.

The term *RAM*, as it is commonly used, is synonymous with main memory, the memory available to programs. For example, a computer with 8*MB RAM* has approximately 8 million bytes of memory that programs can use. In contrast, *ROM* refers to special memory used to store programs that boot the computer and perform diagnostics. Most personal computers have a small amount of *ROM* (a few thousand bytes).

Main memory

Main memory refers to physical memory that is internal to the computer. The word main is used to distinguish it from external mass storage devices such as disk drives. The computer can manipulate only data that is in main memory. Therefore, every program you execute and every file you access must be copied from a storage device into main memory. The amount of main memory on a computer is crucial, because it determines how many programs can be executed at one time and how much data can be readily available to a program. Now, most *PCs* come with a minimum of 32 megabytes of main memory. You can usually increase the amount of memory by inserting extra memory in the form of chips.

Boot

Boot is short for bootstrap, the starting-up of a computer, which involves loading the operating system and other basic software. A cold boot is when you turn the computer on from an off position. A warm boot is when you reset a computer that is already on.

Control Unit

The control unit is responsible for the automatic operation of the computer. It examines the instructions of the program one by one and sends signals to the other computer units to execute them. Each instruction is transferred to the control unit from the memory, is translated and then executed one by one each time, until the program is completed. As each instruction is examined or decoded, the control unit is generating control signals at predetermined time instances, which guide the other computer units to act properly, so that the instructions will be executed. The instructions and, consequently, the control signals can affect all the other computer units.

Arithmetic-Logic Unit

The *Arithmetic-Logic Unit*, *ALU*, is the computer unit that executes many of the operations determined by the program instructions. In particular, the *ALU* executes either arithmetic operations like addition, subtraction, shifting or logic operations like AND, OR or NOT. Thus, if a stored in memory instruction is to execute an addition, the control unit recalls it from the memory, translates (i.e. decodes) it and sends signals to *ALU* that executes the addition of two numbers. As was said above, the control unit and the *ALU* are so closely connected in a digital computer that usually are considered as one unit, called the *Central Processing Unit, CPU*.

Input/Output Unit

The *input-output unit* (*i/o*) of the computer allows the communication of the memory and the *CPU* with the outside world. Through this unit the data are introduced in the computer and are given out from it processed. Thus the input/output unit acts as an interconnecting unit of the computer and the various peripheral instruments. These instruments, the peripherals that are connected through the input/output unit of the computer, are used to display data. Data can be introduced to the computer through a keyboard for example, or data that come out of the computer can be displayed on the screen of a Cathode-Ray Tube (*CRT*) or Liquid-Crystal Display, *LCD*, or be printed by a printer. An external memory disk can be used for extending the memory of the computer while compact disks, *CD* or digital video disks, *DVD*, are now used for extra storing of data and back-up of the computer memory contents and useful program. Another peripheral memory device is the *USB* (*Universal Serial Bus*), an accessory useful for someone to carry stored data in one's pocket.

Apart from the connection of the computer to peripheral units, the *i/o* unit may connect the computer to external elements like transducers for temperature, pressure etc measurements or to actuators, for example relays or display lamps the condition of which should be adjusted.

In both above duties, the *i/o* unit does the interconnection, the matching of the external instruments etc with the *CPU*. This required compatibility is achieved by the *i/o* unit by suitable digital logic circuits.

Digital Sequential Electronics 197

The information from (or to) the external world to (from) the computer can be transmitted in parallel or serially. Since the speed of the peripheral units is relatively low, the most usual transmission is serial. This is economically advantageous, particularly when the transmission should be made to long distances through telephone lines. On top, parallel transmission to long distances has an additional problem due to electrical noise. In serial transmission, data are sent as bytes (of 8 bits).

The computer and the peripheral units compose a complete computer system for data processing or for automatic control applications.

Microcomputers

A microcomputer is a digital computer of very small dimensions. It can be constructed on an *LSI* (*VLSI*) chip (see Section 6.11) or on a printed circuit board, *PCB*, from many integrated circuits.

Nearly all microcomputers have a microprocessor as their basic unit. The microprocessor is an *LSI* (*VLSI*) circuit that includes the control unit and the *ALU*. The microprocessors are usually called *microprocessing units* or *MPU*. Adding more chips for timing, *RAM*, *ROM*, *i/o* circuits and other auxiliary circuits one can build a full computer on a board with the dimensions of a journal page. One such system constitutes a microcomputer, in which the microprocessor is its main component.

As in the *CPU* of a large computer, the purpose of the microprocessor is to receive data in the form of series of bits, to store them for later processing, to execute mathematical operations on these, according to instruction that had been stored earlier and to give out the results to the operator through a printer or a display.

6.11 Digital Integrated Circuits

An introduction to integrated circuits (*IC*) has been presented in Section 2.6. Predominant digital *ICs* can be classified according to the transistor type used in their design, in those using *BJTs*, *TTL* families, and those using *MOSFETs*, *NMOS*, *PMOS* and *CMOS* families. *NMOS* and *PMOS* use *n*-type and *p*-type *MOSFETs* respectively, while in *CMOS* (*Complementary Metal-Oxide Semiconductor*) *ICs* both types of *MOSFETs* are employed.

a. Digital *ICs* based on *BJT* transistors

General purpose *TTL* digital *ICs* are designed and constructed using mostly the circuit of *TTL* NAND gate as basic building block. The basic two-input *TTL* (*Transistor-Transistor Logic*) *IC* circuit is shown in Fig. 6.17(a). In this circuit Q_1 is a multiple-emitter transistor, equivalent to two transistors, as shown in Fig. 6.17(b) operating in parallel. It can be seen, that the two transistors have their bases connected in parallel, their collectors connected in parallel and their emitters (which can be more than 2, up to 8) being the input terminals of the gate. The output part of this circuit is completed with extra components in order to reduce the output resistance of the gate and increase the number of gates that can be driven by this gate. This number is called the *fan-out* of the gate. Other variations of the basic TTL NAND gate provide protection of the circuit from unwanted short-circuit of the output and increase in the speed, i.e. reduce the propagation time of the signal through the gate.

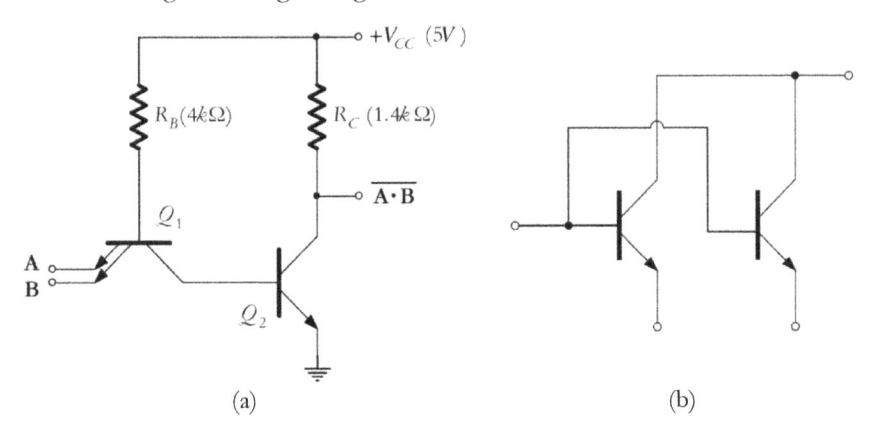

Figure 6.17. (a) Circuit of basic TTL NAND gate and (b) equivalent of transistor Q_1

Another important digital IC family using BJT transistors is available, namely, the ECL (*Emitter-Coupled Logic*) family. The basic gate implements the OR and NOR gates simultaneously (two outputs) being the fastest among all other discrete gates.

b. Digital ICs based on MOSFETs

As was mentioned above three IC families have been emerged using MOSFETs, namely, the PMOS, NMOS and the CMOS. Of these the most useful in practice has been proved to be the CMOS. Most important reason for this is the fact that it consumes much less power for its operation than the other two. It is also most suitable for integration in VLSI (*Very Large-Scale Integration*) form. So the following discussion is concerned with the CMOS IC family. The basic building block in CMOS ICs is the inverter circuit shown in Fig. 6.18.

The *n*-channel MOST and *p*-channel MOST in the CMOS inverter are of the enhancement type and the circuit operates as follows: With zero input voltage, **0** logic state, the *n*-channel transistor is off due to zero voltage between its gate and its source. But the *p*-channel transistor with negative voltage between gate and source is on and thus, the output voltage is V_{DD}, i.e. at logic state **1**. On the other hand, when the input voltage is $v_i = V_{DD}$, i.e.

Digital Sequential Electronics 199

at logic state **1**, the *n*-channel transistor conducts (is on) while the *p*-channel transistor is off, since its gate will be at the same voltage as its source. Thus v_o will be nearly $0V$ and, therefore, the circuit operates as an inverter.

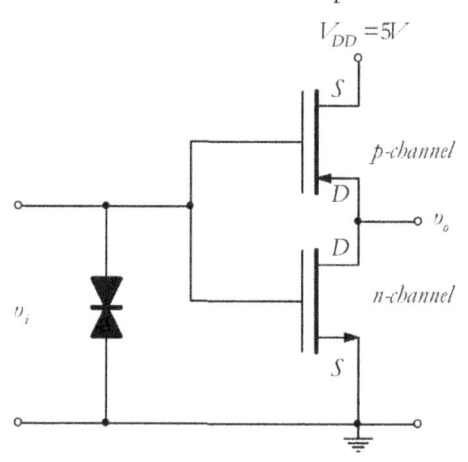

Figure 6.18. *CMOS* inverter

Clearly, the two transistors are not on at the same time apart from the transition time between the two states. Thus, power is consumed only during the transition time, which is very short, and, consequently, the consumed power is very low. This is one of the most important advantages of the *CMOS ICs*.

As mentioned above, the circuit of *CMOS* inverter is the basic building block of all the other *CMOS* gates. The *CMOS ICs* usually have two breakdown diodes connected back-to-back at their inputs, as shown in Fig. 6.18, for protection against excessive static charges and voltages. To prevent any destruction of a *CMOS IC*, any input terminal not connected to the signal source should be connected to the power supply. No signals should be connected to the gate input before the power supply has been connected to the circuit.

An important addition to the *TTL*, *ECL* and *CMOS* gate families is the *BiCMOS* family, *Bi* standing for Bipolar. In this family bipolar transistors are used in combination with the *CMOS* circuits in order to increase the speed of *CMOS* gates.

c. **Very High Speed *ICs*, *VHSIC***

Very High Speed ICs are *LSI* and *VLSI* circuits in which the dimensions of the basic components (transistors) are made very small, smaller than $1\mu m \times 1\mu m$, with their interconnections short and such that any capacitances etc, i.e. strays, are as low as possible. These circuits are produced under names like *gate arrays*, *custom made* and *functional cells* e.g. Programmable Logic Arrays (*PLA*), multibit multipliers etc. Gate arrays are for general purpose use, while the custom made and the functional cells are produced for special purposes and made on the request of the customer. Of course, in some cases gate arrays can be used as parts of the custom made ICs.

In *VHSICs* the semiconductor material is silicon or *Gallium-Arsenide*, GaAs. In the latter technology the components are *MESFET* (*Metal-Semiconductor FET*) of the depletion or enhancement types and Schottky diodes. It should be added though, that the above are results that can be changed, since the progress of silicon *CMOS* and GaAs logic technologies is very fast.

d. Scales of Integration

According to Gordon Moore prediction (1965), which has actually been proved correct at least up to now, the number of transistors/chip will be doubled every 1.5 years, while the dimensions of the transistor will be halved every three years.

The density of components in an integrated circuit package is expressed by the *Scale of Integration* and is given by the number of gates per chip. The up to now created Scales of Integration are as follows:

Small-Scale of Integration, *SSI* for	<	12 gates/chip
Medium-Scale of Integration, *MSI* for	<	100 gates/chip
Large-Scale of Integration, *LSI* for	<	1000 gates/chip
Very Large-Scale of Integration, *VLSI* for	<	10^5 gates/chip
Ultra Large-Scale of Integration, *ULSI* for	<	10^6 gates/chip
Giga-Scale of Integration, *GSI* for	<	10^7 gates/chip
Wafer-Scale of Integration, *WSI* for	<	10^8 gates/chip

e. Gate Characteristics

The most important characteristics of a gate are as follows:

- *Voltage and Current Levels*

In Fig. 6.19 these are described as follows:

V_{OHmin}: The lowest output voltage in the high state

V_{OLmax}: The highest output voltage in the low state

V_{IHmin}: The lowest voltage in the high state that the input recognizes as high

V_{ILmax}: The highest voltage in the low state that the input recognizes as low

Digital Sequential Electronics

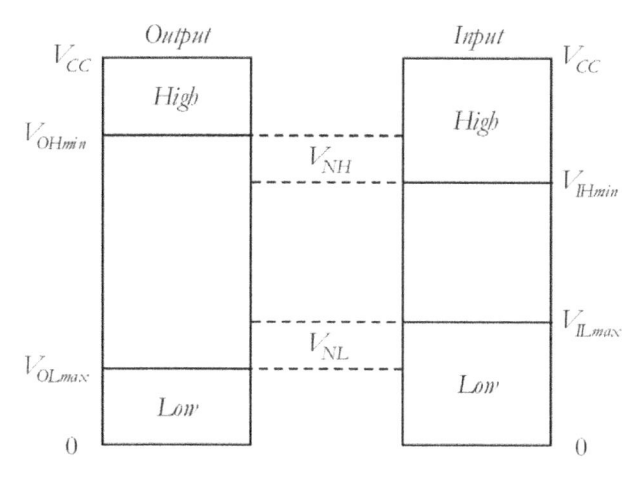

Figure 6.19. (a) Voltage levels and (b) noise margin of a gate

- *Noise Margin*

As described in Chapter 8, noise is any undesired variations of a voltage. *Noise margin* is the highest noise voltage that does not lead to a wrong state. Its value for a gate is given as follows:

Noise Margin at the High Level: $V_{NH} = V_{OHmin} - V_{IHmin}$

Noise Margin at the Low Level: $VNL = V_{ILmax} - V_{OLmax}$

For *TTL*: $\quad V_{NH} = 0.4V \quad V_{NL} = 0.4V$

For *CMOS*: $\quad V_{NH} = 1.4V \quad V_{NL} = 0.9V$

For *ECL*: $\quad V_{NH} = 0.14V \quad V_{NL} = 0.135V$

- *Fan-Out and Fan-In*

Fan-Out is the highest number of gates inputs that can be connected to the gate without change in the signal. The fan-out for *TTL* is about 10, for *CMOS* about 40 and for *ECL* about 25 similar gates. *Fan-In* is the number of inputs that the gate provides.

- *Propagation Delay*

This is defined as the elapsing time from the input state change till the corresponding change occurs at the output. As shown in Fig. 6.20, t_{PLH} is the delay for the output pulse to reach 50% of its final value going from Low to High, measured from the moment the input pulse going from Low to High reaches 50% of its final value. Similarly, t_{PHL} is defined for the pulse going from High to Low. The *propagation delay* of the gate is then the mean of t_{PLH} and t_{PHL}. The propagation delay for a *TTL* is about 10ns, for an *ECL* about 1ns, for a *CMOS* about 40ns and for a *BiCMOS* about 5ns.

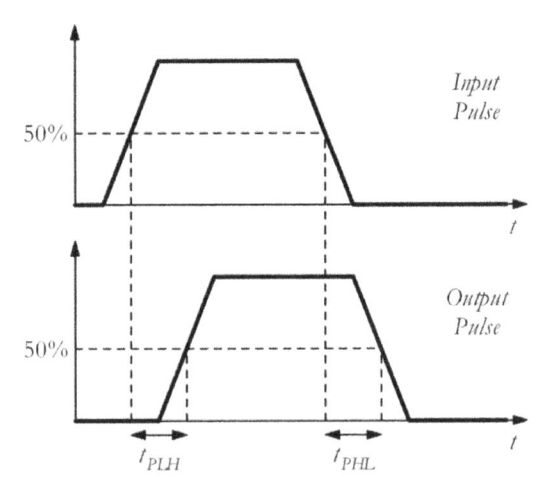

Figure 6.20. Defining the propagation delay

- *Power Consumption*

The power consumption by a *TTL* gate is about $10mW$, by a *CMOS* about $0.3mW$, by an *ECL* about $50mW$ and by a *BiCMOS* about $1mW$.

A figure of merit for a gate is the product of speed times power consumption expressed in pJ.

- *Power Supply*

The power supply voltage depending on the technology can be $5V$ to as low as $1.3V$.

6.12 Design with *HDL*

Hardware Description Languages (HDLs) are programming languages that are designed to program *Field-Programmable Gate Arrays* (FPGAs) and *Complex Programmable Logic Devices* (CPLDs). FPGA is an integrated circuit designed to be configured by the customer or designer after manufacturing. CPLD is a programmable logic device with less complexity than that of a FPGA.

The basic level for *FPGA* design entry is the *Register Transfer Level* (RTL), which represents a digital circuit as a set of connected primitives (adders, counters, multiplexers, registers, etc). There are two basic ways to create an RTL design: schematic entry and HDL entry. Schematic entry describes the circuit connectivity (*netlist languages*) and it is not very convenient to use it for large projects. HDL entry is more convenient, but needs to be processed by a compiler (usually called *synthesizer*) in order to transform the HDL code listing into a physically realizable gate-level netlist.

The gate-level netlists consist of interconnected gate-level macro cells. This gate map can be downloaded onto the programmable device, and run.

The *HDLs* formally define how the device should operate. This is more portable and convenient alternative to creating a circuit from elements. They can describe the circuit's operation, its design and organization, and tests to verify its operation by means of simulation. *HDL* based design has established itself as the modern approach to design of digital systems. Actually, the *HDLs* were initially designed to assist simulation rather than replace schematic design entry. Synthesizers were created much later. Therefore *HDL* languages have two subsets of language constructs: synthesizable (suitable for synthesis and simulation) and non-synthesizable (suitable only for simulation).

The syntax constructs of *HDL* languages are similar to those of conventional programming languages. However, the semantics is quite different. In contrast to most software programming languages, *HDLs* also include an explicit notion of time, which is a primary attribute of hardware.

The first *HDLs* were *Instruction Set Processors* (*ISP*), developed at Carnegie Mellon University, and *KARL*, developed at University of Kaiserslautern, both around 1977. Although many proprietary *HDLs* have been developed, *Verilog* and *VHDL* are the major standards used for simulation and logic synthesis. Both of these languages are widespread. *Verilog* was introduced by Gateway Design Automation in 1985. In 1987 the U.S. Department of Defence and the *IEEE* sponsored the development of *VHDL*.

Verilog is a *HDL* used to model electronic systems. It is most commonly used in the design, verification, and implementation of digital logic chips. It is also used in the verification of analog- and mixed-signal circuits. *Verilog* is more similar to *C*.

VHDL stands for *Very High Speed Integrated Circuits* (*VHSIC*) *HDL*. It has become now one of industry's standard languages used to describe digital systems. It is used in electronic design automation to describe digital and mixed-signal systems such as *FPGA* and integrated circuits. *VHDL* is more *Pascal*-like.

References

[1]. Mano M. M. & Ciletti M. D., *Digital Design*, 4th Edition, Pearson Prentice Hall, 2007.

[2]. Greenfield J. D., *Practical Digital Design Using IC's*, Wiley, New York, 1977.

[3]. Millman J. & Tamb H., *Pulse Digital and Switching Waveforms*, McGraw-Hill, New York, 1973.

[4]. Millman J. & Grabel A., *Microelectronics*, McGraw-Hill, 1987.

[5]. Pascoe R. D., *Solid-State Switching: Discrete and Integrated*, Wiley, New York, 1973.

[6]. Malmstadt H. V. & Enke C. G., *Digital Electronics for Scientists*, W. A. Benjamin, New York, 1969.

[7]. Antoniou A., *Digital Signal Processing*, McGraw-Hill, 2006.

[8]. Hodges D. A., *Microelectronic Memories*, Scientific American, Vol. 237, No. 3, pp. 130-135, September 1977.

[9]. Texas Instruments Inc., *Designing with TTL Integrated Circuits*, McGraw-Hill, 1971.

[10]. Storey N., *Electronics – A Systems Approach*, Addison-Wesley, 1992 (reprinted 1995).

[11]. Mermet J., *Fundamentals and Standards in Hardware Description Languages*, Springer Verlag, 1993.

MATLAB Problems

❖ *The reader is advised first to consult the MATLAB tutorial in Appendix D*

6.M.1. Using Simulink confirm the truth table of a *JK* flip-flop.

6.M.2. Repeat Example 6.M.4 for modulo-16 binary-up and binary-down ripple counters.

6.M.3. Build a Simulink model for the synchronous binary counter of Fig. 6.13 and extract the waveform at the output of each flip-flop. What modulo is the counter?

6.M.4. Build a Simulink model for the shift-register of Fig. 6.14 using *D* flip-flops and extract its output waveforms when the input D is **1**.

Problems

6.1. Show that the circuit in Fig. P.6.1 operates as a *T* flip-flop and comment on the required type of pulses at input *T* for reliable operation.

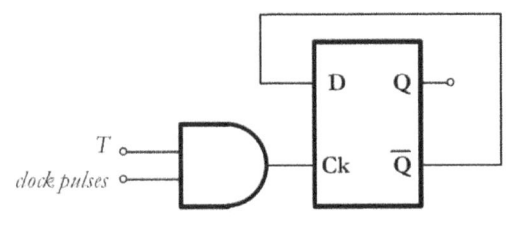

Figure P.6.1.

6.2. The input and clock waveforms given in Fig. P.6.2 are respectively applied to the input and clock terminals of a) a *D* flip-flop and b) a

Digital Sequential Electronics

negative triggered JK flip-flop in which K=1. Draw the output waveforms from each of the two flip-flops.

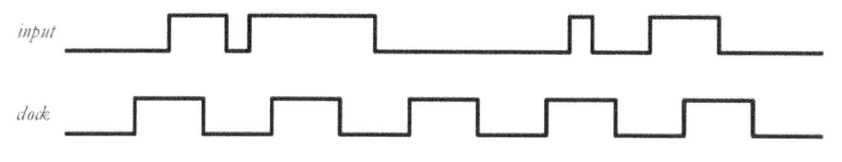

Figure P.6.2.

6.3. Determine the cycle of the counter in Fig. P.6.3. All J and K inputs are kept high.

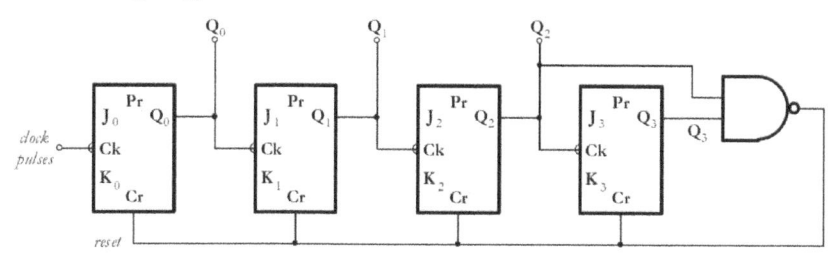

Figure P.6.3.

6.4. Is the counter in Fig.P.6.4 synchronous or asynchronous (ripple)? Explain its operation and draw the voltage waveforms at the outputs Q_1 and Q_2, if the flip-flops are negative-edge triggered and the clock pulses are square waves.

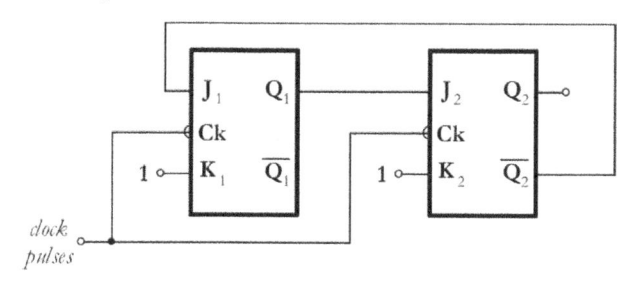

Figure P.6.4.

6.5. Design a modulo-6 counter using a modulo-3 counter and a flip-flop. Next draw the output waveforms relative to the clock pulses for the two cases when the flip-flop a) precedes and b) follows the modulo-3 counter.

6.6. Draw the waveforms of pulses at the output Q_3 relative to the clock pulses and determine the counter modulo for the arrangement in Fig. P.6.5.

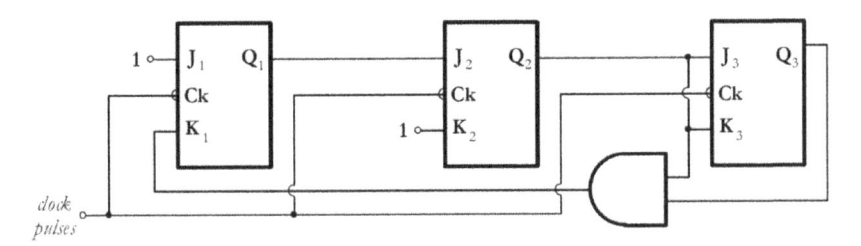

Figure P.6.5.

6.7. Determine suitable values for R and C in the OS in Fig. 6.8 for the time duration T of the output pulses to be 3/4 of the pulse period (or 3/1 marks-to-space ratio).

6.8. Using JK flip-flops and gates design a modulo-9 synchronous counter and show, with the help of a table with the flip-flop states, that it operates properly. If T_F and T_G are the propagation delays of a pulse in a flip-flop and a gate respectively, determine the highest frequency of pulses that your counter can follow reliably.

6.9. Design a circuit that will reduce a pulse rate of 1 *Mpps* at its input to 12.5 *kpps* at its output (pps : pulses per second).

6.10. Design a circuit that will energize 8 *LED*, one after the other with 1 *s* delay in between and stay lit. The circuit will be reset by the operator at will.

Chapter 7

Data Conversion and Processing of Digital Signals

7.1 Introduction

In the last two chapters, digital circuits and systems have been introduced in a concise way. These circuits operate on digital signals however, while in the real world most signals are in analog form (voice, music, continuous measurements of physical quantities etc). In order to make use of digital systems and get the benefit of it, the analog signals have to be converted to digital form. On the other hand, digital signals are not comprehensive by humans unless, they can be converted to analog form.

In this chapter some methods are introduced for converting signals from analog to digital and digital to analog form. The former operation requires the analog signal to be sampled first and this is initially explained. Next, the most suitable in practice circuit for digital-to-analog conversion is presented first, because this is used in some analog-to-digital converters. Then, three methods for analog-to-digital conversion of signals are briefly introduced and explained. Included is a necessary circuit for performing the sample-and-hold operation. Finally, the required digital filters in digital processing are very briefly introduced.

7.2 Conversion of Analog Signals to Digital and Vice Versa

Signal processing can be analog or digital, however, digital processing has been proved to be more accurate and less noisy. If the signal is in digital form, processing will be by digital means. But digital processing of an analog signal requires its conversion first to digital representation, before applying any digital processing on it. A circuit performing this conversion is called *Analog-to-Digital Converter* or *ADC*. On the other hand if the processed signal has to be returned to analog form, a suitable circuit converting the digital signal to analog will be required. Such a circuit is called *Digital-to-Analog Converter* or *DAC*.

The whole operation for digital processing of an analog signal can be better understood by means of Fig. 7.1. In this figure, the transmission path of the analog signal is represented by a single arrow, while that of the digital signal by a wide arrow, since this transmission can be parallel. For reasons to become apparent below, the *ADC* is preceded by a low-pass filter (and a sampler), while the *DAC* is followed by a smoothing simple low-pass filter. Because of its simplicity, the structure and operation of the *DAC* is discussed first.

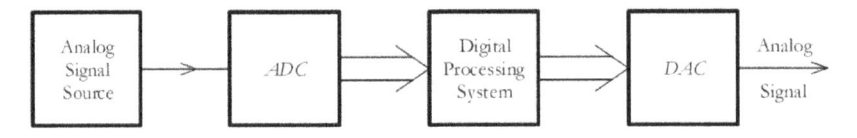

Figure 7.1. Digital processing of an analog signal

7.3 Sampling and Quantization

Sampling is the operation for the conversion of an analog signal to a series of pulses, the amplitude of which follows the variations of the amplitude of the sampled signal (see Fig. 7.2). This operation is effected using the *sampler*, which is an analog switch that remains closed for a small duration of time t_s every T seconds. Time T is the sampling period, whereas $1/T$ is the sampling rate. Time t_s is the sampling time.

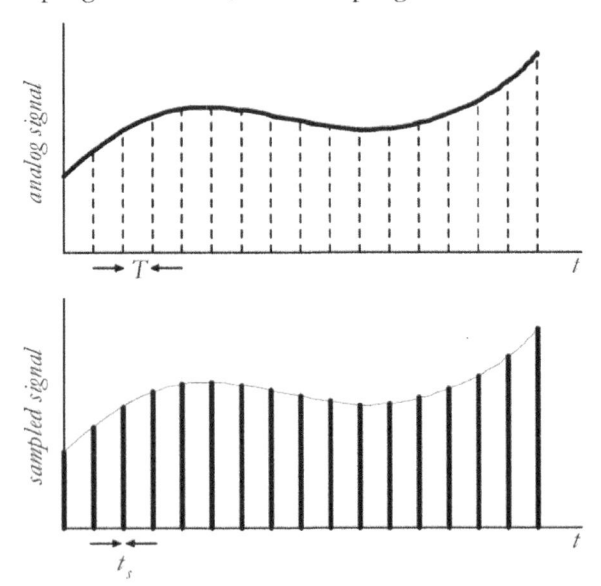

Figure 7.2. Sampling process of an analog signal

It has been shown that satisfactory reconstruction of the analog signal from the sampled, theoretically, can be achieved, if the sampling rate is at least twice the highest frequency in the initial analog signal. Thus, if for an audio signal the highest frequency is $3kHz$, for satisfactory reconstruction of the signal the sampling rate should be at least $6kHz$. Half of this lowest necessary sampling frequency is the, so-called, *Nyquist frequency*.

The conversion relationship usually is not 1-to-1, (i.e. a $10V$ voltage is not represented by the 8-bit word 00001010), but another that corresponds to some scale. For example, full scale of input voltage corresponds to 2^n-1 levels, where n is the number of bits that are used for the digital representation of the analog signal. According to this, each of the input pulses will be represented by one of the 2^n-1 levels, while two consecutive levels will differ by $V_{max} \cdot 2^{-8}$, where V_{max} is the highest value of the input signal. If for example $V_{max}=10V$, this difference will be $10 \cdot 2^{-8}V$, which also represents the *resolution* of the process. Thus, the input signal is *quantized*. The quantization of $V_{max}=10V$ by 8 bits is as follows:

$$00000000 = 0 \times 10 \times 2^{-8}V = 0V$$
$$00000001 = 1 \times 10 \times 2^{-8}V = 0.03906V$$
$$00000010 = 2 \times 10 \times 2^{-8}V = 0.07813V$$
$$\ldots \quad \ldots \quad \ldots$$
$$11111110 = 254 \times 10 \times 2^{-8}V = 9.92188V$$
$$11111111 = 255 \times 10 \times 2^{-8}V = 9.96094V$$

It can be seen from the above, that each signal sample will be represented by a digital number. This number will accurately represent the signal pulse only, if the height of the latter equals one of the voltage levels into which the maximum value of the analog signal is split. When it falls in between two consecutive levels, it will be represented by the digital number corresponding to its nearest level. Thus the analog signal is *quantized*. The difference between two quantization levels is the *quantization step*, which is equal to $V_{max} \cdot 2^{-n}$, where V_{max} is the maximum value of the input signal and n the number of bits in the digital representation.

According to the above argument, an error will arise in the representation, which will be equal, at most, to half the difference between the two consecutive voltage levels, i.e. $\frac{1}{2}V_{max} \cdot 2^{-n}$. This is called *quantization error* and, in actual fact, it creates a type of noise, namely, the *quantization noise*. In practice, the quantization noise and its maximum value depend on the quantization step, which gets smaller as the number of bits gets higher. On the other hand, the maximum number of bits in each digital word that the digital processing system can manipulate is restricted by the system hardware. This means, that each digital word representing a sample may have to be truncated or rounded and this will create the truncation or rounding errors respectively, which will increase the noise of the system. The interested reader is advised to consult a specializing book on digital signal processing, for example Antoniou [7].

Example 7.M.1

Consider a 3-bit analog-to-digital conversion with $V_{max}=10V$ being the highest value of the input signal. Using MATLAB convert the analog voltages $1.85V$ and $5.30V$ to digital.

Code

The following code converts any analog signal between 0 and V_{max} to digital representation with 3 bits. The validity will be checked for signals $1.85V$ and $5.30V$.

The quantization of V_{max} by 3 bits is as follows:

Digital representation				Corresponding analog value x
0 → 000	→	$0 \times V_{max} \times 2^{-3} = 0$	→	$x < V_{max}/16$
1 → 001	→	$1 \times V_{max} \times 2^{-3} = V_{max}/8$	→	$V_{max}/16 \leq x < 3V_{max}/16$
2 → 010	→	$2 \times V_{max} \times 2^{-3} = V_{max}/4$	→	$3V_{max}/16 \leq x < 5V_{max}/16$
3 → 011	→	$3 \times V_{max} \times 2^{-3} = 3V_{max}/8$	→	$5V_{max}/16 \leq x < 7V_{max}/16$
4 → 100	→	$4 \times V_{max} \times 2^{-3} = V_{max}/2$	→	$7V_{max}/16 \leq x < 9V_{max}/16$
5 → 101	→	$5 \times V_{max} \times 2^{-3} = 5V_{max}/8$	→	$9V_{max}/16 \leq x < 11V_{max}/16$
6 → 110	→	$6 \times V_{max} \times 2^{-3} = 3V_{max}/4$	→	$11V_{max}/16 \leq x < 13V_{max}/16$
7 → 111	→	$7 \times V_{max} \times 2^{-3} = 7V_{max}/8$	→	$13V_{max}/16 \leq x$

Then the m-file bit3.m could be written as follows:

```
function digital=bit3(analog,V)
    if analog<V/16
        digital=0;
    elseif analog>=V/16 & analog<3*V/16
        digital=1;
    elseif analog>=3*V/16 & analog<5*V/16
        digital=2;
    elseif analog>=5*V/16 & analog<7*V/16
        digital=3;
    elseif analog>=7*V/16 & analog<9*V/16
```

```
                digital=4;
        elseif analog>=9*V/16 & analog<11*V/16
                digital=5;
        elseif analog>=11*V/16 & analog<13*V/16
                digital=6;
        else
        digital=7;
        end
digital;
end
```

and the code in MATLAB is:

```
>> digital1=bit3(1.85,10)        % V_max=10V and analog input 1.85V

digital1 =

    1

>> digital2=bit3(5.30,10)        % V_max=10V and analog input 5.30V

digital2 =

    4
```

Example 7.M.2

Consider again the 3-bit analog-to-digital conversion of example 7.M.1. Using MATLAB and the m-file bit3.m of example 7.M.1 plot the digital output versus the analog input as well as the quantization error against the analog input.

Code

```
>> analog=0:0.001:10;            % analog input from 0 to V_max
>> n=length(analog);
>> for i=1:n
digital(i)=bit3(analog(i),10);    % digital output for every analog input
error(i)=(digital(i)*10/8)-analog(i);
end
>> plot(analog,digital); axis([0 10 -0.5 7.5]); grid; xlabel('analog input'); ylabel('digital output')
>> figure
>> plot(analog,error); axis([0 10 -1.2 0.8]); grid; xlabel('analog input'); ylabel('quantization error')
```

It is interesting to mention here that Simulink includes an *idealized ADC quantizer* block for converting a continuous-time signal to digital (Fig. 7.M.3).

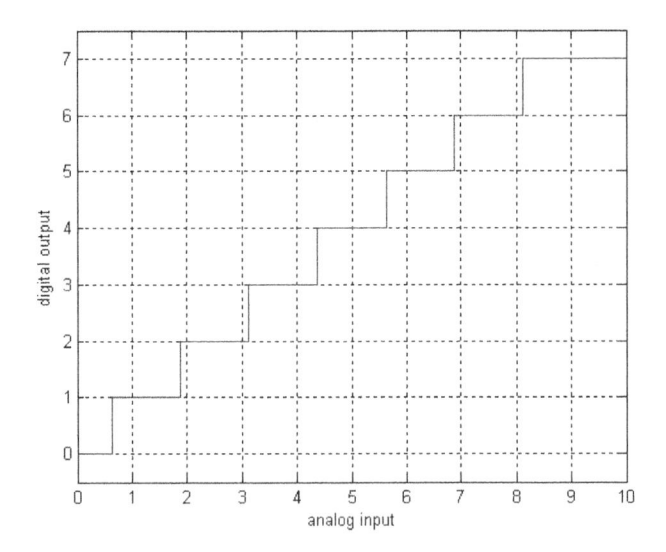

Figure 7.M.1. Digital output against analog input for a 3-bit analog-to-digital converter with $V_{max}=10V$

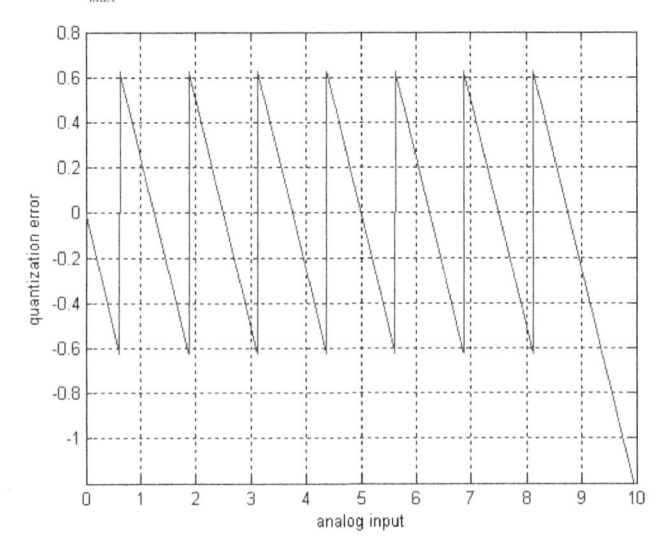

Figure 7.M.2. Quantization error versus analog input for a 3-bit analog-to-digital conversion with $V_{max}=10V$

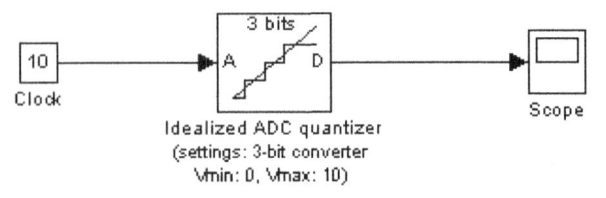

Figure 7.M.3. Simulink model for a 3-bit analog-to-digital converter with $V_{max}=10V$

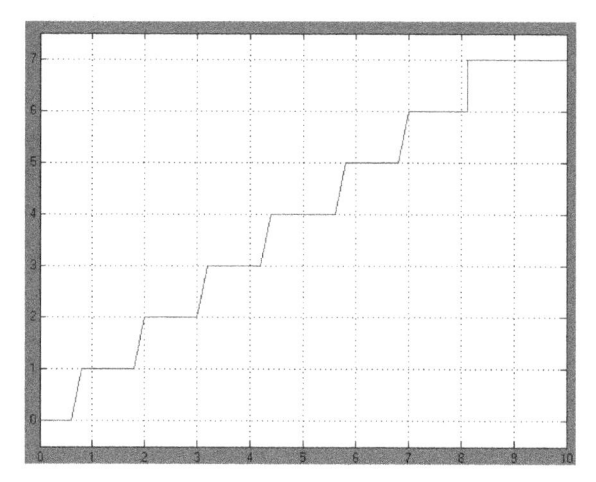

Data Conversion and Processing of Digital Signals   213

Figure 7.M.4. Digital output versus analog input for the Simulink model of Fig. 7.M.3

7.4 Digital-to-Analog Converter

The classical and very useful in practice *Digital-to-Analog Converter*, *DAC*, employs a resistive ladder network combined with an opamp, a *dc* reference source and switches. From all possible schemes, the most successful in practice is shown in Fig. 7.3. Two reasons make this circuit very useful in practice, the following:

a.   The dispersion of resistor values is small (1/2).

b.   The switches, which in the practical circuit will be transistors, usually *FETs*, operate with relatively small voltage values.

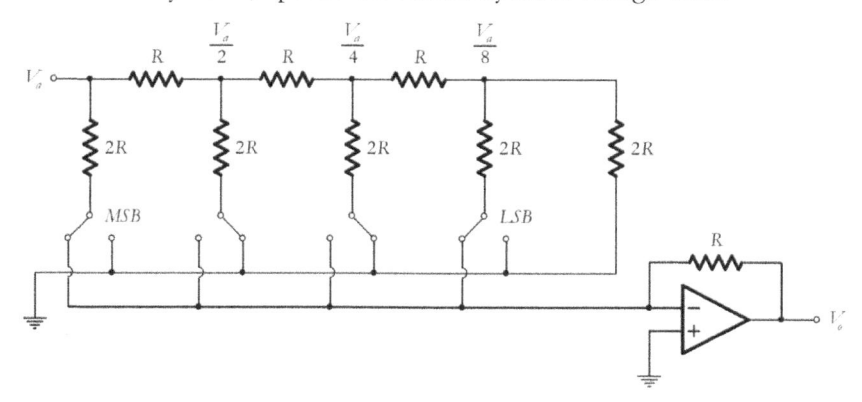

Figure 7.3. A 4 bit Digital-to-Analog Converter

The *DAC* in Fig. 7.3 is suitable for the conversion of a 4 bit digital signal to analog; however, the resistive ladder can be extended toward the reference voltage source V_a by connecting resistors alternately in parallel of value $2R$ and serially of value R.

Each switch in Fig. 7.3 is controlled by the corresponding bit in the number to the under conversion digital signal, thus allowing the proper current to pass through the corresponding resistor 2R. If that bit is 1, the switch connects the resistor terminal 2R to the inverting input of the opamp, whereas if that bit is 0, the switch connects this terminal to ground. The most significant bit (MSB) controls the switch, which is nearest to the reference voltage source.

The output voltage V_o for the random state of the switches, that is defined by the digital number $A_3A_2A_1A_0$, is found to be

$$-V_o = \left(\frac{1}{2}A_3 + \frac{1}{4}A_2 + \frac{1}{8}A_1 + \frac{1}{16}A_0\right)V_a \qquad (7.1)$$

where $A_i=0$ or 1, $i=1,2,3,4$.

Thus for the case in Fig. 7.3, where the number to be converted is 1001, V_o will be

$$-V_o = \left(\frac{1}{2}\times 1 + \frac{1}{4}\times 0 + \frac{1}{8}\times 0 + \frac{1}{16}\times 1\right)V_a$$

or

$$V_o = -\frac{9}{16}V_a$$

For bipolar signals one solution is to use two reference voltage sources, one positive and the other negative, with the sign bit to be used to control a switch which will connect the proper reference voltage source to the circuit. Another solution could be to use the sign bit to connect or disconnect a second opamp, which will be used to reverse the output voltage of the first opamp.

The DAC resolution is the output voltage that corresponds to the LSB of the digital number. For a R-2R ladder with n nodes, the resolution will be $\frac{1}{2^n}V_a$ Volts. In the example in Fig. 7.3 this will be $\frac{1}{2^4}V_a = \frac{1}{16}V_a$ Volts.

It should be noted, that practically, a DAC multiplies the reference voltage by a number. Thus, if the reference voltage is an analog signal, a DAC can be used as an attenuator that is digitally controlled.

Finally, in order to obtain a smooth signal at the output of the DAC, the latter must be followed by a low-pass filter that will reject the frequency band around and beyond the sampling frequency.

7.5 Analog-to-Digital Conversion

As was previously stated, the conversion of an analog signal to digital requires sampling of the signal at regular time instants and the samples that are thus obtained to be converted to digital words. Therefore, the conversion of an analog signal to digital is not a continuous process. This introduces a new frequency band in the spectrum of the analog signal around the sampling frequency, the bandwidth of which is twice the bandwidth of the analog signal. Actually, this phenomenon is repeated theoretically infinite times around any integer multiple of the sampling frequency, as shown in Fig. 7.4. These bands of frequency should be removed from the analog signal, after the digital processing has been performed, by a low-pass filter.

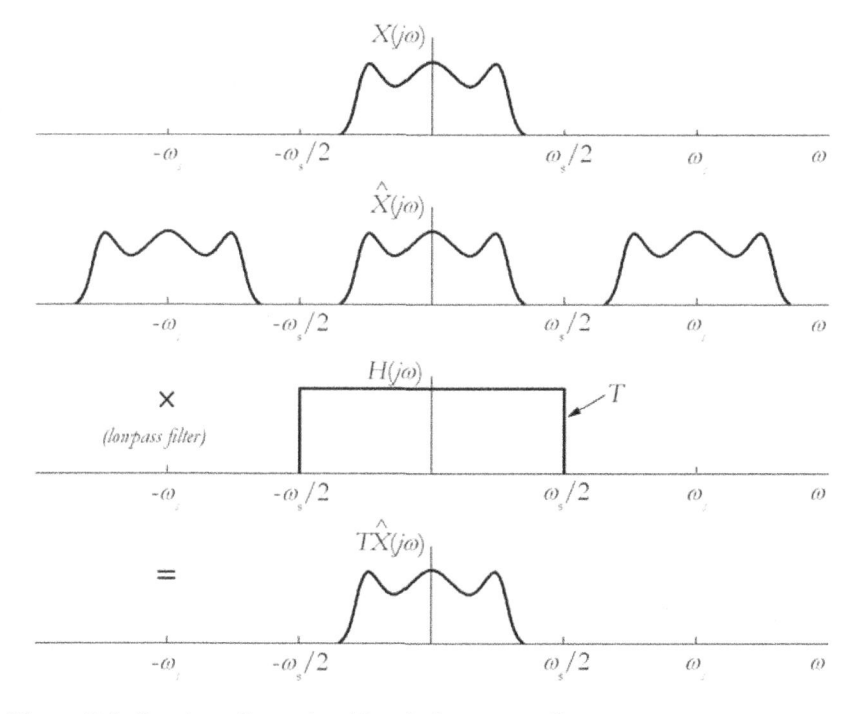

Figure 7.4. Creation of new signal bands due to sampling

However, this new frequency band will create the problem of *aliasing*, if the basic signal frequency band, or baseband, extends beyond half of the sampling frequency. In this case, parts of the new frequency band will enter the baseband, as shown in Fig. 7.5, and will alter the signal waveform, i.e. they will distort it. The problem of aliasing will be avoided by the low-pass filter preceding the sampler, which will reject all frequencies in the signal beyond half of the sampling frequency.

Analog-to-Digital Conversion is usually more involved than Digital-to-Analog Conversion and in some cases, a *DAC* is used for the construction of an *Analog-to-Digital Converter*, *ADC*. A number of conversion methods have been proposed in the past, some of which are more frequently used than others. In what follows, three such *ADC*s are briefly described, namely, the *successive approximation*, the *flash converter* and the *oversampled ADC*.

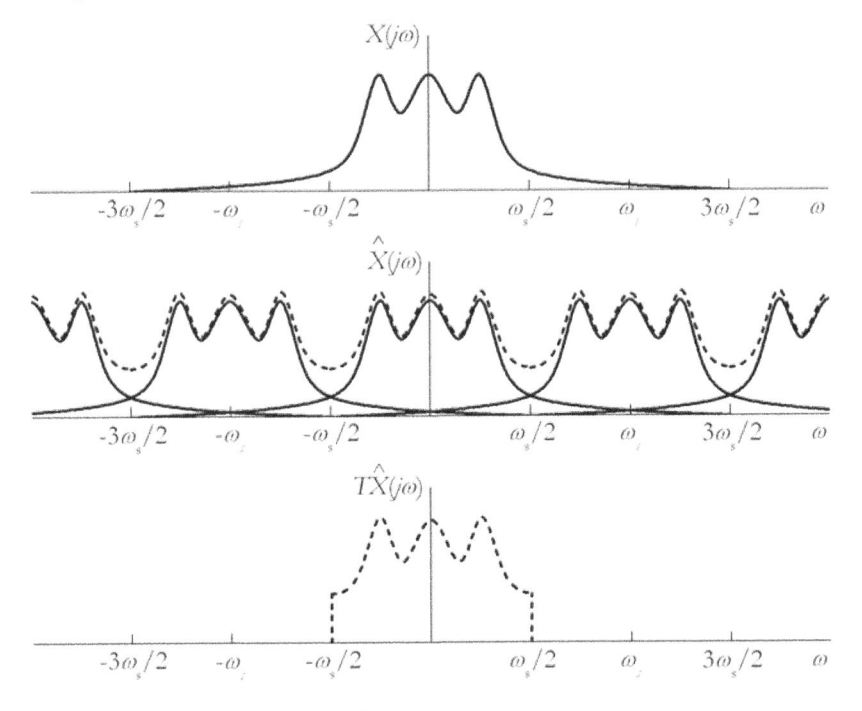

Figure 7.5. The phenomenon of aliasing

7.5.1 Successive Approximation *ADC*

In this method of *successive approximation*, an analog comparator compares the output signal from a *DAC* with the input signal and, when the two signals are equal, it stops the conversion. The analog comparator feeds a special register, which is controlled by a suitable logic circuit and is called the *successive approximation register* (*SAR*). The output (digital) signal of the *SAR* is applied to the input of the *DAC* and also forms the *ADC* output signal. Applying a successive approximation trial and error method finally the right digital word is obtained.

The whole set-up is shown in Fig. 7.6. In the beginning of the conversion the *MSB* in the *SAR* is set to 1 and all the other bits to 0. This word is converted by the *DAC* to the analog signal V_D and this is compared to the input signal by the analog comparator. If V_D is higher than V_{in}, the *MSB* in the *SAR* is set to 0, while if $V_D < V_{in}$, the MSB is kept to 1. Next, the following bit is set to 1. Then, the new output signal from the *DAC* is compared to V_{in} and if $V_D < V_{in}$ the second bit is kept to 1, otherwise it is set to 0 and the process is repeated for determining all the other bits of the digital word.

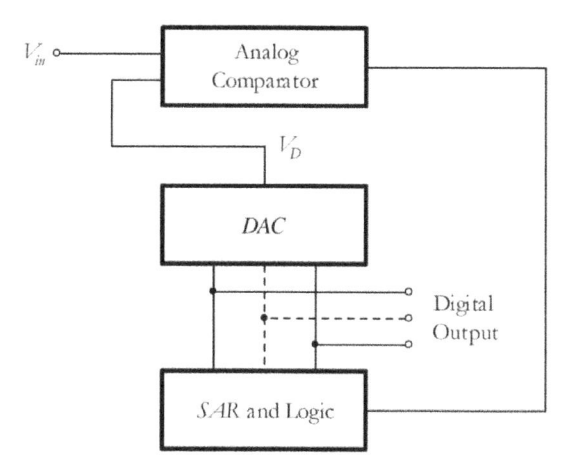

Figure 7.6. Successive approximation *ADC*

Successive approximation *ADC*s are relatively fast and low cost converters and for these reasons are frequently used in practice. Their characteristics are mainly determined by the characteristics of the *DAC* and the analog comparator.

7.5.2 Flash *ADC*

The structure of a flash *ADC* is shown in Fig. 7.7.

The under conversion analog signal V_{in} is compared to different reference signals by a number of analog comparators working in parallel. Then, the outputs of the comparators are decoded by a priority decoder, the output of which is the *ADC* output digital word. This word corresponds to the highest number of 1s in the input of the priority decoder, as given in Table 7.1.

Table 7.1. Truth table for the *Priority Encoder* in Fig. 7.7

W_7	W_6	W_5	W_4	W_3	W_2	W_1	Y_2	Y_1	Y_0
0	0	0	0	0	0	0	0	0	0
0	0	0	0	0	0	1	0	0	1
0	0	0	0	0	1	1	0	1	0
0	0	0	0	1	1	1	0	1	1
0	0	0	1	1	1	1	1	0	0
0	0	1	1	1	1	1	1	0	1
0	1	1	1	1	1	1	1	1	0
1	1	1	1	1	1	1	1	1	1

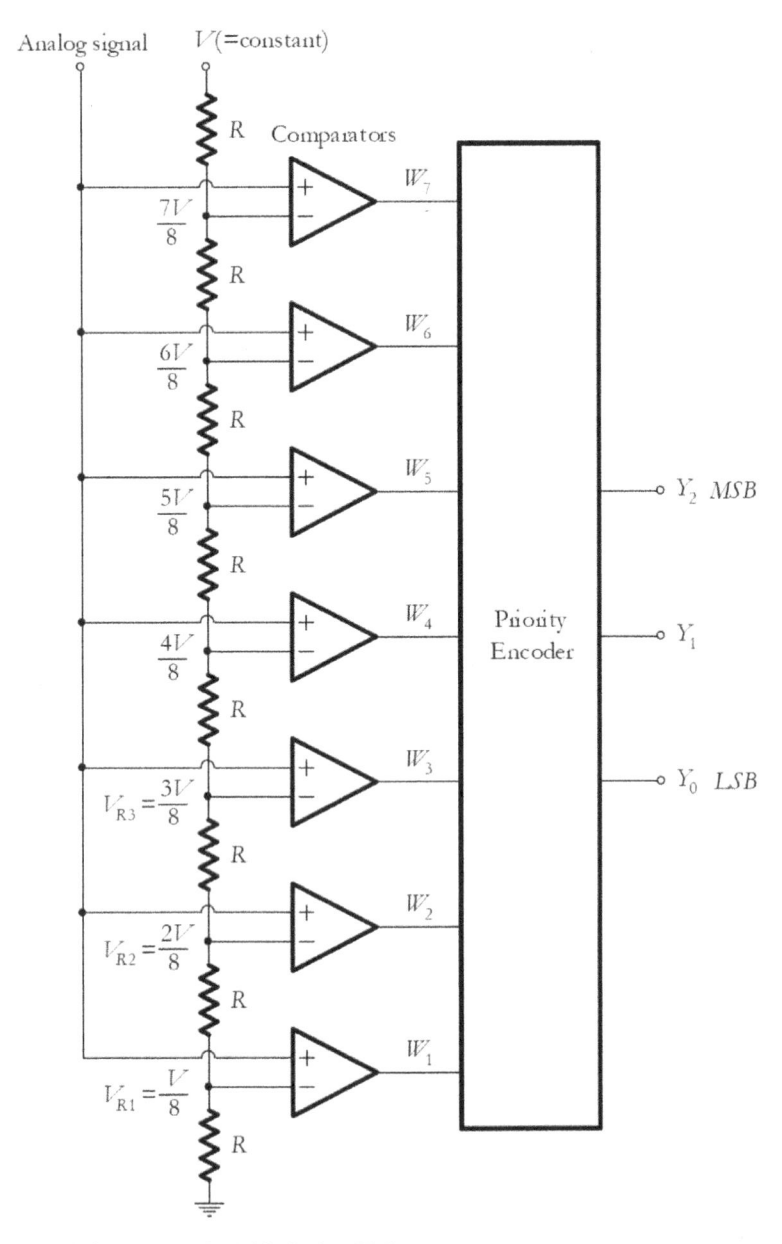

Figure 7.7. Structure of a 3 bit flash *ADC*

Flash *ADCs* are very fast and, consequently, used for the conversion of relatively high frequency analog signals.

7.5.3 Oversampling Converters

Another method for the conversion of a signal from analog-to-digital form and vice-versa is employing the *Delta-Sigma* (ΔΣ) *Modulation*.

Surely, the reader is familiar with terms like *Amplitude Modulation (AM)*, *Frequency Modulation* (FM) used in radio, *Phase Modulation* and of course *Pulse-Code Modulation*, the latter been in effect the conversion of an analog-to-digital form. In ΔΣ modulation the analog signal is first oversampled, i.e. it is sampled at a rate much higher than the Nyquist frequency. Then the integral is taken of the difference between the information-carrying signal and the modulator output signal and this difference is quantized. The result of oversampling is that the power of the quantization error is spread over a wider part of the frequency spectrum, compared to the signal power that remains within the band of the signal frequencies. Using precision digital filters (see Section 7.7) for the removal of the out-of-band noise, *ADCs* of high accuracy and very low noise can be built. Such *ADCs* are very successfully used in *CD* players and other high quality devices. For more about ΔΣ modulators and related *ADCs* and *DACs* the interested reader should be referred to specializing books like [5,6].

The circuit of a ΔΣ modulator is shown in block diagram form in Fig. 7.8.

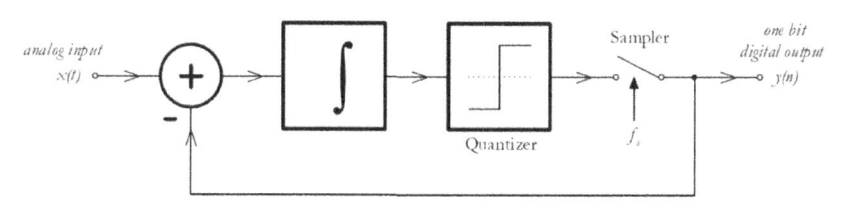

Figure 7.8. The ΔΣ modulator

7.6 Sample and Hold Circuit

When the analog signal varies very fast, in order to be more accurately converted to digital using an *ADC*, it is advisable, that each sample enters the converter having constant amplitude during the sampling period, as shown in Fig. 7.9(c). This way the conversion accuracy will be higher. The circuit by which this can be achieved is called *Sample-and-Hold*.

The basic operation of such a circuit is the following: An analog switch samples the signal and, during the time it is closed, a capacitor is charged up to the corresponding voltage value of the signal. The capacitor retains its voltage during the time the switch is open. The switch is controlled by a digital control signal. Such a circuit is shown in Fig. 7.10(a), which however, due to various problems, is not practical. More successful in practice is the circuit in Fig. 7.10(b), which constitutes the basis for the construction of a hybrid circuit available off-the-shelf.

In both circuits in Fig. 7.10 the opamp is used in its non-inverting mode. In this case the analog input voltage cannot be larger than the common-mode voltage of the opamp. In cases when the input voltage is higher than the permitted value, it is necessary to attenuate the signal. This problem can be solved using the opamp on the left in its inverting mode operating as a summer, as shown in Fig. 7.10(b).

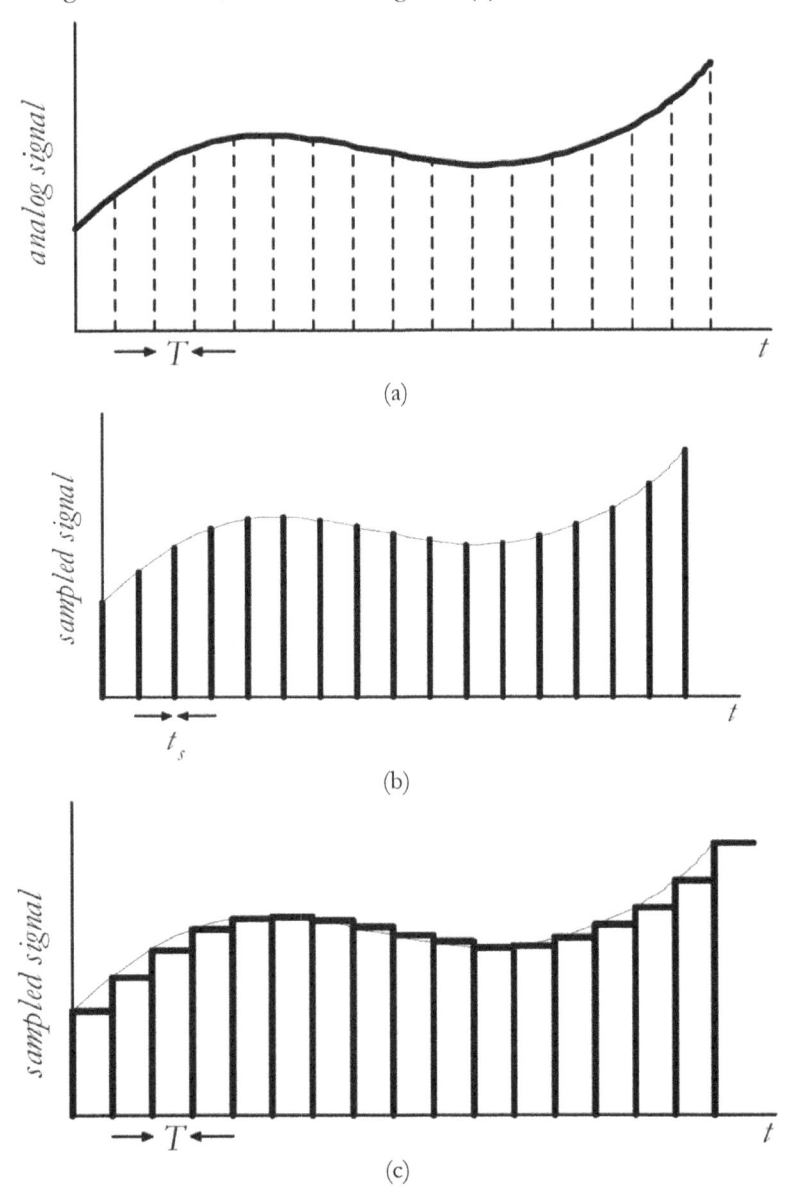

Figure 7.9. Sampling process of an analog signal

The Sample-and-Hold circuit, apart from its use combined with an ADC, can also be used for pulse height-to-pulse width conversion, in measurements of time duration etc.

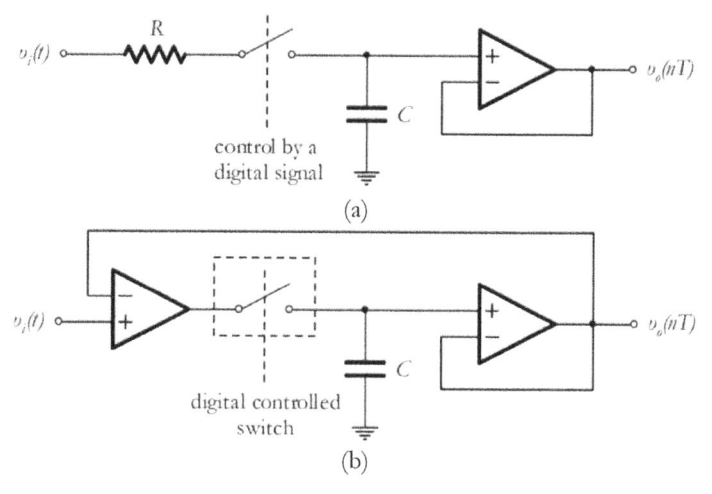

Data Conversion and Processing of Digital Signals 221

Figure 7.10. Sample-and-Hold circuit: (a) basic circuit and (b) practical circuit

Example 7.M.3

Consider a sinusoidal analog signal with frequency $1 rad/sec$ and amplitude $1V$. Using Simulink, sample this analog signal with sampling frequency $2.5Hz$.

Model

In Fig. 7.M.5 a *zero-order hold* block has been used in order to sample and hold the sine wave signal. Inside the block the sample period has been defined at $0.4sec$. The sine wave signal has $1 rad/sec$ frequency and $1V$ amplitude.

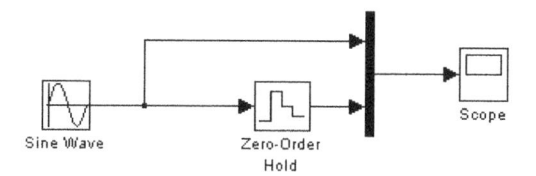

Figure 7.M.5. Simulink model for sampling a sine wave

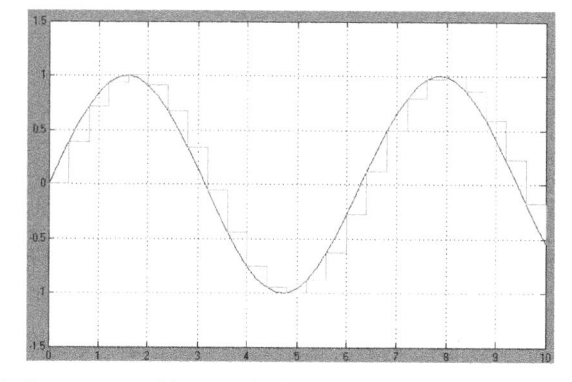

Figure 7.M.6. Sine wave and its sampled version

7.7 Digital Filters

A *digital filter* is a system that, operating on a discrete-time input signal, produces a discrete-time output signal with the additional requirement that the spectrum of the latter be related to that of the former by some rule of correspondence.

Digital filters can process numerical data and simulate dynamic systems and analog filters by means of computer programs. However, apart from this software solution of such problems, the availability of integrated circuits makes possible the same solution using dedicated hardware. There are certain very important properties or characteristics of hardware digital filters, namely:

- Non-critical component tolerances
- High accuracy
- High reliability
- Small physical size, etc

Both software and hardware digital filters can be used to process real-time as well as non-real-time (recorded) signals.

Software digital filters can be implemented on a personal computer or a workstation by means of a high level language or MATLAB or on a general-purpose digital processing chip by using a low-level language. Hardware digital filters relative to analog filters have an additional advantage on top of those stated above, namely that, because the filter coefficients are stored in a computer memory, they can be changed at will during the processing of a signal. This means that a digital filter can be programmable, adaptive and time-varying, if this is required. However, there are disadvantages, namely, the limited range of their use (0 up to ω_{max}), while for processing analog signals, when the result has to be also in analog form, an *ADC* and a *DAC* are required.

Nowadays, digital filters have replaced analog filters in many applications, while new applications have emerged. Such applications are in communication systems, audio systems including *CD* players, instrumentation, processing of seismic and biological signals, speech synthesis, digital radio, high-definition television etc.

There are various types of digital filters but, only two are briefly explained here namely, the *non-recursive* and the *recursive* digital filters.

Non-Recursive Digital Filters (*NRDF*)

Non-recursive digital filters are discrete-time systems the output of which is a function of the present and past samples of the input signal. This refers to a causal system. It can be shown, that such systems have a finite impulse response. Impulse response of a system is the response of the system when the excitation is an impulse. In this case a non-recursive digital filter is a *Finite Impulse Response* system, also referred to as FIR *digital filter*. An FIR digital filter is always stable and can have linear phase.

Recursive Digital Filters

Recursive digital filters are discrete-time systems the output of which at any instant depends on the present and past samples of the input signal as well as on past samples of the output signal. Clearly this definition entails the existence of feedback in the system. Such a system has an *Infinite Impulse Response* and for this reason, such filters are referred to as IIR *digital filters*.

The application of feedback in IIR digital filters means that such systems can be unstable under certain conditions.

Further discussion on digital filters is beyond the scope of this book and the interested reader should refer to specialized texts on the subject, as for example to the excellent book by A. Antoniou [7].

References

[1]. Analog Devices, *Analog Devices Conversion Handbook*, 3rd Edition, 1986.
[2]. Storey N., *Electronics – A Systems Approach*, Addison-Wesley, 1992 (reprinted 1995).
[3]. Millman J. & Grabel A., *Microelectronics*, McGraw-Hill, 1987.
[4]. Price T. E., *Analog Electronics*, Prentice Hall, 1997.
[5]. Norsworthy S., Schreier R. & Temes G., *Delta-Sigma Data Converters, Theory, Design and Simulation*, New York, IEEE Press, 1997.
[6]. Bourdopoulos G., Pneumatikakis A., Anastassopoulos V. & Deliyannis T., *Delta-Sigma Modulators, Modelling, Design and Applications*, Imperial College Press, 2003 (Distributed by World Scientific Publishing Co. Pte. Ltd.).
[7]. Antoniou A., *Digital Signal Processing*, McGraw-Hill, 2006.

MATLAB Problems

❖ *The reader is advised first to consult the MATLAB tutorial in Appendix D*

7.M.1. Consider a 4-bit analog-to-digital converter with $V_{max}=8V$ being the highest value of the input signal. Using MATLAB convert the voltages $1.85V$ and $5.30V$ to digital.

7.M.2. Consider again the 4-bit analog-to-digital conversion of problem 7.M.1. Using MATLAB plot the digital output versus the analog input as well as the quantization error versus the analog input.

7.M.3. Consider a sinusoidal signal with frequency $2 rad/sec$ and amplitude $5V$. Using Simulink sample this analog signal with sampling frequency of $1Hz$.

Problems

7.1. An analog signal varies between $0V$ and $5V$. It is sampled and quantized to four bit words. Calculate the quantization step and the error in Volts corresponding to the digital representation of a sample of $4V$ high.

(Answers: 0.3125, 0.0625V)

7.2. Complete the DAC of Fig. 7.3 to become a 6-bit DAC and show the position of switches, if the digital control word is 011010.

Chapter 8

Electrical Noise

8.1 Introduction

The electrical noise has been mentioned in preceding chapters as an unwanted content of the useful signal. In fact, electrical noise is any signal appearing at the output of a device that is not correlated with the signal at its input. In general it is random, but, in some cases, it may be periodic at the frequency of 50 (60) *Hz* due to interference by the mains voltage or current. Interference can also be caused by any electromagnetic field surrounding the conductor carrying the signal. Noise can be generated inside a device, e.g. an amplifier, through which the signal passes and this is due to various physical reasons. In some cases the useful signal can be buried in noise meaning, that its amplitude can be smaller than that of the noise. In such cases special processing of the signal is required to apply in order to dig out the useful information carried by the signal.

In this chapter the various noise sources are reviewed first and the noise characteristics are examined. Next, ways for the reduction of the noise content of the signal are discussed briefly. Then, a noise model of the opamp is introduced and used as an example to determine the total noise generated in the simple inverting amplifier circuit using the opamp. Finally, some basic points are mentioned, which have to be considered by the designer of a circuit using opamps, in order to avoid the introduction of excessive noise in the signal applied to the circuit.

8.2 Sources of Noise in Analog Signals

There are various reasons for the presence of noise in the signal. The signal picks up noise at the generation point and on its way to the device by which it is measured. The noise sources can be distinguished as *external* or *interference* and as *inherent*.

8.2.1 External or Interference Noise Sources

These sources include

- Electromagnetic radiation inducing noise to the wires carrying the signal
- Capacitor coupling noise from any power lines and transformers in the neighborhood
- Hum due to faulty grounding of the measuring devices

On the other hand, the inherent noise sources refer to the monitoring and measuring devices, for example the amplifiers. The noise added to the signal inside a device can be

- Thermal or Johnson noise
- Shot noise
- $1/f$ noise
- Noise due to drift in *dc* coupled cascade amplifiers, because of small changes in the component characteristics with time and temperature

If the signal is to be converted from analog form to digital for more accurate processing, additional noise can occur due to *quantization error* and *truncation*. We will consider these two noise sources separately in Section 8.7 below.

The effect of the interference noise sources on the signal can be minimized by careful shielding and grounding of the devices used in the measurement and by using coaxial cables for transferring the signal from its generation point to the measurement system.

However, the noise generated internally in an electronic device requires more detailed explanation and this is done separately below.

8.2.2 Inherent Noise

Consider a *dc* current I_{dc} in a conductor. This is defined by the mean number of positive charges passing through a cross section of the conductor in $1s$. In a metal the moving charges are free electrons. Their number passing a cross section of the conductor is not the same at any instant due to the fact that the electrons in the conductor move at high speeds to various directions and at the same time drift along the conductor due to the electric field at much lower speeds. Thus, the instantaneous current is not constant, but its mean value is.

Thermal Noise

The variations in the current at any instant depend on the temperature, which causes changes in the electron agitations and thus to the number passing through the conductor cross-section at any instant. Thus, although these changes do not appear in the *dc* current, they lead to the existence of a *rms* value of current, which is noise. It can be shown, that the noise *rms* voltage at the terminals of a resistor with resistance R is given by

$$E_n = \sqrt{4kTRB} \tag{8.1}$$

where,

k is Boltzmann's constant equal to $1.38 \times 10^{-23} J/K$

T the absolute temperature in degrees Kelvin

B the bandwidth of the instrument measuring the noise voltage

Taking the square of both sides in Eq. (8.1) gives

$$E_n^2 = 4kTRB \tag{8.2}$$

Then dividing both sides by R we get

$$\frac{E_n^2}{R} = 4kTB \tag{8.3}$$

Clearly, this is the power of noise generated in the resistance at temperature T, and it is constant at all frequencies within the bandwidth B. For this reason it is called *white noise*.

In Fig. 8.1(a) the equivalent of a "noisy" resistance is shown as a noise voltage source in series with a 'noiseless" resistance, as in Fig. 8.1(b).

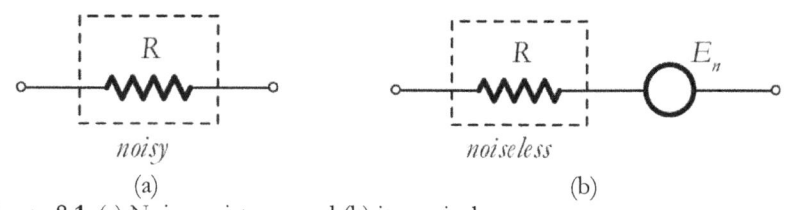

Figure 8.1. (a) Noisy resistance and (b) its equivalent.

Example 8.1

Calculate the thermal noise generated in a $100k\Omega$ resistance at the temperature of $40°C$ in $1MHz$ bandwidth. The Boltzmann's constant is $1.38 \times 10^{-23} J/K$.

Solution

Substituting in Eq. (8.1) gives

$$E_n = \sqrt{4 \times 313 \times 1.38 \times 10^{-23} \times 10^5 \times 10^6} = 41.57 \mu V$$

Shot Noise

Shot noise occurs when charge carriers have to pass over a potential barrier, e.g. electrons in a semiconductor diode. If the mean value of the current is I_{mean}, then the *rms* value I_{shot} of the shot noise is given by

$$I_{shot} = \sqrt{2qI_{mean}B} \tag{8.4}$$

where q is the electronic charge and B as in Eq. (8.1) for thermal noise. This type of noise is also *white*.

1/*f* or Flicker Noise

In the $1/f$ noise the noise power is proportional to $1/f$, where f is the frequency, and it is characteristic of semiconductor materials. Since the noise power is getting lower with increasing frequency this type of noise is called *pink*. It is clear that at lower frequencies in the noise power spectrum the $1/f$ noise dominates but, as this noise gets lower and lower with increasing frequency, the white noise dominates, as shown in Fig. 8.2. The frequency at which the two plots $1/f$ and white noise meet is the so-called *corner frequency*. The lower the corner frequency the better for the device as far as noise is concerned.

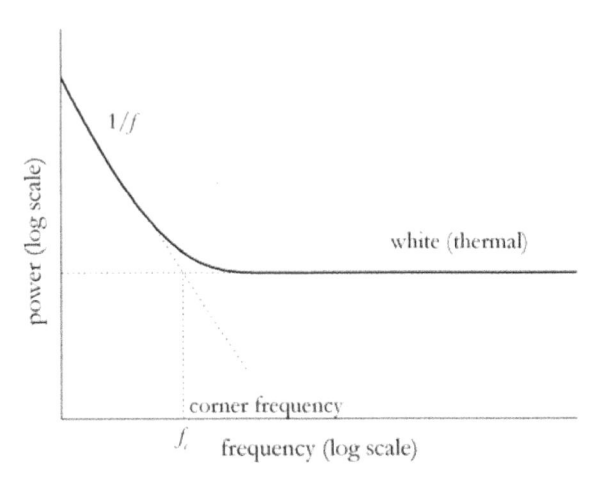

Figure 8.2. Plot of noise power against frequency in a conductor.

Example 8.2

From manufacturer's data the flicker noise power against frequency is given as k_v^2/f, where k_v^2 is in $(\mu V)^2$. Determine the total flicker noise power between frequencies f_L and f_H.

Electrical Noise

Solution

$$V_{ntot}^2 = \int_{f_L}^{f_H} \frac{k_v^2}{f} df = k_v^2 \ln \frac{f_H}{f_L}$$

Example 8.M.1

Using MATLAB plot the thermal noise of example 8.1 versus temperature for $-40°C$ to $140°C$.

Code

```
>> k=1.38e-23;
>> R=1e5;
>> B=1e6;
>> theta=-40:0.1:140;                   % define temperature range in °C
>> En=sqrt(4*k*R*B*(273+theta));
```
% the plots below are derived with the following procedure (the reader is advised first to consult the MATLAB tutorial in Appendix D)
```
>> plot(theta,En); axis([-40 140 3.6e-5 4.8e-5]); grid; xlabel('temperature (degrees Celsius)'); ylabel('thermal noise (volts)')
```

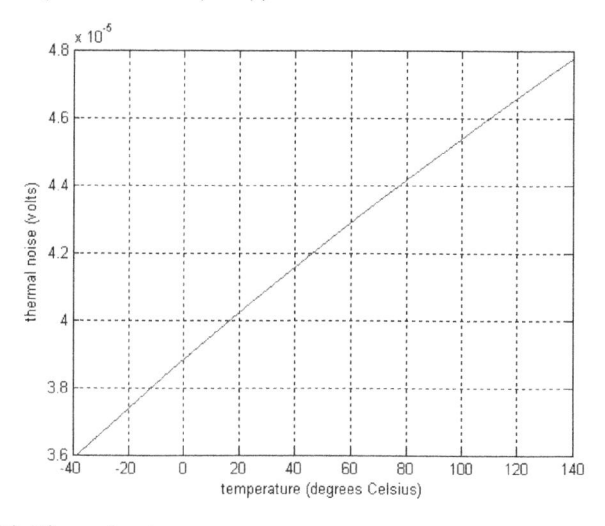

Figure 8.M.1. Thermal noise versus temperature

Example 8.M.2

Using MATLAB find the total flicker noise power of example 8.2 for $k_v^2=10(nV)^2$, $f_L=1Hz$ and $f_H=1MHz$.

Code

```
>> flicker_integral=quad('flicker',1,1e6)      % use flicker.m and take the integral
```

```
flicker_integral =

   138.1551
```

The m-file flicker.m is:

function a=flicker(f)
a=10./f;
end

8.3 Noise Characteristics and Measures

As has already been mentioned, the noise waveform is irregular. Thus one characteristic of noise is its power. The noise power P_n is given by

$$P_n = \frac{E_n^2}{R} = I_n^2 R$$

where, E_n and I_n are the respective *rms* values of the corresponding noise voltage and current generated in a resistance R. Usually, for comparison reasons, the value of R is taken to be 1Ω. With this in mind, the noise power is taken to be the square of the *rms* value of the noise voltage or current.

Let $x(t)$ be the value of noisy voltage or current at any instant t and x_{mean} be the corresponding mean value of this voltage or current. Then the *rms* value of noise x_{rms} from time $t=0$ to $t=T$ will be

$$x_{rms} = \sqrt{\frac{1}{T}\int_0^T [x(t) - x_{mean}]^2 dt}$$

It can be shown that, if E_n has Gaussian distribution, then, as a rule of thumb, the p.t.p. (peak-to-peak) noise amplitude is the *rms* noise value E_n multiplied by 6. The probability that the noise p.t.p. amplitude will not exceed this value is 99.73%.

From Section 8.2 it is clear, that another characteristic of noise is its bandwidth and its distribution in the various frequencies. The corresponding plot gives the noise power spectrum. The manufacturers of various devices (transistors, opamps, etc) give this spectrum as E_n^2/f or I_n^2/f in $(\mu V)^2/Hz$ or $(\mu A)^2/Hz$ respectively or the square root of them, as shown in Fig. 8.3. As was stated above, if the noise power is constant at all frequencies, the noise is called *white*, whereas if it falls off with increasing frequency, like the $1/f$ noise, it is called *pink*.

The relative noise content of a signal is given by the *signal to noise ratio* S/N. If S_p and N_p are the signal power and the noise power respectively, then

Electrical Noise

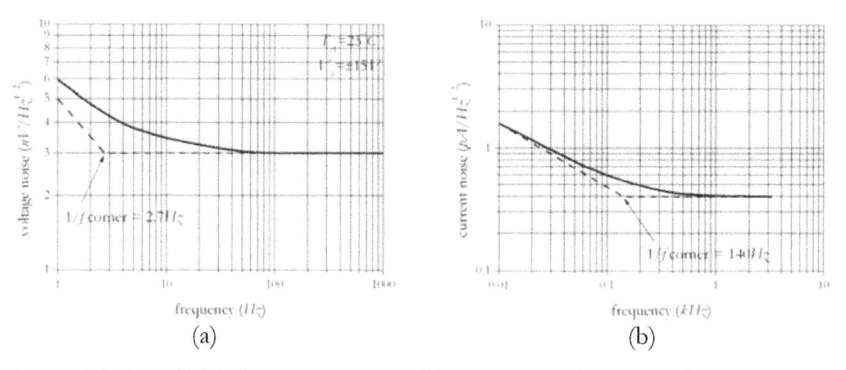

Figure 8.3. (a) OP-27 Noise voltage and (b) current as a function of frequency

$$S/N = \frac{S_p}{N_p}$$

or

$$S/N = 10\log\frac{S_p}{N_p} \quad \text{in } dB$$

Any signal applied to the input of an amplifier is characterized by a signal power S_{pi} and noise power N_{pi}. On the other hand, at the output of the amplifier the resulting signal power is S_{po} and the corresponding noise power is N_{po}. As a measure of how good the amplifier is from the noise point of view, a figure of merit is used called *Noise Figure* (NF). This is defined as follows:

$$NF = 10\log\frac{N_{po}}{G_p N_{pi}} \tag{8.5}$$

where G_p is the power gain of the amplifier. Since G_p is the ratio of the output signal power S_{po} to the input signal power S_{pi}, i.e.

$$G_p = \frac{S_{po}}{S_{pi}}$$

then Eq. (8.5) can be written as follows:

$$NF = 10\log\frac{N_{po} S_{pi}}{S_{po} N_{pi}}$$

or

$$NF = 10\log\frac{S_{pi}/N_{pi}}{S_{po}/N_{po}} \tag{8.6}$$

Thus, the NF (given in dB) is the ratio of the signal to noise ratio at the input divided by the signal to noise ratio at the output of the amplifier. It is clear from Eqs. (8.5) and (8.6), that for a noiseless amplifier the NF=0 and that the smaller the NF the better the amplifier is w.r.t. noise. The NF is a function of frequency, since both N and S depend on frequency. In Fig. 8.4 a coarse plot of this dependence is given for a semiconductor IC amplifier. In the region AB, where the $1/f$ noise dominates, the NF falls off at a rate of 3dB/octave as expected. In the region BC the NF is constant, because the noise is white while in the region CD the NF increases at a rate of 6dB/octave due to the reduction of the amplifier power gain with increasing frequency.

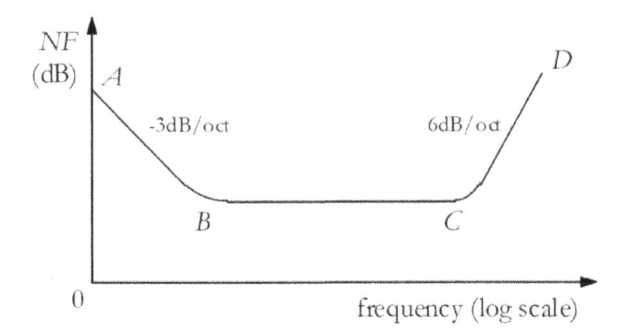

Figure 8.4. Plot of NF against frequency for an amplifier

8.3.1 Total Noise Due to White and Flicker Noise

Consider the white noise power taken from one of the manufacturer's data plot to be given by

$$e_n^2 = 4kTRB = \alpha_v^2 B$$

where B is the bandwidth.

For an elementary bandwidth $dB=df$ we can write

$$de_n^2 = \alpha_v^2 df$$

At the corner frequency f_c the flicker noise and the white noise are equal. For the elementary bandwidth df the flicker noise power will be $(k_v^2/f)df$ and the white noise $4KTRdf = \alpha_v^2 df$. Then at f_c

$$\frac{k_v^2}{f_c} df = \alpha_v^2 df$$

or

$$k_v^2 = \alpha_v^2 f_c \tag{8.7}$$

Using this relationship one can calculate the total noise, flicker plus white, for a bandwidth f_H-f_L, as follows:

$$E_{ntot}^2 = \int_{f_L}^{f_H} \frac{k_v^2}{f} df + \int_{f_L}^{f_H} \alpha_v^2 df = k_v^2 \ln \frac{f_H}{f_L} + \alpha_v^2 (f_H - f_L)$$

Substituting for k_v from Eq. (8.7) gives

$$E_{ntot}^2 = \alpha_v^2 \left(f_c \ln \frac{f_H}{f_L} + f_H - f_L \right) \qquad (8.8)$$

In the above discussion and indeed, in all cases in this Chapter, the various types of noise are considered to be statistically independent (uncorrelated).

8.4 Reduction of Noise

As has been mentioned in Section 8.2, noise can be added to any signal, by interference from various sources as well as inherently in a device like an amplifier.

Interference noise can be prevented by proper shielding and effective grounding of the measuring system. Once this has been achieved, the remaining noise is inherent and as such it is random. It can be white or pink. It has an infinite bandwidth and a way to reduce its power is to use an appropriate filter to restrict its bandwidth in the bandwidth of the signal. In Section 1.9 a simple RC lowpass filter was presented and this can be quite adequate in many cases. More advanced filters are introduced and discussed in Chapter 4.

Another method to reduce the inherent noise is to take advantage of its random nature and employ averaging of the noisy waveform. According to this method, successive records of the signal are added and the result is divided by the number of records. Because of the randomness of the noise, the value of noise at each point of the record is not the same at the corresponding points at the repeated records, taking any random value higher, lower, zero even negative. Thus, the result of averaging leads to noise reduction, which can be quite substantial, if a large number of records is averaged. Obviously, the method requires storing the records, adding and averaging, which can be easily achieved using a computer.

8.5 Noise Model of An Opamp

Any amplifier is built up from active and passive devices, like transistors, resistors, etc. During the operation of the amplifier, each of these elementary devices produces some kind of noise. As a result the amplifier produces internally some noise, which appears at its output and influences the amplified signal. In noise studies of the amplifier, all internal noise sources of the amplifier are referred to the input as equivalent noise sources, i.e. noise sources that would produce the same noise power at the output as the noise power produced by all the internal noise sources of the amplifier.

It has been accepted that the opamp noise model includes three noise sources, a voltage E_n and two current noise sources I_{n1} and I_{n2} connected to the input, as shown in Fig. 8.5. These noise sources combine the generation of all white and $1/f$ noise.

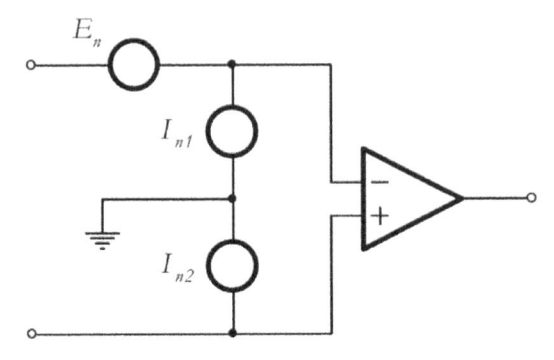

Figure 8.5. Opamp noise model

Values of E_n, I_{n1} and I_{n2} and their variation with frequency can be found in the manufacturer's data sheets for each opamp.

As an example of applying the opamp noise model, consider the simple inverting circuit shown in Fig. 8.6(a). The corresponding noise model of the circuit is shown in Fig. 8.6(b). In the latter, R_s is equal to $R_1//R_f$ producing the noise voltage E_{ts}, while E_{f2} is the noise voltage produced by R_2 according to Eq. (8.1). Following this, the total noise power E_{nT}, which will appear at the opamp input will be the following:

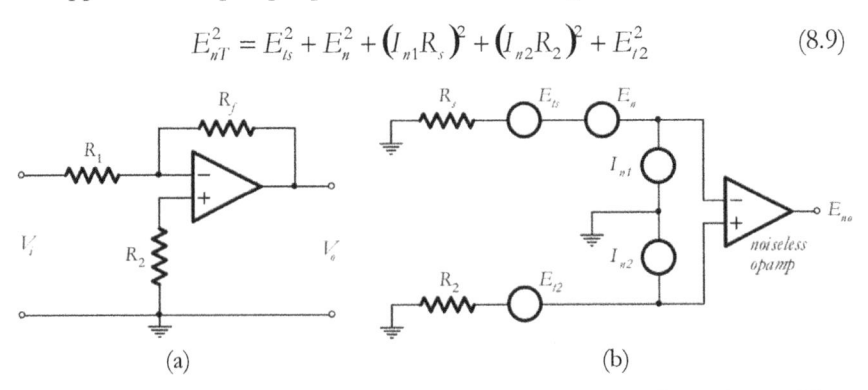

$$E_{nT}^2 = E_{ts}^2 + E_n^2 + (I_{n1}R_s)^2 + (I_{n2}R_2)^2 + E_{f2}^2 \tag{8.9}$$

Figure 8.6. (a) Inverting amplifier circuit and (b) the corresponding noise model

Electrical Noise

Note that the signal source has been taken to be an ideal voltage source. Any non-zero source resistance can be included in R_s.

8.5.1 Effect of Negative Feedback on Noise

The noise appearing at the output of an amplifier can be due to two reasons:

a. Noise accompanying the signal before entering the amplifier
b. Noise produced inside the amplifier

In case (a) the noise is treated as signal and is amplified accordingly. However, in case (b) application of negative feedback may reduce the internally produced noise. To show this, the presence of the internally produced noise in the amplifier is modeled by the noise voltage source V_n as shown in Fig. 8.7(a). Applying negative feedback, the circuit will become as shown in Fig. 8.7(b).

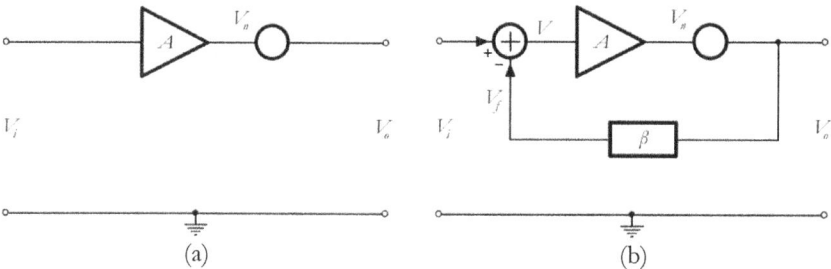

Figure 8.7. (a) Model of noisy amplifier and (b) noise reduction using negative feedback

From Fig. 8.7 we can write

i. Without feedback (Fig. 8.7(a))
$$V_o = AV_i + V_n \tag{8.10}$$

ii. With negative feedback (Fig. 8.7(b))
$$V_o = AV + V_n \tag{8.11}$$

where
$$V = V_i - V_f = V_i - \beta V_o \tag{8.12}$$

Substituting for V from Eq. (8.12) into Eq. (8.11) gives
$$V_o = AV_i - \beta AV_o + V_n$$

and solving for V_o

$$V_o = \frac{AV_i}{1+\beta A} + \frac{V_n}{1+\beta A}$$

It can be seen that the internally produced noise voltage has been reduced by a factor of $(1+\beta A)$. However, this reduction has been followed by the same reduction of the signal. Assuming that the amplifier noise is independent of the size of the input signal, one can increase the amplitude of the latter in order to increase the signal to noise ratio.

It should be noted that the additions of the noise voltage V_n and the other voltages are purely symbolic and it is not correct to add these quantities arithmetically.

8.6 Isolation Amplifiers

In the isolation amplifiers the output is completely isolated from its input. This means that the input common terminal and the output common terminal are completely isolated from each other, i.e. the leakage resistance between these is of the order of $10^{13}\Omega$. Also, the corresponding dielectric breakdown voltage is high. This isolation applies equally well in the case of the power supply and the ground terminals.

The isolation amplifier is an integrated circuit (IC), which structurally is more complicated than an opamp or an instrumentation amplifier. Its input stage is differential, but the signal is transferred to the output stage via a transformer or an optical coupler. One important difference from the usual opamp and the IA is that the isolation amplifier can withstand high common mode voltages, a situation that arises in measurements where high voltage devices are encountered, or when two grounded points are involved. One of the symbols used for the isolation amplifier is shown in Fig. 8.8.

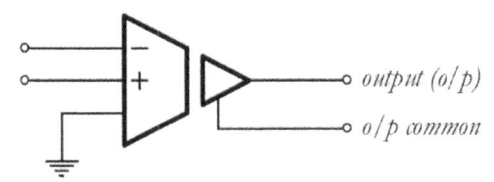

Figure 8.8. A symbol of the isolation amplifier

A typical IC isolation amplifier is the BB3450. Usually, the output common terminal must be connected to the power supply common terminal. No connection would be normally made from input common to the power supply common or to the output common terminals. The amplifier can be used as an opamp to serve as impedance buffer and for the amplification of low-level signals. Special care should be taken when using the isolation amplifier in medical applications regarding the patient's safety and the protection of the amplifier against defibrillator voltages.

8.7 Noise in Digital Systems

When an analog signal is to be processed by a digital system, it has first to be converted to digital form. Depending on the number of bits of the *ADC*, which will be used in the conversion, an error will occur called *quantization error* (see Section 7.3). The greater the number of bits in the binary word the smaller this error will be. Due to this error the resulting digital signal is considered noisy, i.e. it contains some kind of noise, which, quite appropriately here, is called *quantization noise*.

The output digital signal from the *ADC* will be passed to digital processing hardware, which can process binary words of limited length. If this word length is shorter than that of the samples to be processed, the latter should be brought to the right length, i.e. be truncated. The resulting error, called *truncation error*, will be considered a source of noise for the system.

Consider that the processing unit is implementing a digital filter. The digital filter is described mathematically by a function with numerical coefficients, which are decimal numbers. When the digital filter is realized in digital hardware, the coefficients will be represented in binary form. Depending on the word length of the processing hardware the filter coefficients may have to be rounded and truncated to a certain number of bits. Also, the product of each coefficient times each sample of the signal will have to be truncated after rounding, since the product will normally have a higher number of bits than each of the product terms. Clearly, this truncation of the product will be another source of noise.

Other types of noise, having to do with the distortion of pulses due to actual components of the hardware and spurious pulses, can be eliminated by proper design and use of Schmitt trigger circuits.

In spite of all the above noise sources though, a digital system is less noisy than a corresponding analog, and this is an important advantage, among others, of digital systems over analog. Contemporary technology of integrated circuits and its foreseen future advance ensures the above advantage of digital systems over analog.

Example 8.M.3

Consider a sinusoidal signal with frequency $3 rad/sec$ and amplitude $3V$. Using Simulink add noise into the signal. Using the *digital filter design* block of Signal Processing Blockset, filter the noise with a lowpass *FIR* digital filter.

Model

In Fig. 8.M.2 the Simulink model has been implemented using a *sine wave* block for the input signal, a *random source* block for the noise, an *add* block for incorporating noise into the input signal and a *digital filter design* block for filtering the noisy signal. The sine wave is of $3V$ amplitude and $3rad/sec$ frequency. In the random source block the noise has been defined as Gaussian type with sample time $0.01sec$. The Signal Processing Blockset has to be installed in order to use the digital filter design block (see Appendix D). Into the digital filter design block the filter has been defined to be a lowpass equiripple FIR filter. The original input signal, the Gaussian noise as well as the noisy signal together with the filtered one are shown in Fig. 8.M.3.

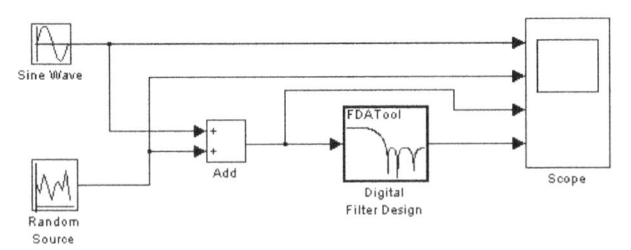

Figure 8.M.2. Simulink model for filtering noise from a sine wave signal

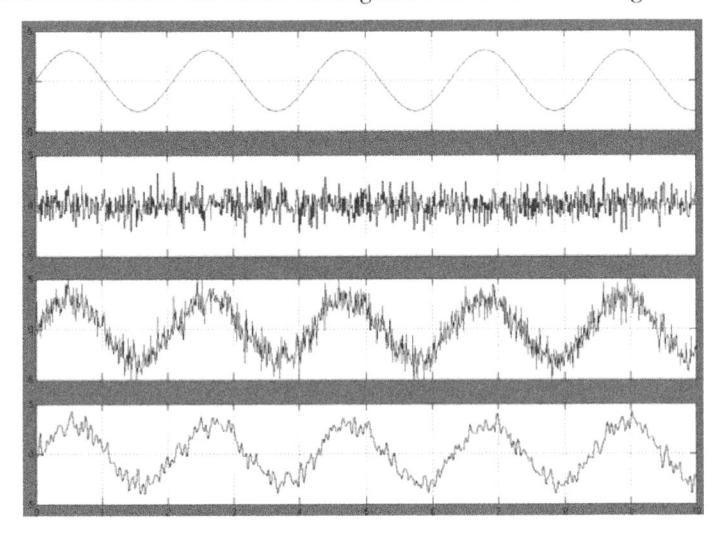

Figure 8.M.3. Sine wave signal, Gaussian noise, noisy signal and filtered signal

8.8 Basic Points To Consider For Low Noise Measurements

When designing a measurement system, special care must be taken so that the noise in the signal should be as little as possible. Some basic points to consider are as follows:

a. Minimization of externally generated noise

Starting from the electrodes, which act as the signal source, care should be taken that any artifacts are eliminated. Then shielded cables should be used to carry the signal from the electrodes to the measuring system and from device to device within the system. Proper shielding of the system itself is necessary, while all individual devices should be carefully grounded. One grounding point, preferably at the input terminal, must be used, when several electronic devices are connected together. Multiple grounding points can lead to large ground-loop currents.

b.   Choice of amplifier

The amplifier to be used, opamp or instrumentation, must have very low noise characteristics. This requires that the $1/f$ corner frequencies as well as the white noise are very low.

c.   Amplifier bandwidth

The amplifier bandwidth should be limited to the signal bandwidth to avoid amplification of noise with frequencies beyond those of the signal.

d.   Resistance connected to the Input

Since each resistance generates a noise voltage proportional to the square root of its value, excessive resistance connected mainly to the input should be avoided. This resistance at the input is most important, since the noise it generates will be amplified by the system gain.

References

[1]. Storey N., *Electronics – A Systems Approach*, Addison-Wesley, 1992 (reprinted 1995).
[2]. Normann R. A., *Principles of Bioinstrumentation*, Wiley, 1988.
[3]. Pallás-Areny R. & Webster J. G., *Sensors and Signal Conditioning*, Wiley, 1991.
[4]. King R., *Electrical Noise*, Chapman and Hall, 1966.
[5]. Delaney C. F. G., *Electronics for the Physicist*, Ellis Horwood Publishers, 1980.
[6]. Karoubalos K., *Introduction to Noise Theory and Applications*, Athens, 1979 (in Greek).

MATLAB Problems

❖ *The reader is advised first to consult the MATLAB tutorial in Appendix D*

8.M.1. Using MATLAB plot the shot noise of Eq. (8.4) versus I_{mean} for $I_{mean}=0$ to $10mA$. It is given that $q=1.6\times10^{-19}C$ and $B=1MHz$.

8.M.2. Consider a sinusoidal signal with frequency $3rad/sec$ and amplitude $3V$ and another one with frequency $15rad/sec$ and amplitude $2V$. Using Simulink add these two signals as well as Gaussian noise. Using the *digital filter design* block of Signal Processing Blockset filter the noise with a Butterworth lowpass *IIR* digital filter as well as with a lowpass *FIR* filter with Kaiser window.

Problems

8.1. Two resistances R_1 and R_2 are connected in series. Determine the total noise voltage.

8.2. Two resistances R_1 and R_2 are connected in parallel. Determine the noise voltage of the combination.

8.3. Draw the current equivalent of a noisy resistance and then apply this in the results of problems 8.1 and 8.2.

8.4. Derive the total noise power due to flicker and white noise using noise current data.

8.5. Derive the total noise at the output of the non-inverting amplifier in Fig. P.8.1 using the opamp noise model in Fig. 8.5.

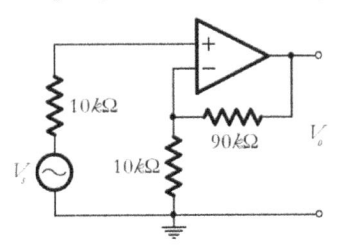

Figure P.8.1.

8.6. The noise model of the amplifier in Fig. P.8.2(a) is as shown in Fig. P.8.2(b). Determine the signal to noise ratio (S/N) referred to the input of the amplifier.

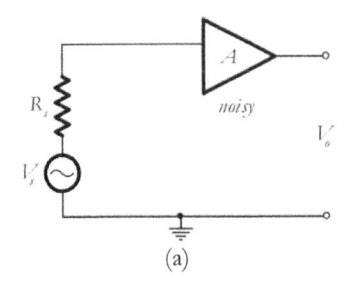

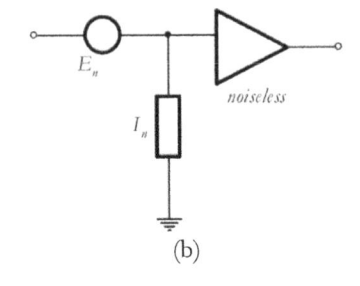

Figure P.8.2.

(Answers: $\dfrac{V_s^2}{4KTR_s B + I_n^2 R_s^2 + E_n^2}$)

8.7. The output voltage of an amplifier with voltage gain A contains, apart from the amplified sinusoidal input signal, also its second harmonic, which distorts it. Show that application of negative feedback to the amplifier will reduce the amplitude of the second harmonic of the signal by $1+\beta A$ times. (Hint: Follow a procedure similar to that presented in Section 8.5.1)

PART II

MEASUREMENTS

Chapter 9

Measurement of Physical Quantities

9.1 Introduction

In previous chapters, electronic devices, circuits and systems have been introduced and analyzed in some detail. They are useful to the reader in order to understand the various techniques of detection and measurement of some physical quantities. These quantities have been selected in this book to be related mainly to bioscience, since this field is of great interest to the welfare of all human beings.

In this chapter, first the general measuring system in its simplest form is presented and some important items, related to the quality of measurement of physical quantities, are briefly reviewed. Next, the Wheatstone resistance and capacitance bridge is discussed, as this is of great importance for the measurement of many physical quantities. The various *transducers* and *sensors* are described, which are required for the measurement of pertinent physical quantities. Such quantities are the temperature, pressure, displacement, liquid and gas flow, which are important in the field of medicine and bioscience in general.

9.2 Electronic Measuring System

Any electronic measuring system in its simplest form consists, in general, of a transducer or sensor, a signal conditioning and/or a processing stage and a recording or displaying device, as shown in Fig. 9.1.

A *transducer* is a device that converts one type of energy to another for purposes of measurement, information transfer etc. It can be electromechanical, electromagnetic, photovoltaic, piezoelectric, electronic etc.

```
[Transducer or Sensor] → [Signal Conditioner and/or Signal Processor] → [Recording or Displaying Device]
```

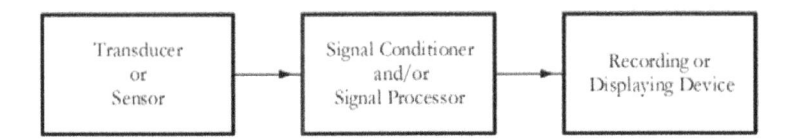

Figure 9.1. An electronic measuring system

On the other hand, a *sensor* is a device that detects (senses, measures) a physical quantity and converts it into an electrical signal. This signal, next, can be processed and read by an observer or an instrument.

The difference between a transducer and a sensor is that the transducer converts one form of energy into another, whereas a sensor receives information coming from a property of a physical quantity and converts it into an electrical signal only.

It is important to realize, that the presence of the transducer or sensor should not affect the quantity to be measured.

The electrical signal appearing at the output of the transducer/sensor is modified by the signal conditioner or processor, i.e. it may be amplified, filtered etc and, when it is put in the proper form, is sent to the recording system, which can be a counter, a printer or a visual imager. To take advantage of the digital technology, the signal in the processor can be turned into digital form using an *ADC* and be processed by digital filters.

9.2.1 Review of Some Definitions Concerning Measurements

Some definitions concerning the quality of measurement of physical quantities are as follows:

- **Accuracy** of a measurement: It is the measure of how close the obtained value is to the true or accepted value of the measured quantity. Calibration of the measuring instrument can partly improve poor accuracy.

- **Precision**: It refers to how close together the readings of the measuring instrument are, when measuring the same quantity repetitively under the same conditions. It is quite possible for a measurement to be very precise but not accurate, because precision does not refer to the true or accepted value of the measurement.

- **Sensitivity** of an instrument is the smallest amount it can measure of the quantity it is built to measure. The instrument does not notice anything smaller than its sensitivity.

- **Repeatability**: It refers to how close the results agree with each other, when they have been obtained under the same conditions and in short time interval.

- **Reproducibility**: It refers to how close to each other successive readings are, when the same quantity is measured in the long run with the same method, or with measurements carried out by different people, or made using different instruments, or in different laboratories.

- **Linearity** of an instrument is the amount of deviation from an instrument's ideal straight-line performance.

- **Resolution** of an instrument is the smallest amount of input signal change that the instrument can detect reliably.

9.3 Bridges

The Wheatstone Bridge, shown in Fig. 9.2, is a circuit that is used in measurements of resistances, capacitances, inductances and impedances in general.

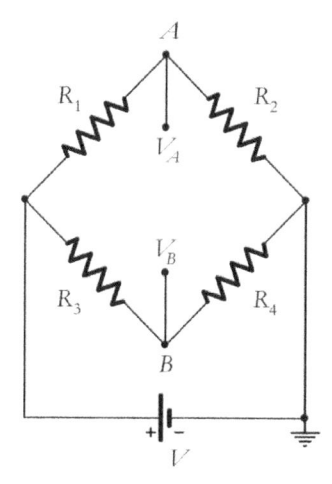

Figure 9.2. The Wheatstone resistance bridge

Referring to Fig. 9.2 we can write for V_A and V_B

$$V_A = \frac{R_2}{R_1 + R_2} V \qquad V_B = \frac{R_4}{R_3 + R_4} V$$

Then

$$V_{AB} = V_A - V_B = \left(\frac{R_2}{R_1 + R_2} - \frac{R_4}{R_3 + R_4} \right) V \qquad (9.1)$$

The bridge is said to be balanced when $V_{AB}=0$ and this, as found from Eq. (9.1), happens when

$$R_2 R_3 = R_1 R_4 \tag{9.2}$$

Thus, when three of the resistances are known the fourth, say R_4, can be found from Eq. (9.2) by balancing the bridge. This can be achieved by varying one of the other three, e.g. R_3, which can be a standard varying resistor. Condition (9.2) is reached independently of the power supply voltage or current and its possible variations.

The balance condition is monitored by placing a galvanometer or other null indicating instrument between nodes A and B.

This measurement method can also be used as a polarity detector, because, starting from balance, the indication of the instrument positive or negative depends on the changes in the value of R_4, if this is used as a sensor. In this case the readings of the indicating instrument are not independent of the supply voltage and its variations. Also, the readings are not exactly linear and some compensation method should be devised.

Such a sensor resistance bridge with analog linearization is shown in Fig. 9.3. The output voltage V_o is found to be

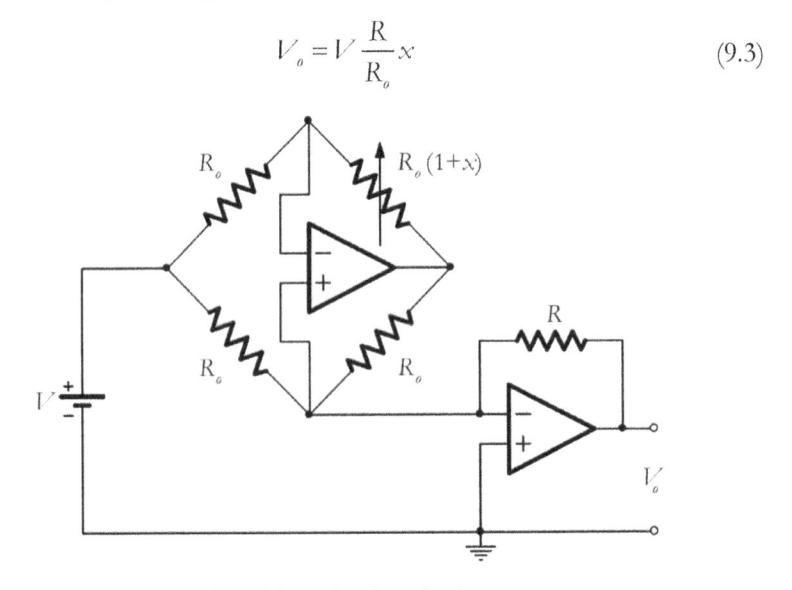

$$V_o = V \frac{R}{R_o} x \tag{9.3}$$

Figure 9.3. A resistance bridge with analog linearization

In case the unknown component is a capacitor, the circuit in Fig. 9.2 can be turned to a simple capacitance bridge as shown in Fig. 9.4. Notice that, in this case, the applied voltage V is ac, not dc.

Again, it can be shown, that for the capacitance bridge to be balanced the condition is

$$C_x R_2 = C_1 R_1 \tag{9.4}$$

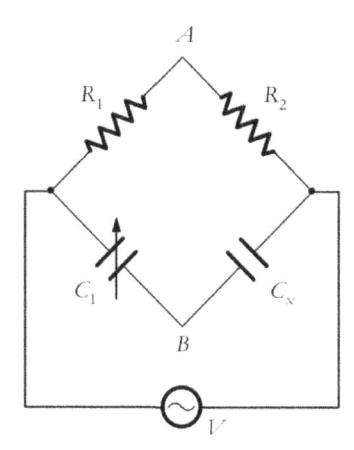

Figure 9.4. Simple capacitance bridge

Finally, when bridge branches are in general impedances Z_1, Z_2, Z_3, Z_4, then the balance condition is

$$Z_1 Z_4 = Z_2 Z_3 \tag{9.5}$$

Usually, Z_1 and Z_2 are made equal, so that the impedances in the opposite arms Z_3 and Z_4 must be equal to satisfy the balance condition in Eq. (9.5.). Again, in this case, the applied voltage is *ac*.

9.4 Temperature Measurement Transducers

Temperature is measured every day in nearly all patients in a hospital, since it constitutes the parameter that is correlated with the patient's health condition. Temperature is measured at various parts of the human body for clinical diagnosis and patient monitoring.

There are many types of *temperature sensors* available, namely:

- Fluid in glass
- Thermoresistive
- Thermocouple
- Thermistor
- *pn* junction diodes
- Quartz resonance
- Radiation

The common thermometer used at every home is an example of *fluid in glass*, in which the liquid is mercury. The measurement of temperature is based on the expansion of the volume of the liquid with an increase in temperature. While other liquids are also used in this category of thermometers, the range of temperature measurements is limited by the freezing and boiling points of the liquid and the softening point of the glass.

9.4.1 Thermoresistive Sensors

Thermoresistive sensors are based on the variation of the resistivity of a metal with changes of temperature. This variation, although limited in range, is approximately linear and is given as follows:

$$\rho = \rho_0 \left[1 + \alpha(T - T_0)\right] \quad (9.6)$$

Here, ρ and ρ_0 is the *resistivity* at temperatures T and T_0 respectively and α is the *temperature coefficient of resistivity*.

Since resistance is proportional to resistivity, we have

$$R = R_0 \left[1 + \alpha(T - T_0)\right] \quad (9.7)$$

where R and R_0 is the resistance at temperatures T and T_0 respectively and α the temperature coefficient of resistivity. Platinum is the most common material used as thermoresistor due to its resistance to corrosion and the fact that the coefficient α is relatively constant over a wide temperature range.

9.4.2 Thermocouples

A *thermocouple* is created when two different types of conductors are joined at one end. Because of the *Seebeck effect*, a voltage appears between the other ends held at different temperature, as shown in Fig. 9.5.

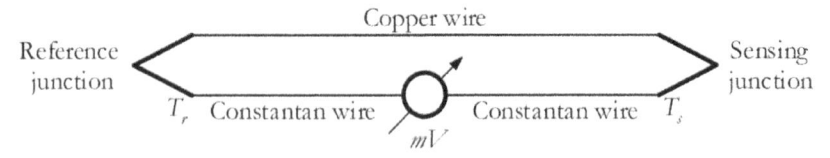

Figure 9.5. Copper/constantan thermocouple temperature measurement

The measured voltage difference ΔV is given by

$$\Delta V = \alpha(T_s - T_r)$$

where, T_s is the temperature to be measured, T_r the reference temperature and α is the sensitivity $\delta V / \delta T$ of the pair of metals used in the thermocouple. For the copper/constantan pair $\alpha = 40.28 \mu V/°C$. The voltage depends on the thermoelectric properties of the two conductors and the temperature difference of the junction. Thermocouple junctions may be placed in series to increase the voltage, thus creating a thermopile. A most common thermocouple is that consisting of copper and constantan (-270° to 400°C). Other thermocouples are chromel/constantan (-270° to 1000°C), iron/constantan (-210° to 760°C) etc. The choice of the thermocouple should be based on the temperature range, sensitivity and corrosion resistance. Other characteristics of thermocouples are their accuracy, usually 0.1K, and their resolution approximately 0.001K.

In very small size the thermocouple can be a very fast temperature sensor being thus suitable for biological applications. Commercially available thermocouples can be obtained with diameters of the order of 25μm and very low time constants, about 2ms.

The thermocouples have very low resistance and, due to this, the interference as well as the internally produced noise are very low. However, also due to the low voltage produced, particularly for low voltage differences, special amplifying circuitry is required, which is, however, commercially available.

As was said above, the thermocouple temperature measurement requires the use of a temperature reference. Traditionally an ice bath was used for this, but it is common now, to provide this by compensating circuitry.

9.4.3 Thermistor

A *thermistor* is a semiconductor resistive temperature sensor made of sintered metal oxides. It has a negative temperature coefficient, as can be seen from its resistivity, which is given by

$$\rho \propto \exp\left(\frac{E_g}{2kT}\right) \tag{9.8}$$

where, E_g is the *energy band-gap*, k the Boltzmann constant and T the absolute temperature. For germanium, $E_g=0.72eV$ while $k=8.67\cdot 10^{-5} eV/K$. Thus the resistance of the thermistor is:

$$R = B\exp\left(\frac{E_g}{2kT}\right) \tag{9.9}$$

B is a constant that can be determined as follows: Measuring R at a certain temperature T and knowing the energy gap for the semiconductor the thermistor is made of (e.g. germanium, silicon etc) B can be obtained from Eq. (9.9). On the other hand, if at temperature T_0 the thermistor resistance is R_0, then at temperature T the resistance R will be

$$R(T) = R_0 \exp\left[\left(\frac{1}{T}-\frac{1}{T_0}\right)\frac{E_g}{2k}\right] \tag{9.10}$$

It can be used for measuring temperature in the range $-190°C$ to $300°C$ with high precision, about $0.1mK$. When encapsulated in glass the thermistors become robust and corrosion resistant. Thermistors find a variety of applications as temperature sensors. Some of them are the following:

- Fluid flow measurements (cardiatic output monitoring)
- Gas flow measurements (pneumotachometer)
- Thermometry
- Calorimeters
- Vacuum gauges
- Fluid level detectors
- Temperature compensation circuitry
- Temperature control, time delays

Due to their small size, they have a very fast response to variations in air or fluid temperature. Thermistors can be used as thermal velocity probes. The velocity-sensitive thermistor R_u is exposed to the velocity stream, as shown in Fig. 9.6(a), with a second thermistor R_t placed inside the probe. Then both thermistors are connected as branches of a resistance bridge, as shown in Fig. 9.6(b).

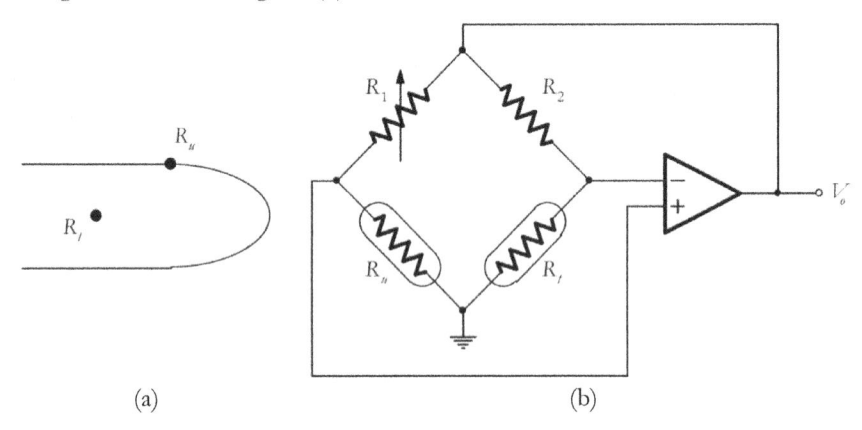

Figure 9.6. Thermal velocity sensor using two thermistors

Initially the bridge is balanced. Then the output voltage of the op-amp is zero. When the thermistor R_u is exposed to the velocity stream, it gets cooler and, subsequently, its resistance increases. This leads to increased voltage at the non-inverting input of the op-amp leading to increased voltage V_o at its output. This voltage heats thermistor R_u however, the presence of thermistor R_t provides temperature compensation.

When using a thermistor as temperature sensor, care should be taken how this is fed. Thus, if it is fed from a constant voltage source, the current will be V/R and the dissipated power in the thermistor V^2/R. This power, if it gets above a certain value, will heat the thermistor increasing its temperature. Then its resistance will drop with a subsequent reduction of the thermistor resistance, which will lead to higher current and so on and so forth. Thus the *V-I* characteristic of the thermistor will be as shown in Fig. 9.7. This phenomenon is called thermal runaway and will lead to the destruction of the thermistor. It will not happen, if the current is very low or, if it is kept constant, fed from a constant current source.

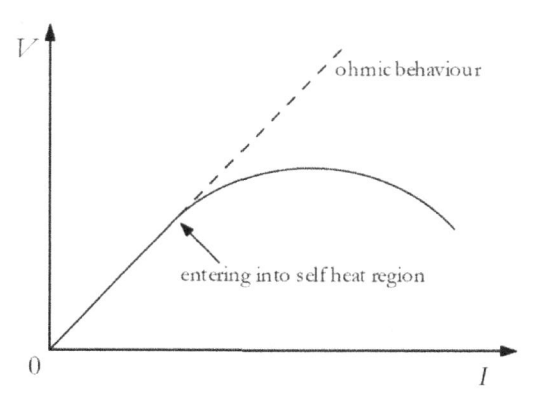

Figure 9.7. Thermistor *V-I* characteristic

The thermistors have very high temperature coefficient of resistance of 4%/°C and the advantage of being available in a variety of sizes and shapes. Their major drawback is their high nonlinearity, particularly so, if the temperature range under measurement is very wide. For use of the thermistor over a relatively small temperature range it has to be linearized. This can be achieved to a first-order approximation by using a current-limiting resistance R_L in series with the thermistor and a voltage source. The value of R_L is found to be

$$R_L = \frac{R_m(\beta - 2T_m)}{\beta + 2T_m}$$

where R_m is the resistance of the thermistor at the midpoint temperature T_m in the range over which the temperature is to be measured and $\beta = E_g/2k$. With this resistance the current in the thermistor becomes linear and so, the measured voltage becomes a linear function of temperature.

Thermistors can be of various sizes. Very small size devices can be built into probes for sensing in the human body, being relatively fast, because of their small size. They are quite sensitive to variations in temperature, quite stable over long periods of time and relatively inexpensive sensors.

Example 9.M.1

Using MATLAB plot the resistance of a thermistor versus temperature for −190°C to 300°C. It is given that $E_g=0.72eV$, $k=8.67 \cdot 10^{-5} eV/K$ and $R=10k\Omega$ at 25°C.

Code

```
>> Ro=10;
>> Eg=0.72;
>> k=8.67e-5;
>> To=298;
>> theta=0:0.01:100;                          % define temperature range
>> R=Ro*exp((Eg/(2*k))*((1./(273+theta))-(1/To)));
>> plot(theta,R); axis([0 100 0 40]); grid; xlabel('temperature (degrees Celsius)');
ylabel('thermistor resistance (kilohms)')
```

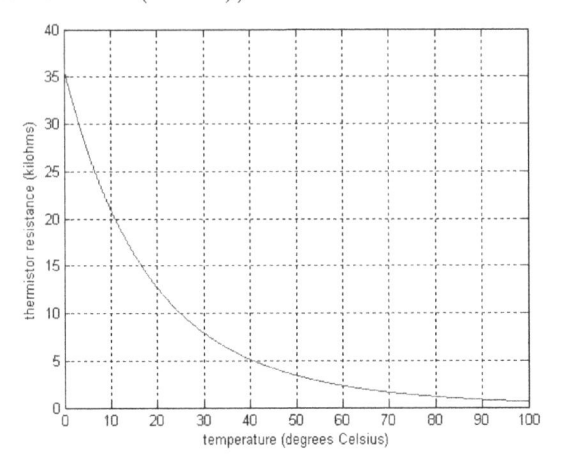

Figure 9.M.1. Thermistor resistance versus temperature

9.4.4 Other Temperature Sensors

a. The use of a *quartz crystal* to build a thermometer is based on the *piezoelectric effect* according to which, if the crystal is pressed in one direction, it develops an electric voltage in another direction. Electrically, this crystal has an equivalent circuit as shown in Fig. 9.8(a). This circuit exhibits one series resonant and one parallel resonant frequency. Placed in the feedback path of an amplifier one can build a high quality resonance circuit, see Fig. 9.8(b), taking advantage of one or the other resonant frequency of the quartz crystal. A resonator of this short exists in all quartz wristwatches. The resonant frequency of the quartz crystal is temperature dependent and this makes it suitable to be used as a thermometer. The temperature coefficient is of the order of $10^{-4} K^{-1}$ providing a temperature resolution around $10^{-4} K$.

b. In *semiconductor pn diodes*, when forward biased, the V-I characteristic is given by

$$I = A \exp\left(\frac{qV - E_g}{kT}\right) \tag{9.11}$$

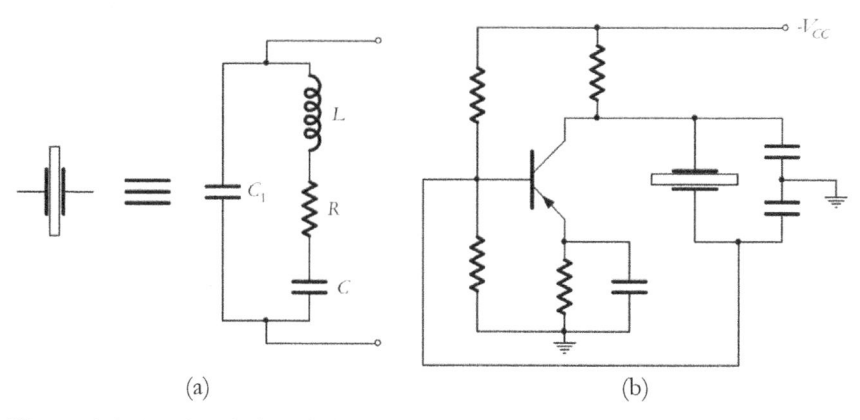

Figure 9.8. (a) Electrical equivalent of a quartz crystal and (b) an oscillator using a quartz crystal

where, I is the forward biased current, A is a constant dependent on the geometry of the junction, E_g the energy band-gap, q the charge of the electron, V the voltage across the junction, k the Boltzmann constant and T the absolute temperature. It can be seen from Eq. (9.11) that, if the current is kept constant the voltage developed across the junction is a linear function of the absolute temperature. Solving for V, Eq. (9.6) gives

$$V = \frac{E_g}{q} + \frac{kT}{q}\ln\left(\frac{I}{A}\right) \qquad (9.12)$$

If this junction is driven by two constant currents I_1 and I_2, a voltage difference V_1-V_2 is developed according to Eq. (9.12)

$$V_1 - V_2 = \frac{kT}{q}\ln\left(\frac{I_1}{I_2}\right) \qquad (9.13)$$

Therefore the difference in voltages is proportional to the absolute temperature. Based on this a thermometer can be built. It is constructed on an integrated circuit with two matched (identical) *pn* diodes, which are driven by two different currents. It measures the voltage difference developed across the *pn* junctions. The achieved accuracy of temperature measurement by such a thermometer is around $0.1K$ for a temperature range of $10K$.

c. The human body, as any object, being at any temperature emits *radiation* with a peak in the far infrared region. Treating the skin as a black body, Stefan's law can be applied, which states that the radiation power is proportional to T^4 with T the temperature of the body. At low temperatures this radiation power, is low, which implies that a radiation thermometer will be more suitable for high temperature measurements. The radiation thermometer does not contact the skin and thus also absorbs ambient thermal power. Because of this the temperature reading has to be corrected.

9.5 Displacement Measurement Transducers

Two types of displacement measurement sensors are examined below: *capacitive* and *inductive*.

9.5.1 Capacitive Sensor

A capacitor consists of two conducting surfaces separated by a dielectric, which is an insulator. In the case of the parallel plates capacitor, shown in Fig. 9.9(a), the capacitance C is given by (see Eq. (1.20))

$$C = \varepsilon \frac{A}{l} = \frac{\varepsilon_o K A}{l} \qquad (9.14)$$

Here, $\varepsilon = K\varepsilon_o$, ε_o is the permittivity of vacuum, K the dielectric constant of the dielectric, A the area of the conducting parallel plates and ℓ the distance between the plates. For vacuum $\varepsilon_o = 8.8542 \times 10^{-12} F \cdot m^{-1}$ and $K=1$ (1.00054 for air). In other types of capacitors the capacitance is also a function of the geometry.

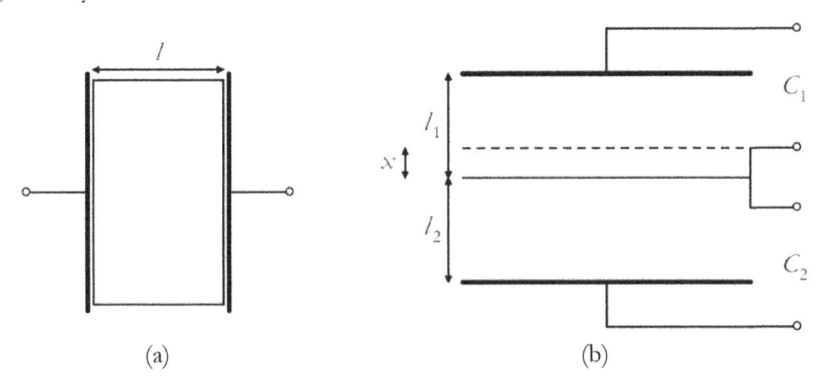

Figure 9.9. (a) Construction of a capacitor and (b) construction of a differential capacitor

Differentiating Eq. (9.14) w.r.t ℓ gives

$$\frac{dC}{dl} = -\frac{\varepsilon_o K A}{l^2} = -\frac{C}{l}$$

Thus

$$\frac{dC}{C} = -\frac{dl}{l}$$

which can be written also as

$$\frac{\Delta C}{C} = -\frac{\Delta l}{l} \quad \text{or} \quad \Delta C = -\frac{C}{l}\Delta l \qquad (9.15)$$

This is a linear relationship between the change in C and the change in ℓ that caused it.

A differential capacitor can also be built, as shown in Fig. 9.9(b), where the middle plate (a conducting membrane for example) is movable. Initially the two capacitances are balanced, i.e.

$$C_1 = \frac{\varepsilon_o KA}{l_1} \qquad\qquad C_2 = \frac{\varepsilon_o KA}{l_2}$$

and for $\ell_1 = \ell_2$, $C_1 = C_2$.

If the middle plate is moved as shown in the figure, ℓ_1 becomes $\ell_1 - x$ and ℓ_2 becomes $\ell_2 + x$. Then

$$C_1' = \frac{\varepsilon_o KA}{l_1 - x} \qquad\qquad C_2' = \frac{\varepsilon_o KA}{l_2 + x}$$

and C_1 increases while C_2 decreases. Subtracting C'_2 from C'_1, next adding C'_1 and C'_2 and finally dividing $C'_1 - C'_2$ by $C'_1 + C'_2$ gives

$$\frac{C_1' - C_2'}{C_1' + C_2'} = \frac{x}{l}$$

where, $\ell = \ell_1 = \ell_2$. If C_1 and C_2 are two branches of a capacitance bridge, as shown in Fig. 9.10 with $C_3 = C_4$, then, with $x = 0$, the bridge is balanced and $V_o = 0$. When $x \neq 0$ then, it can easily be shown, that

$$V_o = \frac{1}{2} \cdot \frac{C_2 - C_1}{C_1 + C_2} V_i = -\frac{1}{2l} x V_i$$

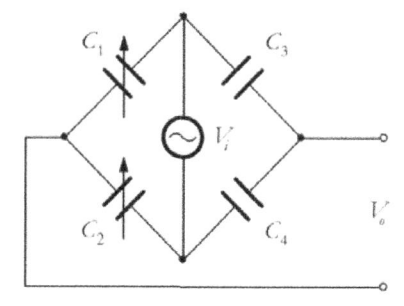

Figure 9.10. Capacitance bridge for displacement measurement

Therefore, V_o is proportional to the displacement of the plate. In the practical sensor the dielectric between the top plate and the middle plate

(C_1) is air, while between the bottom plate and the middle plate (C_2) is some dielectric material.

9.5.2 Inductive Sensor. Linear Variable Differential Transformer, *LVDT*

In Section 2.4 the transformer was considered ideal, meaning that the total power entering the primary coil was transferred to the secondary. In practice, this depends on the degree of coupling between the two coils, which is determined by geometrical considerations and the material within the coils. This material is ferromagnetic taking in such cases the name *ferrite*. In a differential transformer the coupling can vary, if the ferrite is moving within the coils. In Fig. 9.11 a *LVDT* is shown, consisting of a primary and two secondaries symmetrically spaced from the primary. The secondaries are identical, of length ℓ and connected in series opposition. Within the coils is placed a movable ferrite of length ℓ_c and is movable axially within the coils. The primary of the *LVDT* is excited by an alternating current with frequency between $50Hz$ and $50kHz$. When the ferromagnetic core is not displaced from the zero position, the voltage induced to one secondary is equal in magnitude but opposite in phase to the voltage induced in the other secondary so that the total voltage V_o is zero. Now, if the ferrite core is moved axially in one direction, the coupling will not be the same for the two secondary coils and thus the two voltages induced will not be the same in magnitude. Consequently, the voltage V_o will be nonzero. It can be shown that, for a range of core displacements, V_o is approximately linearly dependent on the length of the displacement.

ferrite core

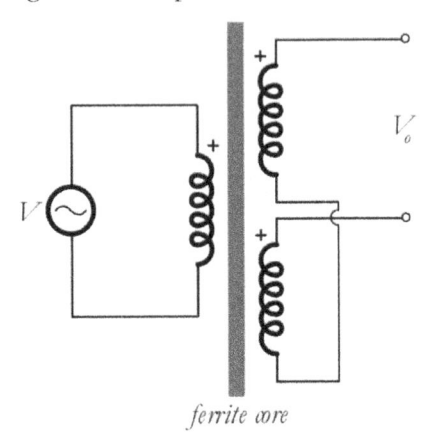

Figure 9.11. A Linear Variable Differential Transformer

*LVDT*s are made for a variety of length scales and are relatively insensitive to temperature and aging.

Other displacement sensors include the following:

Measurement of Physical Quantities

- Variable resistor
- Foil strain gauge
- Silicon strain gauge
- Liquid metal strain gauge
- Parallel plate capacitors
- Sonic/ultrasonic
- Optical stereography
- Interferometer

It can be seen that these are based on various technologies giving them their characteristics, which make them more suitable than the others for each particular displacement measurement.

9.6 Pressure Measurement Transducers

Pressure is important in many medical fields and measurements of blood pressure, foot pressure etc, are of particular interest and importance. Pressure in the cardiovascular system may indicate the health of the heart, vessels, kidneys and other organs of the human body. Using sphygmomanometers the arterial blood pressure is measured at the elbow.

Many pressure sensors use the fact that the resistance of a material changes when this is deformed. Consider for example a metal rod of length L and cross section area A. Its resistance R will be

$$R = \rho \frac{L}{A}$$

where ρ is the resistivity of the material. When the rod is deformed under pressure between its ends, its length will decrease, its cross section area will increase, while, depending on the material, its resistivity will change. The latter is then called *piezoresistivity*. To compare different materials based on resistivity changes under pressure, a gage factor G is used, which is defined as follows:

$$G = \frac{\Delta R / R}{\Delta L / L} \qquad (9.16)$$

For metals, G is approximately 2, while for semiconductors is around 100. Thus, metals will deform but their piezoresistivity will be low, whereas the semiconductors will deform little but their piezoresistivity will be high. Therefore, the semiconductor G is more sensitive than that for the metals.

Since normally ΔR is small, to detect changes four piezoresistors are placed in the four arms of a Wheatstone Bridge, in a way that the R of two elements will increase and the R of the other two will under pressure decrease, as shown in Fig. 9.12.

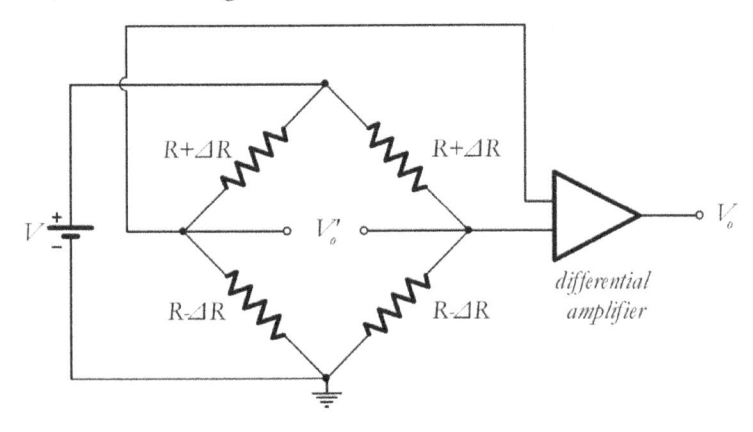

Figure 9.12. Arrangement for pressure measurements

In the past, high quality pressure transducers were very expensive and suffered from temperature sensitivity. The introduction of semiconductor pressure transducers has greatly reduced the cost, enhanced the range of transducer devices available, but these transducers have increased temperature sensitivity. Clearly, for accurate measurements at different temperatures such pressure sensors have to be temperature compensated.

More suitable for miniaturization are the capacitor pressure sensors. As explained in Section 9.5.1, the capacitance C of two parallel metal plates of area A and distance d between them is

$$C = \varepsilon \frac{A}{d} \tag{9.17}$$

When a force is applied, say on one plate, the distance between the plates will decrease and the corresponding change in the capacitance C will depend on the pressure. Capacitor sensors can be very small and thus many can be built and used in parallel, as it is required in foot pressure recordings for example.

Apart from the piezoresistive and capacitive sensors, fiber optics can also be used as pressure sensors.

9.7 Flow Measurements

The flow of a liquid or gas is measured by devices called *flow meters*. Its knowledge is of high importance in industrial processes, in hospitals, in consumers' life etc. There is a great variety of devices used for measurements of flow rate or quantity of a moving liquid or gas. These

Measurement of Physical Quantities 259

measurements are based on various principles and techniques, e.g. in obtaining differential pressure, velocity of moving liquid etc. In general, flow sensors and meters differ in terms of features, applications and operating performance.

9.7.1 Gas Flow Measurements

Air velocity is measured by inserting an air velocity flow sensor into a duct or pipe through an access hole. On the other hand, volumetric flow measurements are obtained by using balometers or flow hoods, which measure air volume flow at the air supply or exhaust outlets.

An air velocity flow sensor is using a thermal anemometer, which is heated up to a fixed temperature and then exposed to the air velocity. By measuring how much more air is required to maintain the original temperature gives an indication of the air speed. The higher the air speed, the more energy is required to keep the temperature at the set level. Vane anemometers have a proximity switch that counts the revolutions of the vane and supplies a pulse sequence that is converted by the measuring instrument to a flow rate. This is based on the conversion of rotation into electrical signal. Other air flow meters are based on differential pressure measurements.

Output for air velocity flow sensors can be analog voltage, analog current or frequency. They can also be connected to computers for data collection and programming. Displays for these sensors can be analog, digital or video terminals.

9.7.2 Liquid Flow Measurements

Liquid flow meters are used for measuring the flow rate of a moving fluid. With most liquid flow measurement instruments, the flow rate is determined inferentially by measuring the liquid's velocity on the change in kinetic energy. Velocity depends on the pressure differential that is forcing the liquid through a pipe or a conduit. With known of the pipe's cross-sectional area A, if it remains constant, the average velocity v is an indication of the flow rate Q, which then is given as follows:

$$Q = v \cdot A$$

Factors that also affect the liquid flow rate include the viscosity and the density of the liquid. Liquid flow meters are based on four metering technologies, the following:

- Measuring the pressure differential and extracting the square root
- Dividing the liquid into specific increments, which are counted by mechanical or electronic techniques

- Measuring the velocity of the liquid
- Measuring directly the mass rate of flow (mass flow meters)

Electrical outputs can be analog current, analog voltage, frequency and switch. Some liquid flow meters provide signal outputs in serial, parallel or other digital formats.

Liquid flow meters are commonly available that can measure temperature, density or level.

Two types of flow meters may be of interest to the readers of this book, namely: a) the *electromagnetic flow meters* and b) the *ultrasonic flow meters*. The principles behind their operation are explained here briefly.

a. The *Electromagnetic flow meters* (*EM*) are based on Faraday's law of electromagnetic induction, which states, that a voltage is induced on a conductor when it moves in a magnetic field. The liquid serves as the conductor while the magnetic field is created by energized coils outside the flow tube. These flow meters can be mounted in-line, or be of the insertion or non-invasive type. In-line flow meters are installed directly in the process line and typically require a straight line or pipe. By contrast, insertion type devices are inserted perpendicular to the flow path. Non-invasive *EM* flow meters do not require mounting in the process flow and can be used in closed piping systems. Outputs can be an analog current, analog voltage, frequency and switch. Interface option is also possible. Also *EM* flow meters suitable for use in sanitary environments, e.g. in hospitals, are also available.

b. *Ultrasonic flow meters* use sound waves to determine flow rates. They can be either Doppler effect meters or Time-of-Flight meters. *Doppler effect meters* measure the frequency shifts caused by fluid flow. This frequency shift is proportional to fluid velocity. In order for Doppler effect flow meters to work properly with a liquid, particles or bubbles must be present in the flow. Ultrasonic flow meters will generally not work with distilled water or drinking water. Aerations would be required in the clean liquid applications. *Time-of-Flight meters* use the speed of a signal traveling between two transducers. This signal increases or decreases with the direction of transmission and the velocity of the fluid being measured. Output types of ultrasonic flow meters are analog voltage, analog current, frequency or pulse and switch. Interface options for ultrasonic flow meters include serial and parallel interfaces.

9.8 Hall Effect Sensors

Consider a conductor or a semiconductor plate of thickness t, inside a magnetic field of flux B at right angles to the direction of its length, as shown in Fig. 9.13(a). If a current I flows along its length, a voltage V_H is generated across the breadth of the plate. This phenomenon was first observed by Hall in 1879 and is referred to as the *Hall Effect*. Quantitatively the *Hall voltage* V_H is given by

$$V_H = R_H \frac{BI}{t} \tag{9.18}$$

where B is the magnetic field and R_H the Hall coefficient. For V_H in *Volts*, I in A, B in *Tesla* and t in m, R_H is given in $m^3/Coulomb$. If R_H is known, t, V_H and I can be measured then, B can be obtained from Eq. (9.18).

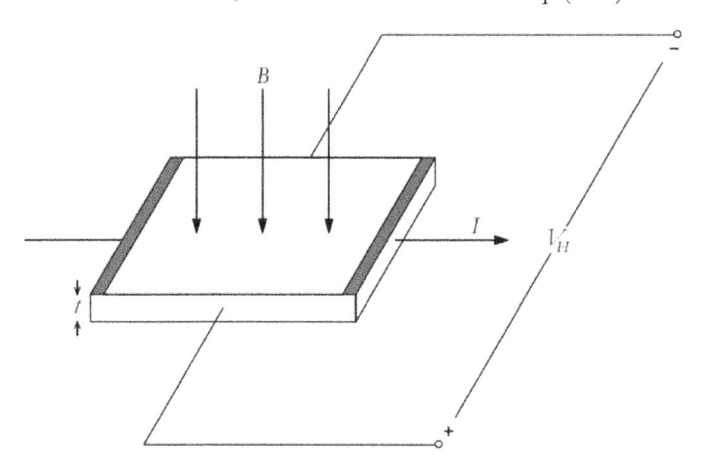

Figure 9.13. The Hall effect device

The generation of the Hall voltage V_H can be explained as follows: The charge carriers in the plate, moving at right angles to the magnetic field are subjected to Lorenz force, which causes the accumulation of charge on one side of the plate depending on the sign, positive or negative, of the charge carriers. This charge accumulation leads to the generation of an electric field, consequently to voltage V_H that its force on the charge carriers balances that exerted by the magnetic field. In a semiconductor the polarity of the Hall voltage depends on the majority carriers and, thus, it is opposite for n and p materials.

In semiconductors the value of R_H is much higher than in metals and thus, Hall effect sensors are based on semiconductors rather than on metals. Due to various reasons, e.g. controllable mobility by adding impurities, Hall effect sensors have been proved to be highly reliable. Some semiconductor materials used for Hall effect sensors are InSb, InAs, Ge, GaAs and Si with silicon having the advantage that circuits for signal conditioning can be integrated on the same chip.

Hall effect sensors are commonly used for the measurement of magnetic fields and for movement measurements with perhaps the most common application in computer keyboards.

9.9 Optical Measurements

According to the discussion in Section 2.7 two types of devices can be used as photodetectors namely, *photoconductors* (or *photoresistors*) and *photovoltaic cells*.

a. Photoconductor Sensors

Photoconductors are intrinsic semiconductor devices, the conductivity of which changes when light of proper wavelength falls on them. The reason is that then, new electron-hole pairs are generated with the electrons moving freely in the conduction band and the holes in the valence band. Thus with larger numbers of free charge carriers (current can be higher for the same voltage) the conductivity will be higher and, consequently, the resistance of the photoconductor will drop. With the photoconductor being part of a resistance bridge followed by a differential amplifier a useful photosensor can be obtained.

Since electron-hole pairs in the semiconductor can be also generated by increasing the temperature, optical measurements using photodetectors can be accurate provided that the temperature of the photoconductor is kept constant during the illumination. This is particularly necessary in the case of photoconductors for measurements in the red and near infrared region of wavelengths. In the visible range of the spectrum (400-700nm) and the nearest infrared (700-1400nm) cadmium-based materials are used (CdS, CdSe, CdTe), while in the near infrared lead-based materials are used (PbS, PbSe, PbTe).

b. Photovoltaic Sensors

Light detection can be achieved by taking advantage of the photovoltaic effect in a *pn* junction (see Section 2.7). A photovoltaic sensor is a *pn* junction that produces an open circuit voltage across its terminals, which increases with the intensity of the incident light on the junction until a saturation point is reached. Placing a resistor across the terminals a current is obtained and the voltage drop in the resistor can be measured. On the other hand, by short-circuiting the terminals of this photodetector the resulting current can be measured, and this is proportional to the intensity of light.

It should be noted, that in the first case the linearity of the measurement decreases as the load resistance increases and the response time also increases.

Also, it should be reminded, that *pn* junction photodetectors, depending on the semiconductors, have different efficiencies in the different wavelengths of the incident light (see Section 2.7). Although their output voltage or current requires amplification, *pn* junction photodetectors (sensors) compared to photoconductors are faster, produce lower noise and offer better linearity.

References

[1]. Wedlock B. D. & Roberge J. K., *Electronic Components and Measurements*, Prentice Hall, 1969.
[2]. Edwards D. F. A., *Electronic Measurement Techniques*, Butterworths, 1971.
[3]. Pallás-Areny R. & Webster J. G., *Sensors and Signal Conditioning*, Wiley, 1991.
[4]. GlobalSpec Inc, *Products and Services / Sensors, Transducers and Detectors*, 07/06/2010 (Internet).

MATLAB Problems

❖ *The reader is advised first to consult the MATLAB tutorial in Appendix D*

9.M.1. Using MATLAB find the resistance of a thermistor at 40°C. It is given that $E_g=0.72 eV$, $k=8.67 \cdot 10^{-5} eV/K$ and $R=20 k\Omega$ at 20°C.

Chapter 10

Biopotentials - Biosignals

10.1 Introduction

Biological signals or biomedical signals, also referred to as bioelectric signals or biosignals in short, are electrical signals generated by some kind of activity in the living cell, which may be chemical, electrical, mechanical or thermal. The generation of all bioelectric signals is due to

a. differences in the concentration of ions like Na^+, K^+, Cl^-, etc in the living cell and the extra - cellular or interstitial fluid

b. transient changes in the permeability of the cell membrane to these ions

The nervous system plays a fundamental role in nearly every function of the body. The propagation of the biosignals inside the living body is mainly carried out by the nervous system. Because of its importance, we examine in this Chapter the nervous system and the electrical characteristics of the nerves first. Next we examine the conduction or propagation of biosignals along the nerves. Then we look at the most frequently recorded biosignals for diagnostic purposes like the electromyogram *(EMG)*, the electrocardiogram *(ECG)*, the electroencephalogram *(EEG)*, the electroretinogram *(ERG)*, and the electrooculogram *(EOG)*. Finally, we refer briefly to the evoked potentials *(EP)*.

10.2 The Nervous System

The nervous system can be distinguished in the *central nervous system, CNS,* and the *autonomic nervous system, ANS*.

The central nervous system includes the brain, the spinal cord and the peripheral nerves. The brain is connected to the spinal cord and acts in a way somewhat similar to a computer, i.e. it receives information from the environment, internal or external, through the peripheral nerves (afferent nerves), and passes information orders through the peripheral nerves (efferent nerves) to striated muscles and glands (see Fig. 10.1(a)).

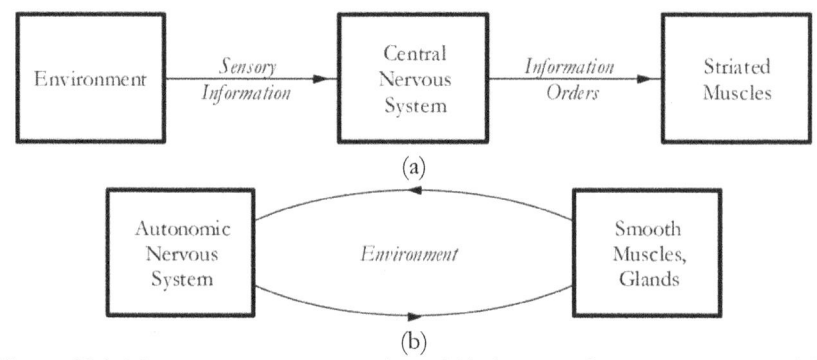

Figure 10.1. The nervous system consists of (a) the central nervous system and (b) the autonomic nervous system

The *autonomic nervous system* controls, basically involuntarily, various internal organs like the heart, the intestines and glands (Fig. 10.1(b)). It has components in both the central nervous system and the peripheral nervous system, *PNS*.

The basic structural unit of the nervous system is the neuron and this is what we describe next.

10.2.1 The Neuron

The neuron is the basic structural unit of the nervous system with the nerves being bundles of neurons. It is the nerve cell that receives signals from a *receptor* or another neuron through a *synapse* and transmits it further to the *CNS*, to another neuron or to a muscle. The structure of the neuron is shown schematically in Fig. 10.2. This nerve cell is separated from its surrounding fluid, the *interstitial fluid*, by a membrane, which is semipermeable to certain ions (as explained later). Attached to the *cell body* are the *dendrites* and the *axon*. The length of the axon can be up to $1m$ with its diameter being $1\text{-}20\mu m$. The axon serves the purpose of conducting *action potentials* to the *synapses*, points at which they make functional contact with other cells. Parts of some axons along their length, typically $1mm$ long, are covered by layers of myelin, which form a sheath that changes the electrical properties of the axon at these parts. They are called *Schwann cells*. Consecutive *myelinated* parts of the axon are separated by short *unmyelinated* parts. The latter are called *nodes of Ranvier* and are important for the regeneration and thus conduction of an electric pulse to long distances in the axon up to its *endings*. Neurons attached to striated muscles are called *motor neurons*.

Biopotentials – Biosignals 267

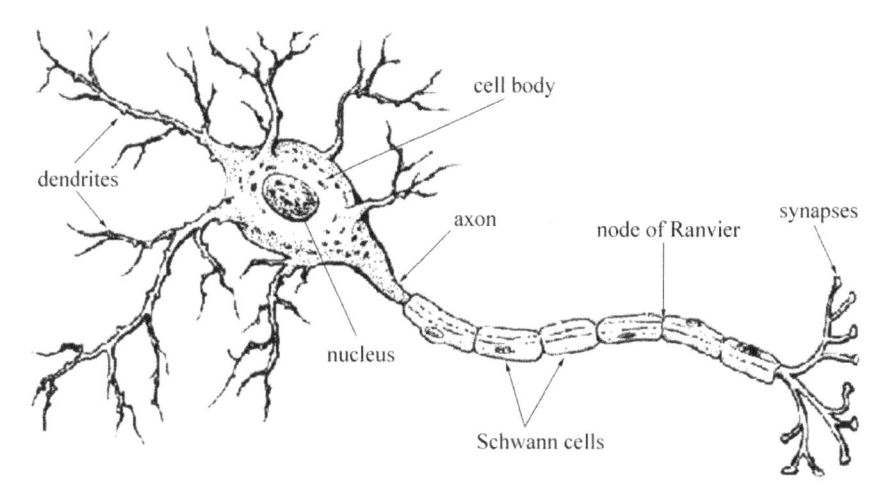

Figure 10.2. The structure of the neuron

Without any stimulus, there is a constant potential across the membrane of the cell, which is called the *resting potential*. When an electric stimulus is applied to the nerve cell, it disturbs the resting potential, and, if it is adequately depolarizing, an action potential is created which is transmitted along the axon up to its endings. This is the *action potential*. Both the resting and the action potentials are examined separately in some detail below.

10.3 The Resting Potential

The difference in ion concentrations C_i inside the axon and C_o outside the axon as well as differences in the membrane permeabilities of the various ions cause the generation of the resting potential. These ions are mainly the sodium (Na^+), the potassium (K^+), and the chlorine (Cl^-), and are the main ions that can pass through the membrane. Other heavy ions either inside or outside the axon cannot pass through the membrane. The concentration of Na^+ outside the membrane is much higher than that inside, while that of the K^+ is much higher inside than outside the axon. Also, the concentration of the Cl^- outside the axon is much higher than inside the axon. Due to their chemical gradient, the different ions tend to pass through the membrane from the side of higher concentration to the side of lower concentration; i.e. Na^+ and Cl^- from the outside to the inside of the axon and K^+ from inside to outside the axon. As they move passing the membrane, they form an electric field, which opposes the movement of more ions across the membrane. Eventually, equilibrium is reached and a constant potential is established. It should be stressed that in equilibrium the movement of ions across the membrane is balanced. Taking the potential outside the axon to be zero, by convention, the potential inside the axon is found to be about $-90mV$. Thus polarization across the membrane is created.

Using the *Nernst equation* one can calculate the contribution of each kind of ions in the total resting potential. Consider, for example, the case of Na⁺ ions, the concentrations of which are $C_o=145 moles/m^3$ and $C_i=12 moles/m^3$ for a typical mammalian axon. According to the Nernst equation

$$q(V_i - V_o) = kT \ln \frac{C_o}{C_i} \quad (10.1)$$

where

q	is the charge of the electron (electronic charge) $1.6 \cdot 10^{-19} C$
V_i	the potential inside the axon
V_o	the potential outside the axon, conventionally zero
k	Boltzmann's constant $1.38 \cdot 10^{-23} JK^{-1}$
T	the temperature in degrees Kelvin
$\ln \frac{C_o}{C_i}$	the natural logarithm (to the base e) of the ratio of the concentration of the particular ion outside and inside the axon

Substituting values in Eq. (10.1) with $T=310K$ and solving for V_i it is found that this is $+66mV$ (equilibrium potential for Na⁺).

Using the same equation for the cases of K⁺ ($C_o=4 moles/m^3$, $C_i=155 moles/m^3$) and Cl⁻ ($C_o=120 moles/m^3$, $C_i=4 moles/m^3$) it is found that the potentials V_i are $-98mV$ and $-90mV$, respectively.

However, the final value $-90mV$ of the resting potential is formed by the contribution of all ions acting simultaneously taking into consideration the corresponding membrane permeabilities P for these ions.

Two important points should be mentioned here. First, the permeability of the membrane for the K⁺ ions is much higher than that for the Na⁺ ions. So at rest, the positive K⁺ ions can pass across the membrane from inside to outside the cell due to the difference in concentrations, while the positive Na⁺ ions can pass from outside to inside the cell due to the difference in concentrations and the electric field. No energy is required for these flows, which are called *passive flows*. Second, in order to maintain the concentration gradients, K⁺ and Na⁺ ions should cross the membrane, the K⁺ from outside to the inside of the cell, and the Na⁺ from inside to outside the cell. These flows are taking place against differences in concentration and, additionally for the sodium ions, against the electric field. This is the active Na-K *transport* or the Na-K *pump* and works on expenditure of metabolic energy.

10.4 Electrical Properties Of The Axon

Electrically the axon behaves as a poor conductor. It has an external membrane resistance R', an intra-axonal resistance R along its length that is leaky and a capacitance due to the potential differences across the membrane.

The resistance of the axon is that of the fluid inside the membrane, called *axoplasm*, which is conductive. If we consider the axon cylindrical of length ℓ and cross-sectional area $S=\pi r^2$ (where r is the radius), the axon resistance R will be

$$R = \rho \frac{l}{S} \tag{10.2}$$

where ρ is the resistivity of the axoplasm.

The fact that the axon is leaky can be simulated by a parallel resistance for each very small segment of the axon. This resistance evidently will depend on the area of the segment of the axon, as any change in the potential will leak in all directions. So, if R_m is the leakage resistance per unit surface area, then the resistance R' of the axon piece considered cylindrical of length ℓ will be

$$R' = \frac{R_m}{2\pi r l} \tag{10.3}$$

(with r being again the radius of the cross-section of the cylinder).

According to these considerations, the current inside the axon progressively will be reduced, and at a certain distance from its starting point, the current along the axon and the leakage current will become equal. Considering that the axon is uniform, let this distance be λ. Thus, at distance $\ell=\lambda$, called the *space parameter*, the resistances R and R' will be equal, i.e.

$$R = R'$$

Substituting in this equation the values of R and R' from Eqs. (10.2) and (10.3) gives

$$\frac{\rho \lambda}{\pi r^2} = \frac{R_m}{2\pi r \lambda}$$

or

$$\lambda = \sqrt{\frac{R_m r}{2\rho}} \tag{10.4}$$

Therefore, the space parameter λ indicates how far the axon current can reach before most of it has leaked out through the membrane. For an unmyelinated axon the value of λ is much smaller than that for a myelinated axon, because the myelin sheath in the latter acts as an insulator, thus reducing the leakage of the axon current.

Apart from the two resistances R and R', the fact that there is a potential difference across the membrane corresponds to the presence of a capacitance between the axoplasm and the interstitial fluid. If C_m is the capacitance per unit area of the axon surface then the capacitance C for the cylindrical segment of length ℓ of the axon will be

$$C = C_m (2\pi r \ell) \tag{10.5}$$

Note that the presence of the myelin sheath increases the distance between the two conductors, i.e. between the *plates* of the capacitor. Accordingly, the corresponding capacitance of geometrically equal (same ℓ, same r) axon segments will be much smaller for the myelinated axon than for the unmyelinated axon.

Following the above discussion, we may represent the electrical behaviour of a small axon segment by the equivalent circuit shown in Fig. 10.3(a).

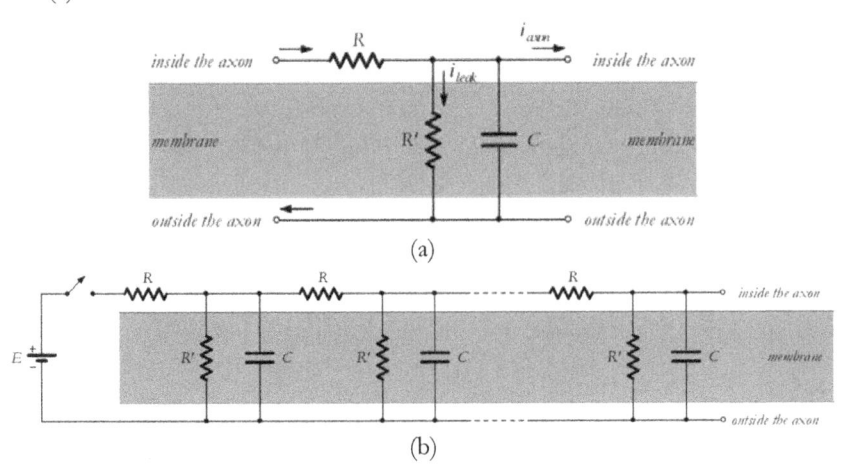

Figure 10.3. (a) The equivalent circuit of a very small segment of the axon and (b) the equivalent circuit of a large part of the axon

As the axon can be considered to be made up of small segments connected in series, the equivalent of a part of the axon can be considered to be made up of the equivalent circuits of all segments connected in cascade, as shown in Fig. 10.3(b). Note that in this equivalent circuit we have ignored the existence of the resting potential, since this is constant and is not going to affect the transmission of the signal (varying voltage or current) along the axon.

Biopotentials – Biosignals

The circuit in the last figure operates both as an *attenuator* due to the presence of resistance R' in the parallel branches and as a *low-pass filter* (see Section 1.9) due to the capacitances also in the parallel branches. Generally speaking, it is a *distributed parameter circuit* due to the uniformity of its structure, which is made up of similar subcircuits corresponding to very small segments of the axon.

Now let us assume that the value of the depolarization caused by a stimulus E is not higher than a certain value, say not higher than $30mV$. This is the case of a *weak stimulus*. If the switch in the circuit in Fig. 10.3(b) is closed, the application of the voltage E will result in a current through the first resistance R, which then will be divided into the current in the first R' and the current in the second R, while the capacitor will tend to charge towards the voltage across R'. Thus, the current in the second R will be smaller than that in the first R. This procedure will be repeated in the second R, R', C- section, and so on. Thus, the current in the series branch will be getting smaller and smaller until it becomes zero. On the other hand, the voltage in each node will become smaller and smaller as the distance from the point of the application of the *stimulus* increases. This description applies equally well for the myelinated as well as for the unmyelinated axon with the difference that the current propagates further in the former than in the latter. One may calculate the voltage V_x at a distance x from the point the stimulus is applied using the following equation:

$$V_x = V_d e^{-x/\lambda} \qquad (10.6)$$

where, V_d is the voltage at $x=0$ and λ the space parameter.

The situation is different in the case that the stimulus is not weak and this is what we discuss next.

Example 10.M.1

Refer to the circuit in Fig. 10.3(b), in which $E=1V$ dc, $R=100\Omega$, $R'=1k\Omega$ and $C=2nF$. When only one R, R', C- section is present (the rest have been removed), using MATLAB, plot the voltage across C against time after the switch has been closed at $t=0$. Repeat when two R, R', C- sections are present for the voltage across the two capacitors. In both cases consider that the initial voltage in the capacitor(s) is zero.

Code

Analysis of the circuit gives

$$E = v_R + v_C$$

where

$$v_R = iR = R\left(C\frac{dv_C}{dt} + \frac{v_C}{R'}\right)$$

Then

$$\frac{dv_C}{dt} = \frac{1}{RC}\left(E - \frac{R+R'}{R'}v_C\right)$$

which becomes

$$\frac{dv_C}{dt} = 5\cdot 10^6 - 5.5\cdot 10^6 v_C$$

Using numerical solution (see Appendix D) and considering $v_C(t)=0$ for $t=0$ we have

```
>> [t,vc]=ode23('section_1',[0 3e-6],0);
```
% numerical solution using section_1.m till $t=5\mu s$ considering $v_C(t)=0$ for $t=0$
```
>> plot(t,vc); axis([0 3e-6 0 1]); grid; xlabel('time (sec)'); ylabel('vc(t) (volts)'); title('numerical solution for vc(t)')
```

The m-file section_1.m is:

```
function dx=section_1(t,x)
dx=5e6-5.5e6*x;
end
```

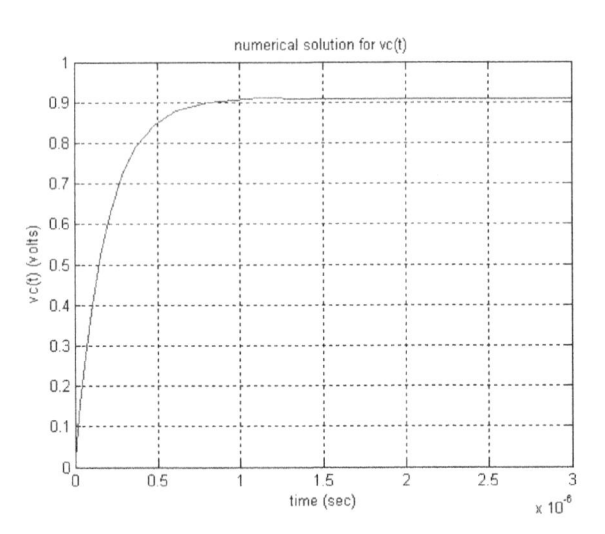

Figure 10.M.1. Voltage across capacitor C against time

Considering now two R, R', C- sections we get

$$E = v_R + v_{C1} = v_{C1} + R\left(\frac{v_{C1}}{R'} + C\frac{dv_{C1}}{dt} + \frac{v_{C2}}{R'} + C\frac{dv_{C2}}{dt}\right)$$

where

$$v_{C1} = v_{C2} + R\left(\frac{v_{C2}}{R'} + C\frac{dv_{C2}}{dt}\right)$$

C_1 is the capacitor of the first R, R', C- section and C_2 the capacitor of the second section. Solving the above system we get

$$\frac{dv_{C1}}{dt} = \frac{1}{RC}\left(E + v_{C2} - \frac{R+2R'}{R'}v_{C1}\right) \qquad \frac{dv_{C2}}{dt} = \frac{1}{RC}\left(v_{C1} - \frac{R+R'}{R'}v_{C2}\right)$$

which becomes

$$\frac{dv_{C1}}{dt} = 5\cdot 10^6 \left(1 + v_{C2} - 2.1 v_{C1}\right) \qquad \frac{dv_{C2}}{dt} = 5\cdot 10^6 \left(v_{C1} - 1.1 v_{C2}\right)$$

Using numerical solution (see Appendix D) and considering $v_{C1}(t) = v_{C2}(t) = 0$ at $t=0$ we have

```
>> [t2,vcnew]=ode23('section_2',[0 3e-6],[0 0]);
% solution using m-file section_2.m till t=5µs, v_{C1}(t)=v_{C2}(t)=0 for t=0
>> plot(t2,vc2(:,1),t2,vc2(:,2),t,vc); axis([0 3e-6 0 1]); grid; xlabel('time (sec)');
ylabel('vc(t) (volts)'); title('numerical solution for vc1(t) and vc2(t)');text(2.6e-
6,0.8,'vc1(t)');text(2.6e-6,0.73,'vc2(t)');text(2.03e-6,0.94,'vc(t) for one RC section');
```

The m-file section_2.m is:

```
function dx=section_2(t,x)
dx=zeros(2,1);
dx(1)=5e6*(1+x(2)-2.1*x(1));
dx(2)=5e6*(x(1)-1.1*x(2));
end
```

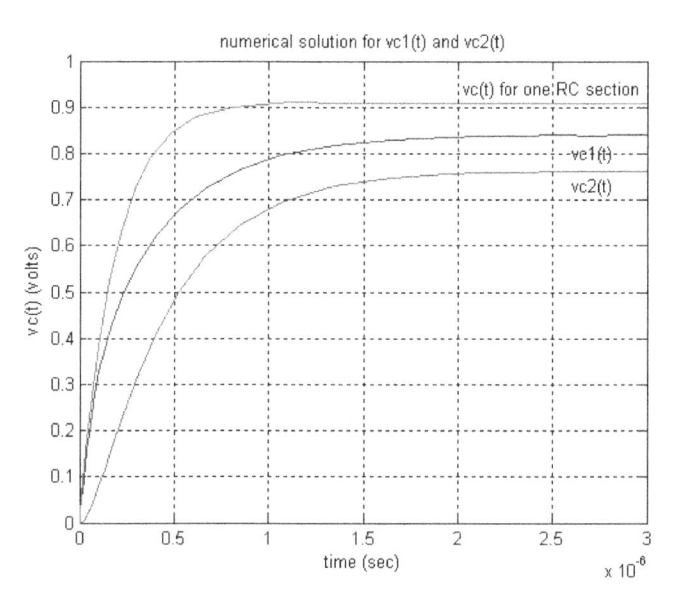

Figure 10.M.2. Voltage across capacitors C_1 and C_2 against time

Fig. 10.M.2 shows the voltage across capacitors C_1 and C_2 together with the voltage across capacitor C of the case of one R, R', C- section (see Fig. 10.M.1) for reasons of comparison. Clearly, the voltage in capacitors C_1 and C_2 rises more slowly than in C, while it approaches a successively lower final value.

10.5 The Action Potential

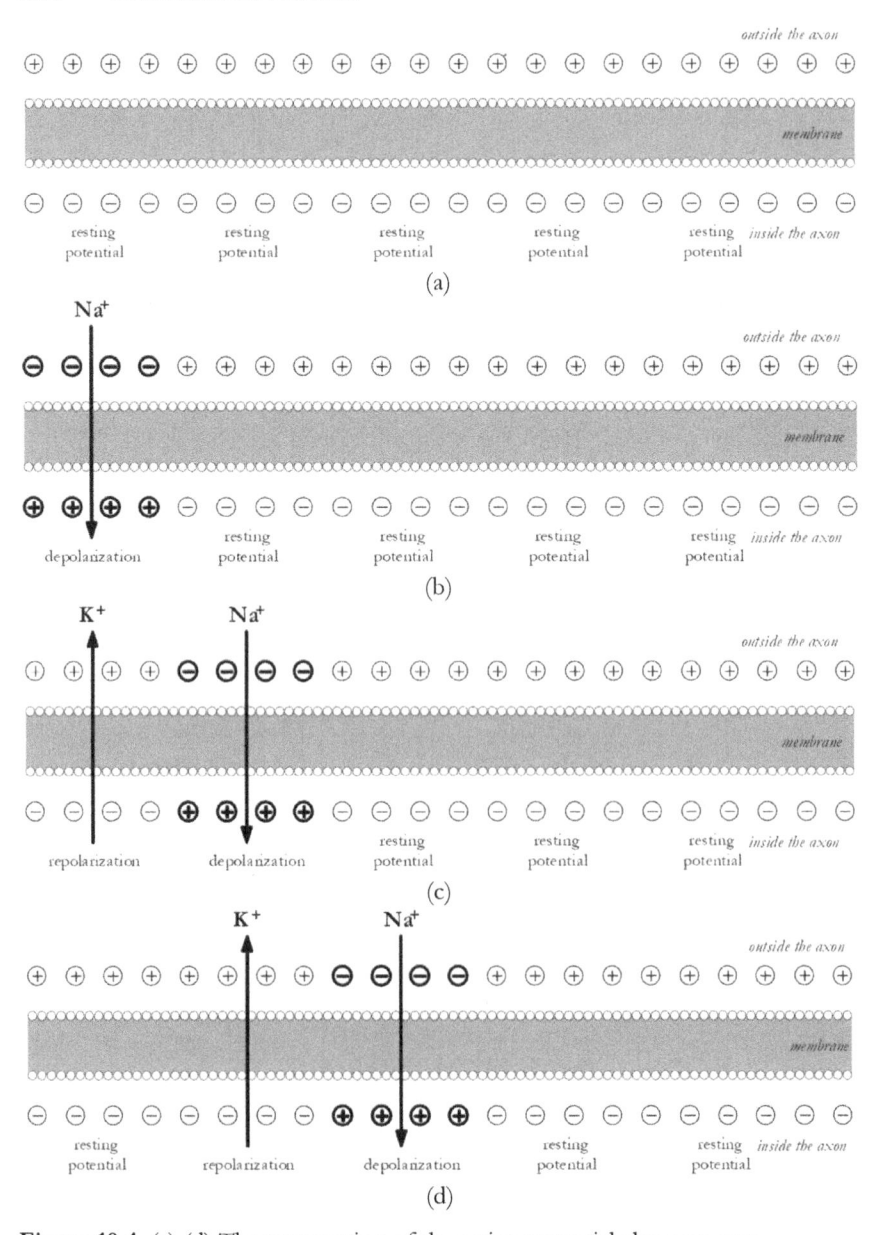

Figure 10.4. (a)-(d) The propagation of the action potential along an axon

Consider now the case when the stimulating voltage at the point $x=0$ of the axon causes the potential at this point to increase above the threshold potential (which is about $-50mV$), say, to $-40mV$. Such a stimulus may be caused, for example, by sensation coming from heat, cold, sound, light etc. Then a large change in the resting potential occurs momentarily at the point of stimulation. This potential change propagates along the axon and is called the *action potential*. The mechanism for the propagation of the action potential along the axon can be schematically represented as in Fig. 10.4 and is described as follows:

With the potential of the axon at $x=0$ being above the threshold, an increase in the permeability of the membrane to sodium ions Na^+ occurs (about 1000 times), which causes an influx of Na^+ ions into the axon causing the potential there to take positive values (about $+50mV$). Thus, the axon there gets initially *depolarized*, Fig. 10.4(b), and then its polarity is reversed. However, the neighboring region is still polarized at −90mV (inside the cell). This causes a flow of positive ions both inside and outside the membrane, Fig. 10.4b, between the two regions. This way the potential inside the membrane in the neighboring region to the right increases above threshold and the previous process is repeated, i.e. the permeability of the membrane to Na^+ ions increases dramatically there and an influx of those ions depolarize this region reversing its polarity as shown in Fig. 10.4c. To avoid the depolarization of the whole axon the permeability to potassium ions K^+, about 1ms later, increases and such ions, helped also by the electric field, pass to the outside of the membrane thus *repolarizing* this region of the axon. Eventually the resting potential and the restoration of the steady ionic concentrations are restored over a much longer interval by the action of the Na^+ - K^+ pump. Thus, the depolarizing edge is followed by a repolarizing one with the action potential propagating further along the axon causing the depolarization of the next region and the process is repeated, as shown sequentially in Figs. 10.4(a) to (d).

Fig. 10.5(a) shows the development in time of the action potential at a small region inside the axon caused by the depolarization and repolarization. The time it takes for this procedure in a cell depends upon the type of cell, i.e. whether it is about a nerve axon, a cardiac cell, etc.

The mechanism of the action potential propagation along an unmyelinated axon, as was described above, will be regenerated very often, because of the simultaneous leakage of the current in such an axon. Evidently this causes reduction in the speed of the signal propagation. But the situation is different in a myelinated axon. Because of the myelin sheath the leakage that takes place attenuates the current pulse much less. At the same time, this pulse propagates faster along the myelinated axon than in an unmyelinated one, since the axon capacitance is lower, as was explained in Section 10.4. The reduced signal has the possibility, when reaching the

Ranvier node, to restore the action potential to its original height and shape, following the procedure described in Fig. 10.4(a) to (d). Clearly, between two neighboring Ranvier nodes the signal propagates as in the case of weak signals, however when reaching the Ranvier node is still strong enough to cause regeneration. This regeneration of the action potential at the Ranvier nodes means the amplification of the signal, thus giving it the capability to propagate along the whole myelinated axon independently of its length. During this way of propagation of the action potential in the myelinated axon the signal seems to jump from node to node, thus forming what has been termed *saltatory conduction*, as shown in Fig. 10.6.

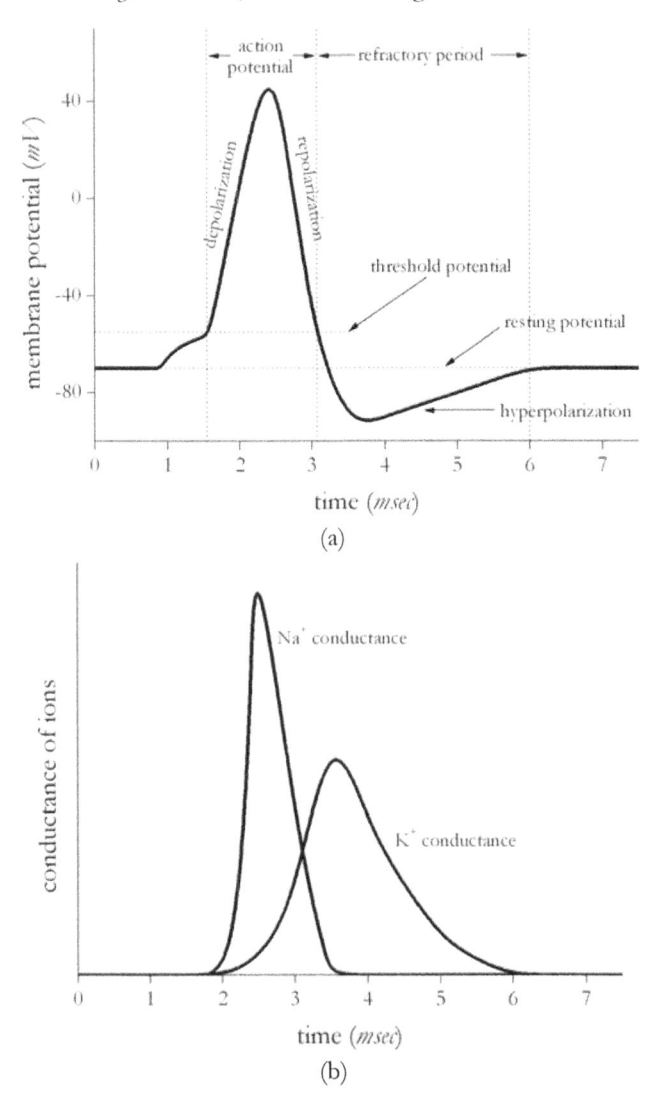

Figure 10.5. (a) Development in time of the action potential and (b) Relative contribution of Na^+ and K^+ ions conductance to the axon excitation

Biopotentials – Biosignals 277

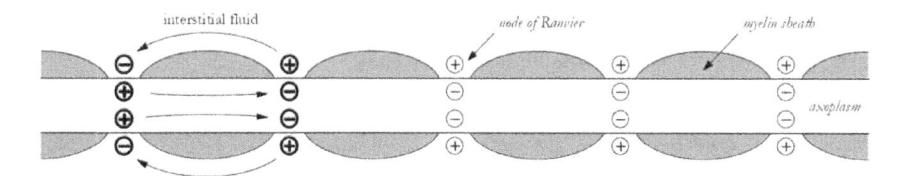

Figure 10.6. Saltatory conduction of action potential in a myelinated axon

10.6 The Electromyogram, *EMG*

The *electromyogram* or *EMG* is the recording of the electrical activity in a muscle. Muscle contraction is the result of the transmission of the action potential from the axon into the neuromuscular synapses resulting in excitation of the muscle. A muscle is made up of many *motor units*. Each motor unit is the set of a single branching neuron from the brain stem or spinal cord and a small or large number (up to 2000) of muscle fibers (cells) to which it is connected through the motor end plates. In the absence of action potential, the resting potential across the membrane of a muscle fiber is similar to the resting potential across a nerve fiber. Muscle action can be triggered by an action potential that moves along an axon and is transmitted across the motor end plates into the muscle fibers causing them to contract.

EMG electrodes (see Sections 11.3 and 11.7) attached to the skin monitor the electrical signals from many motor units. However, on special occasions, a concentric needle electrode inserted under the skin is used to measure the electrical activity of single motor units.

Muscle contraction can be stimulated voluntarily but also electrically by the examiner and this method is often preferred to the former. The reason is that with the electrical stimulation all muscle fibers fire at nearly the same time, while with the voluntary stimulation they do not, and it may occur that each motor unit produces several action potentials depending upon the signals sent from the central nervous system.

The arrangement for monitoring the *EMG* provides also the same signal either directly or integrated (Fig. 10.7(a)). Both signals appear on the screen of a *CRT* or another display unit. The integrated signal gives a measure of the quantity of electricity associated with the muscle action potentials. This is accordingly interpreted in conjunction with the *EMG* record by the clinician doctor.

It should be mentioned that there is a delay in the time of the appearance of the *EMG* signal after the stimulation has been applied, and this can be used to compare the *EMG* signals from symmetrical muscles.

By applying two stimuli at different locations, one can determine the velocity of action potential in motor axons (Fig. 10.7(b)). Calculating the difference in latency periods gives the time taken by the action potential to cover the distance between them. Then dividing this distance by the time taken gives the velocity of the action potential.

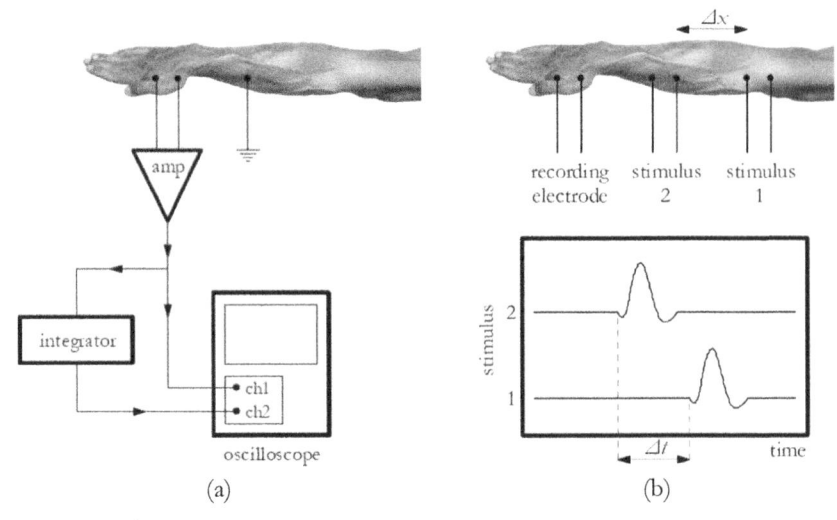

Figure 10.7. (a) Instrument arrangement for obtaining an *EMG* and (b) Compound method for measuring the motor nerve conduction velocity. The electrical response of a muscle is recorded upon successive proximal (upper 2) and distal (1) stimulation of motor axons in cardial nerve.

More about monitoring the *EMG* is given in Chapter 11.

10.7 The Electrocardiogram, *ECG*

The recording of the electrical signals that have been created by the changing and transmission of biopotentials in the heart is the *electrocardiogram*, or *ECG* for short.

In order to get a better understanding of the creation and be able to explain the shape of the waveform of the *ECG*, it is useful to know the construction and the operation of the heart. So this is what we explain first.

The heart consists of four chambers, as shown in Fig. 10.8, two upper ones called the *left atrium* and the *right atrium* and two lower ones called the *left and right ventricles*, all four acting as pumps. The two atria contract in synchronism and so do the two ventricles. The rhythm for the heart operation is kept by the *sinoatrial node* (*SA*), which is made of special muscle cells, is housed in the right atrium and acts spontaneously. This rhythm is normally about 72*ppm* (pulses per minute), but it may be changed by nerves external to the heart that respond to other stimuli. The initial electrical pulse causes the simultaneous depolarization of both atria, which contract and pump blood to the corresponding ventricles. Then the electrical signal reaches the *atrioventricular node* (*AV*), which initiates the depolarization of the two, left and right, ventricles. This causes the contraction of the ventricles,

Biopotentials – Biosignals 279

which, in their turn, pump blood the right ventricle to the lungs and the left ventricle to the general circulation. When this is done, the nerves and muscles of the ventricles repolarize and the whole sequence is repeated.

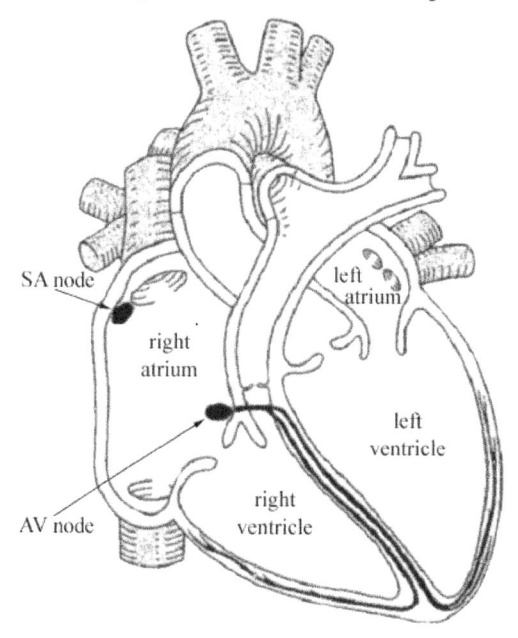

Figure 10.8. The human heart

The electrical activity in the heart cannot be monitored using electrodes on direct contact with the heart muscles and nerves every time the *ECG* is taken. However, by means of ionic currents inside the body, this is transferred to the skin and it can be monitored by attaching electrodes to various parts of the skin. We may consider that any instant electrical fields and associated currents in the cardiatic muscle are the constituents of an electric dipole vector in the three dimensional space, which changes with time in size and orientation. Therefore, the voltage we measure on the skin will be the projection of this vector on the plane formed by the positions of the electrodes, while it will also depend on the position of the electrodes on the skin and will be valid only for the instant the measurement is taken.

The electrodes are commonly located on the left arm (*LA*), right arm (*RA*), and the left leg (*LL*), as shown in Fig. 10.9(a). It is considered that these electrode positions lie on a plane, called the frontal plane, and it is the projections of the cardiatic electrical vector on this plane that are monitored. Of course, projections on other planes, like the transverse or sagittal, which are perpendicular to the frontal plane, can be monitored placing the electrodes at other positions of the skin (Fig. 10.9(b)). In the case of the projections on the frontal plane when taking the *ECG*, we can measure the voltage between the *LA* and the *RA* (Lead *I*), or between the

RA and the LL (Lead II), or between the LA and the LL (Lead III), as shown in Fig. 10.9(a). Schematically, a triangular is formed, the Einthoven triangle, shown in Fig. 10.9(a) and it is the projections of the electric dipole vector on its sides that are given by the ECG. In clinical examination, all these three lead voltages are usually monitored.

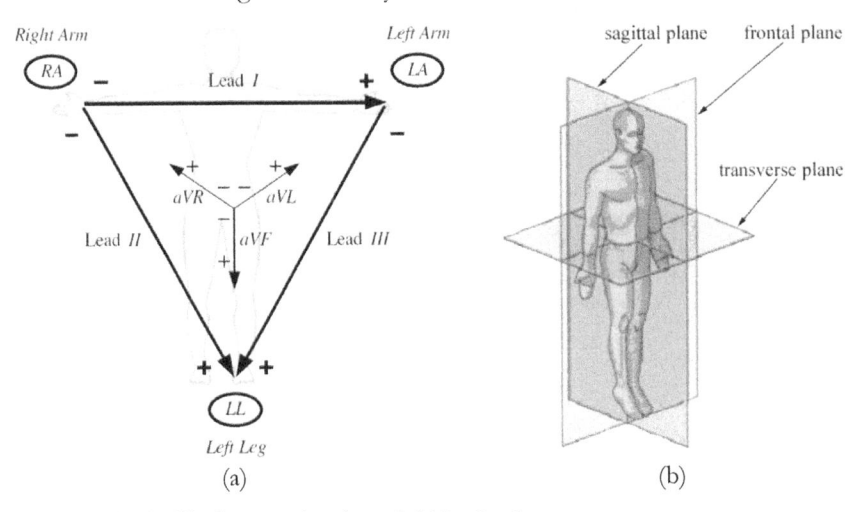

Figure 10.9. (a) Einthoven triangle and (b) body planes

In Fig. 10.10 a typical normal ECG is shown corresponding to Lead II. This can be recorded on a chart or appear on the screen of a CRT or another display device. It is customary to use the letters appearing in the trace to indicate the various parts of the ECG. Each peak in the trace corresponds to a certain electrical activity of the heart at the particular moment. Thus, the P peak corresponds to the depolarization of the atria, while the part between P and Q to the repolarization of the atria, although in most cases no peak appears there. Then comes the QRS part, which corresponds to the depolarization of the ventricles, while their repolarization occurs between S and T. The ECG is examined by the cardiologist who can establish abnormalities due to any heart malfunctioning.

The electrical connections for Leads I, II and III, with the usual polarities of the recording instrument indicated for each lead are shown in Fig. 10.11(a). In an examination, three augmented lead configurations are also recorded. For the connection aVR, one side of the recorder is connected to RA and the other side to the center of two resistors connected to LL and LA, as shown in Fig. 10.11(b). Similarly for the aVL lead, one side of the ECG recorder is connected to LA and the other to the center of the two resistors, which are connected to RA and LL. Finally, for the aVF lead, one side of the recorder is connected to LL and the other to the center of the two resistors, which are connected to RA and to LA.

Biopotentials – Biosignals

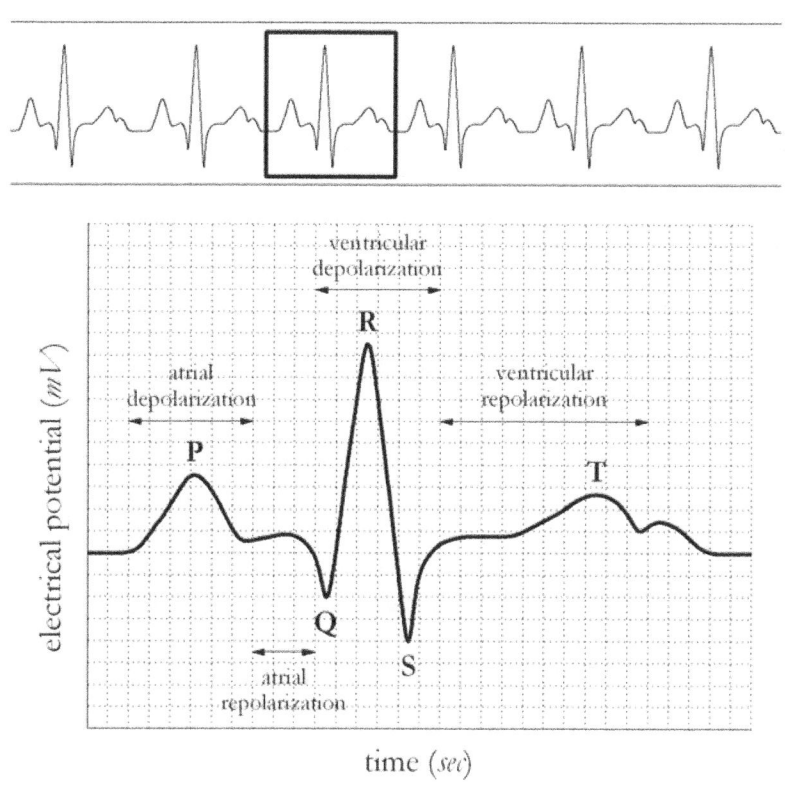

Figure 10.10. Typical *ECG* from Lead *II* position

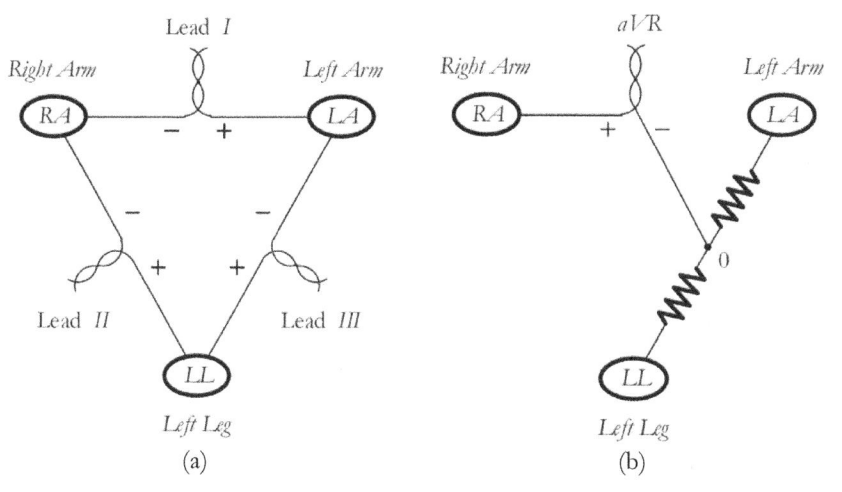

Figure 10.11. (a) Electrical connections for Leads *I*, *II* and *III* and (b) an augmented lead connection for the *aVR* recording

All these six frontal plane leads, i.e. Leads *I*, *II*, *III*, and the augmented *aVR*, *aVL*, and *aVF* are used in taking the *ECG* recordings in a usual clinical examination. However, in addition to the six frontal plane

ECG recordings, another six transverse plane ECG recordings are also taken in clinical examination.

One final word about the electrodes that are attached to the skin. They are flat at the end in the form of a disk in order to get a strong signal, while they are made of a highly conductive metal, such as silver interfaced to silver chloride, and connected to the human body via an electrolytic gel. More details on the electrodes are given in Chapter 11.

10.8 The Electroencephalogram, *EEG*

The *electroencephalogram*, *EEG*, is the recording of electrical activity of the brain. It is obtained by attaching electrodes on the scalp. The obtained signals are very weak, their amplitude being about $50\mu V$, very low compared to *EMG* or *ECG*. They are due to the electrical activity mainly of the neurons in the cortex of the brain, principally that of postsynaptic potentials of pyramidal neurons. It is thought that the potentials are produced through an intermittent synchronization process involving the neurons in the cortex with different groups of neurons becoming synchronized at different instants of time. According to this hypothesis, the signals consist of consecutive short segments of electrical activity from groups of neurons located at various places on the cortex.

The electrodes for recording the *EEG* signals are often small discs of silver/ silver chloride and are attached to the head at locations, which depend on the part of the brain that is to be studied. The international standard 10-20 system is used and in routine tests 16 channels are recorded simultaneously, with the reference electrode usually attached to the ear.

Because of the low amplitude of the *EEG* signals, these are subjected to interference from external signals making their processing very difficult. Apart from the external noise, potentials of muscle activity such as eye movement can cause artifacts in the record. More specifically, interference may be due to the following:

- **a.** Muscle activity (e.g. heart, eyes)
- **b.** Noise from electrodes (e.g. movement, no secure attachment)
- **c.** Interference from electric or magnetic fields
- **d.** Interference from high frequency fields

These problems are examined in some detail in Section 11.6.

The *EEG* signal is composed of different bands of frequencies of the brain, meaning that they are lower (8-$13Hz$) for a relaxed state than for an alert state (above $13Hz$). It is used as an aid to diagnose diseases of the brain such as epilepsy, tumors, and even to indicate the anesthesia level of the patient during surgery. Apart from the spontaneous recordings of

EEGs, it is useful to record the response of the brain to external stimuli when the so called evoked potentials are generated, which are examined next.

10.8.1 Evoked Potentials

The *evoked potential* (EP) is the response recorded from the brain, spinal cord or peripheral nerve evoked by external stimuli. The external stimulation may be visual, auditory or somatosensory. The amplitudes of EP are very low of the order of few microvolts or less, much lower than the EEG, ECG or EMG signals. To extract the EP signals from an EEG, in which all other spontaneous biological signals and ambient noise coexist, a technique called *averaging* is applied. This is achieved by getting a large number (one to two thousand) of EEG recordings when the stimulus is applied at regular instances and using a suitable computer program take the average. All other potentials, spontaneous or noise, are random, and on taking the average cancel out, leaving the EP to emerge amplified.

The EP recording electrodes are placed over the scalp, neck, or spine surface depending on the type of stimulus and the test under consideration. Major types of EP test are the following:

a. **Brainstem Auditory EP (BAEP)**

This test examines the integrity of auditory pathway through the brainstem.

b. **Visual EP (VEP)**

The visual evoked potential (VEP) test examines the integrity of visual pathway from retina to occipital cortex where visual input is perceived by the brain.

c. **Somatosensory EP (SEP)**

This test examines the sensory system from the peripheral nerve to the sensory cortex of the brain.

10.9 The Electroretinogram (ERG) and the Electrooculogram (EOG)

The *electroretinogram* (ERG) is the recording of potential changes produced by the eye when it is exposed to a flash of light, while the *electrooculogram* (EOG) is the recording of potential changes due to eye movement.

To obtain the ERG, one electrode is placed on a contact lens that fits over the cornea and the other is attached to the ear or forehead. On the other hand, to obtain the EOG, a pair of electrodes is attached near the eye.

Clinically, the *ERG* is useful to detect any inflammation of the retina, whereas the *EOG*, which is not frequently used in routine practice, is useful in measurements of the orientation of the eye, its angular velocity and its angular acceleration.

10.10 The Magnetocardiogram (*MCG*) and the Magnetoencephalogram (*MEG*)

Magnetocardiograms (MCG) and *magnetoencephalograms* (MEG) are recordings of the magnetic fields surrounding the heart and the brain, respectively. These fields are produced by the electric currents in the heart and the brain and are very weak, which implies that special equipment and magnetically shielded rooms are required for their measurements.

To measure the weak magnetic fields around the heart ($\sim 5 \cdot 10^{-13} T$) and surrounding the brain, very sensitive detectors of magnetic fields are required. Such a detector is the *SQUID* (Superconductor Quantum Interference Device), which operates at very low temperature (5K) and can detect both steady and alternating magnetic fields of the order of $10^{-14} T$.

One advantage over the corresponding electrical signals is that the magnetic signals give information concerning the direct electric currents, which does not appear in the *ECG* and *EEG*. Nowadays *EEG* amplifiers can detect oscillations as slow as $0.01 Hz$, very close to *dc*. Another difference is that the measurement of the magnetic fields is carried out without any electrodes touching the patient's body.

10.11 Characteristics of the Biosignals

Table 10.1 summarizes the main characteristics of the five biopotentials we refer to here. These are the range of size and frequency of the biosignals. Clearly, the sizes of these biosignals are quite low and the same is true for their frequencies. The small size makes them difficult to detect as, in most cases, the induced noise from the environment and other artifacts have amplitudes larger than the signal amplitude. Special care then should be taken for the detection and the measurement of biosignals. These are the objects of the next Chapter.

Biosignals are usually slowly varying voltages or currents of low amplitudes so low that on many occasions, they are buried in noise in a way that they cannot be recognized. *Noise* is a random fluctuation of the voltage or current or any other signal uncorrelated with the signal of interest that appears together with the latter at the output of the detecting apparatus. Sometimes its amplitude is higher than that of the signal. The concept of noise is treated to some detail in Chapter 8.

The basic characteristics of biosignals are the following:

a. Low amplitudes ($10\mu V$ to $10 mV$)
b. Low frequency range (dc up to several hundred Hz)
c. Noise content sometimes of comparable amplitude

Table 10.1. Characteristics of biopotentials

Biopotential	Range of signal size	Range of frequency
Electromyogram (EMG)	$0.1 - 5\ mV$	$dc - 10\ kHz$
Electrocardiogram (ECG)	$0.5 - 4\ mV$	$0.01 - 250\ Hz$*
Electroencephalogram (EEG)	$5 - 300\ \mu V$	$dc - 150\ Hz$
Electroretinogram (ERG)	$0 - 900\ \mu V$	$dc - 50\ Hz$
Electrooculogram (EOG)	$50 - 3500\ \mu V$	$dc - 50\ Hz$

* Even $600 Hz$ with intracranial electrodes

The biosignals, as they are detected by the electrodes, are in analog form. This means that the function $v(t)$ or $i(t)$ has a value at any time instant t, which is proportional to the corresponding value (at the same time instant) of the physical quantity that it represents. In that respect, $v(t)$ and $i(t)$ are continuous functions of time. In most cases nowadays, when studying biosignals, digital technology is used more advantageously than analog technology. In such cases, the biosignal should be converted from its analog form into digital form. The digital signal, as was explained in Chapters 5 and 7, represents the biosignal at discrete time instants by combinations of a number of pulses of equal height. Each such combination of pulses corresponds to the value of the biosignal the certain time instant. Thus, the digital biosignal is not a continuous function of time as the analog biosignal is.

In the case of EP, since there are many things going on in the brain at any instant, it is very difficult to determine when the EP from a particular stimulus appears from just one recording. So, the technique is to apply the certain stimulus many times, say more than one thousand times, and using a suitable computer program take the average of all these recordings, as was explained above (Section 10.8.1).

References

[1]. Jennings D., Flint A., Turton B. C. H. & Nokes L. D. M., *Introduction to Medical Electronics Applications*, Edward Arnold, 1995.
[2]. Northrop R. B., *Analysis and Application of Analog Electronic Circuits to Biomedical Instrumentation*, CRC Press, 2004.

[3]. Company-Bosch E., *ECG Front-End Design is Simplified with MicroConverter*, Analog Devices, Analog Dialogue, vol. 37, November 2003.

[4]. Semmlow J., *Circuits, Signals, and Systems for Bioengineers (A MATLAB-Based Introduction)*, Elsevier, 2005.

[5]. Webster J. G., *The Measurement, Instrumentation, and Sensors Handbook*, CRC Press, 1999.

[6]. Norman R. A., *Principles of Bioinstrumentation*, John Wiley & Sons, New York, 1988.

[7]. Sansen W., *Microelectronics and Biosensors*, Notes for the PG course on Biomedical Engineering and Medical Physics, University of Patras, Medical Physics Laboratory, 1994.

[8]. Kane J. W. & Sternheim M. M., *Physics SI version*, John Wiley & Sons Inc., 1980.

[9]. Cameron J. R., Skofronic J.G. & Grant R.M., *Physics of the Body*, A Wiley Interscience Publication, 1999.

[10]. Proimos V., *Electricity in Medicine*, Lecture Notes, European Course on Biomedical Engineering and Medical Physics, Department of Medicine, University of Patras, Patras, Greece.

[11]. Kostopoulos G., *Physiology*, Lecture Notes, European Course on Biomedical Engineering and Medical Physics, Department of Medicine, University of Patras, Patras, Greece.

[12]. Despotopoulos A. & Silbernagl S., *Color Atlas of Physiology*, Thieme, 5th Edition, Stuttgart, New York, 2003.

MATLAB Problems

❖ *The reader is advised first to consult the MATLAB tutorial in Appendix D*

10.M.1. Using MATLAB confirm that the potentials inside the axon V_i for the cases of K^+ and Cl^-, which are given in Section 10.3, are indeed $-98 mV$ and $-90 mV$, respectively.

10.M.2. Repeat example 10.M.1 when three R, R′, C- sections are present and the voltage concerns that in the last capacitor on the right. Compare this result to those in example 10.M.1. Initial voltage in all capacitors is zero.

Problems

The data in Tables P.10.1 and P.10.2 should be used in solving the problems below. Also, it is given:

Electron charge = 1.6×10^{-19} C

Boltzmann's constant = 1.38×10^{-38} JK^{-1}

Biopotentials – Biosignals

Table P.10.1. Concentrations of fluids in *moles/m³*

Fluid inside the axon C_i		Fluid outside the axon C_o	
Na⁺	12	Na⁺	145
K⁺	155	K⁺	4
Cl⁻	4	Cl⁻	120
(Others)⁻	163	(Others)⁻	29

Table P.10.2. Axon parameters

Quantity	Myelinated	Unmyelinated
Axonplasm resistivity, ρ^*	$2\,\Omega m$	$2\,\Omega m$
Capacitance per unit area of membrane, C_m	$3 \times 10^{-5}\,F m^{-2}$	$10^{-2}\,F m^{-2}$
Leakage resistance of unit area of membrane, R_m	$40\,\Omega m^2$	$0.2\,\Omega m^2$
Radius of axon, r	$5 \times 10^{-6}\,m$	$5 \times 10^{-6}\,m$

* The resistivity of the interstitial fluid is considered negligible

10.1. Using the data in Table P.10.1, calculate the equilibrium potential difference between the inside and outside an axon at 310K due to the sodium ions Na⁺ ignoring the presence of the other ions.

10.2. Repeat problem 10.1 for the potassium ions K⁺ at the same temperature.

10.3. Repeat problem 10.1 for the chlorine ions Cl⁻ at the same temperature.

10.4. Determine the resistances, axonal and leakage, of an axon of length $\ell = 1\,cm$ when the axon is

a. myelinated
b. unmyelinated

10.5. Calculate the required length of the thinnest copper wire ($r = 0.04\,mm$), the resistivity of which is $1.725 \times 10^{-8}\,\Omega m$, to have the same resistance as the 1 cm length of the myelinated axon in Problem 10.4.

(Answers: $72.8 \cdot 10^3\,km$)

10.6. Using the data in Table P.10.2, determine the space parameter λ for

a. a myelinated axon
b. an unmyelinated axon

(Answers: 0.7 cm, 0.05 cm)

10.7. Determine the capacitance C of $1cm$ axon length when the axon is

   a. unmyelinated
   b. myelinated

Hence, explain why the propagation of an electric signal is faster in a myelinated axon than in an unmyelinated.

10.8. A weak stimulus is applied at one end ($x=0$) of an axon bringing the voltage V_i (with $V_o=0$) from $-90mV$ to $-60mV$. Calculate the value of the potential difference at a distance of $1cm$ from $x=0$ if the axon is

   a. myelinated
   b. unmyelinated

10.9. The switch in Fig. P.10.1 closes at $t=0$ having been open for a long time. Using MATLAB or otherwise plot the voltage v_o against time t.

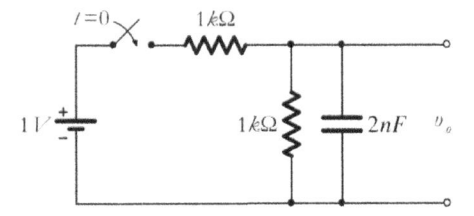

Figure P.10.1.

10.10. Repeat problem 10.9 for the circuit in Fig. P.10.2 and compare the result to that of problem 10.9.

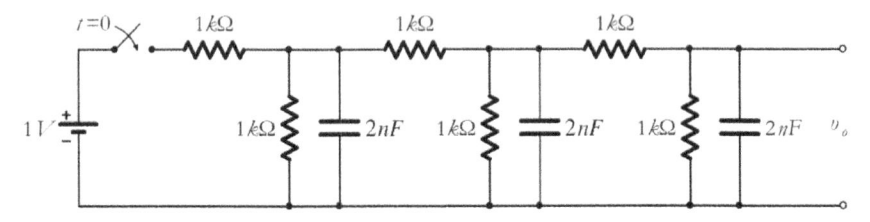

Figure P.10.2.

Chapter 11

Detection and Measurement of Biosignals

11.1 Introduction

In Chapter 10 we discussed the generation of biosignals and their transmission inside the body, either volume conducted or by conduction of action potentials along the nerves. These signals must be detected, processed and displayed or measured in order to reveal the information they carry.

The transmission of biosignals inside the living body is effected by moving ions. However, the currents in metal wires are carried by moving electrons. Therefore, the ionic form of the biosignals has to be converted to electronic form outside the body before being processed and displayed or measured. The conversion is achieved by a special type of sensors, the *electrodes*, as these are referred to in biomedical instrumentation. Specially constructed electrodes are properly attached to the living body on the skin in specific numbers and ways. The detected (sensed) voltage or current can then be amplified and processed further.

In this chapter, the general electronic measurement system is presented in block diagram form and explained briefly in Section 11.2. In Section 11.3 the electrodes used in detecting biopotentials are discussed. Specifically, phenomena arising at the electrode – electrolyte interface and electrode polarization are briefly examined. Their electrical equivalent as well as an electrode – tissue model are introduced. Then, in Section 11.4 the most important characteristics required from an amplifier for the amplification of biosignals are discussed and, based on these, the *Instrumentation Amplifier* (*IA*) is emerged as the most suitable one for this purpose. The detection of *ECG* signals is discussed in Section 11.5 and the importance of the right-leg drive in reducing common-mode signals is explained. Problems when detecting *EEG* and *EP* signals are discussed in Section 11.6, while in Section 11.7 the detection of *EMG* signals is examined.

11.2 A General Electronic Measuring System

To get the information out of a biosignal the following two general steps should be followed:

a.   The biosignal has to be detected or sensed and turned into an electronic signal.

b.   The resulting electronic signal should be processed.

Detection of the biosignal is achieved by a sensor, which is properly selected out of a large variety, to suit best the case of interest each time. For example, in the case of taking an *ECG*, there are various types of metal electrodes available made from silver, gold or steel with different characteristics, shapes and costs and one has to choose among these, depending on the application at hand, i.e., whether it is to be used for medical practice or for research. The electrical signal from the electrode is not in the proper form for the specialist to extract out of it the information he seeks. For this reason, the second step is to process the signal.

Processing of the electrical signal has the meaning of conditioning it in order that the information content can be more accurately obtained from it. This involves increasing its amplitude, reducing its noise content, if not deleting it, and revealing the characteristic frequency spectrum that mainly constitutes the information content of the signal. Increase of the amplitude of the signal is achieved using an amplifier, which in many cases, is of the differential type and, more specifically, the instrumentation amplifier (see Chapter 3). Signal processing can be achieved by using filtering (see Chapters 4 and 7) which can be analog or digital.

Analog filtering is effected by analog filters (see Chapter 4), if the signal is a continuous function of time, which is usually the case with biosignals. However, for higher accuracy and presentation of the results, the analog signal may be converted into digital form and be processed using digital filters (see Chapter 7). The device that will perform the conversion of the signal from analog to digital form is called *analog-to-digital converter*, *ADC*, as was explained in Chapter 7. Next, digital filtering will be applied and the filtered signal will be converted back to analog form by the device called *digital-to-analog converter*, *DAC* (Section 7.4). Finally, a simple smoothing filter will be necessary to follow the *DAC* for rejecting the high frequency components of the resulting analog signal.

In the end, the result will have to be recorded somehow by a printer, a chart recorder or as a visual reading, etc, when the signal should be in its analog or digital form. The above procedure can be presented in the general block diagram form shown in Fig. 11.1, which was also given in Chapter 9 but is repeated here for convenience.

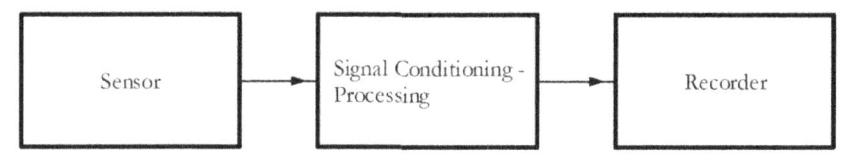

Figure 11.1. An electronic measuring system for biosignals

11.3 Electrodes

Electrodes should obtain the biosignal with as little as possible picking up of any other unwanted signals such as noise, contact potential, etc. They should be designed in shapes that will be securely attached to the body to avoid any noise production due to the electrode movements. In its classic form, the electrode is made of a highly conductive metal, such as silver interfaced to silver chloride, and connected to the human body via an electrolytic gel. The electrode is secured to the skin by means of a non allergenic adhesive tape. Such electrodes are reusable and best suited for serious studies or basic research investigations.

11.3.1 Electrode – Electrolyte Interface

Consider the case of placing a piece of metal M into a solution containing ions of the metal, e.g. silver (Ag) in silver-chloride (AgCl) gel. In the solution apart from the cations, there exist anions in order to maintain charge neutrality. When the piece of metal is inserted into the solution, the following reaction takes place immediately:

$$M \rightleftharpoons M^{n+} + ne^- \tag{11.1}$$

where n is the valence of the metal. Depending on the concentration of the cations in the solution, the reaction initially goes to one direction or the other. This affects the concentration of anions in the vicinity of the metal and thus, the charge neutrality at this region does not hold anymore. Therefore, at equilibrium, a constant potential difference will be developed between the interface of the metal solution and other parts of the solution far from the interface. This potential difference is called *half-cell potential* and is dependent mainly on the metal, its concentration in the solution and the temperature.

In Table 11.1, the half-cell potentials for various electrode materials including the corresponding chemical reactions are given. Clearly, this half-cell potential cannot be measured, because another piece of a different metal should be placed into the solution far from the first. But then, a similar reaction will take place and the measured potential difference will be the difference of the half-cell potentials of the two electrodes. The electrochemists have agreed the second electrode to be that of hydrogen the half-cell of which is taken, by convention, to be zero (see Table 11.1). The voltmeter used for this measurement is considered to have infinite resistance in order not to disturb the zero current condition for the development of the half-cell potential, which is defined at equilibrium.

Table 11.1. Half-cell potentials

Material	Symbol	Reactions	Half-cell potential (V)
Aluminum	Al	$Al \rightarrow Al^{3+} + 3e^-$	-1.706
Iron	Fe	$Fe \rightarrow Fe^{2+} + 2e^-$	-1.706
Hydrogen	H	$H_2 \rightarrow 2H^+ + 2e^-$	0.000
Silver	Ag	$Ag \rightarrow Ag^+ + e^-$	+0.779
Silver/Silver Chloride	Ag/AgCl	$Ag + Cl^- \rightarrow AgCl + e^-$	+0.223
Gold	Au	$Au \rightarrow Au^+ + e^-$	+1.680

11.3.2 Polarization of Electrodes

Let us consider the situation in the interface metal-electrolyte under the passage of current. The passage of current modifies the situation at equilibrium. Apart from Eq. (11.1), a second equation holds referring to the anions A^- of the electrolyte, i.e.

$$A^{m-} \rightleftarrows A + me^- \qquad (11.2)$$

When the reaction goes from left to right, *oxidation* takes place, while when the reaction goes from right to left, it is said, that *reduction* takes place. The situation is demonstrated in Fig. 11.2, where the conventional direction of current I is also shown.

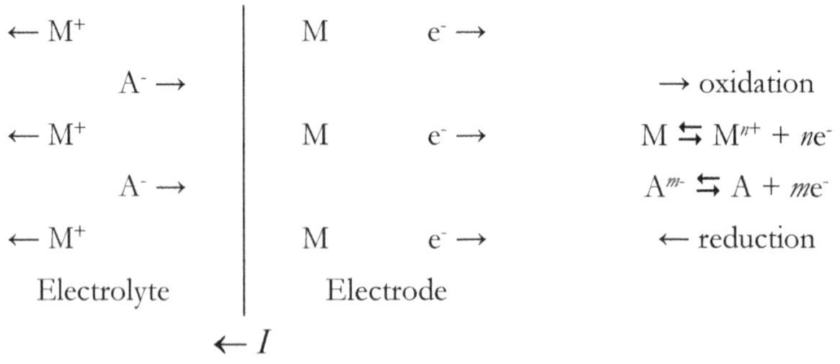

```
← M⁺              M     e⁻ →
    A⁻ →                              → oxidation
← M⁺              M     e⁻ →          M ⇆ M^{n+} + ne⁻
    A⁻ →                              A^{m-} ⇆ A + me⁻
← M⁺              M     e⁻ →          ← reduction
  Electrolyte    Electrode
              ← I
```

Figure 11.2. The metal-electrolyte interface under passage of current I and conventional direction of current

The passage of current in the metal-electrolyte interface leads to the *polarization* of the electrode, meaning that an *overpotential* is added to the half-cell potential. The situation is demonstrated in Fig. 11.3.

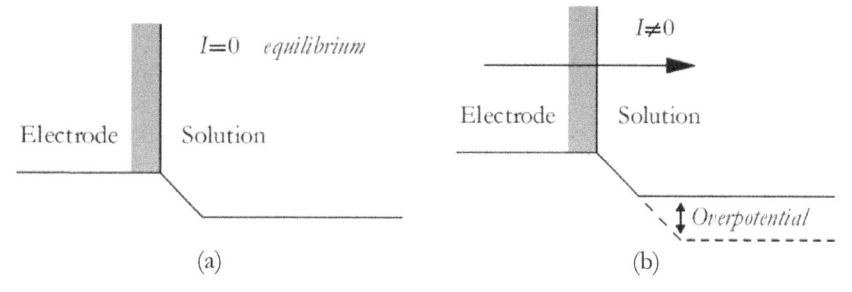

Figure 11.3. The situation (a) at equilibrium (b) under passage of current I

The value of the overpotential can be determined by subtracting the half-cell potential at equilibrium from the total potential measured under the passage of current. The following mechanisms are considered to be responsible for the development of the overpotential in the electrode.

a. Ohmic overpotential due to ohmic resistance of the electrolyte.
b. Concentration overpotential due to changes in the ion distribution in the electrolyte in the vicinity of the electrode-electrolyte interface.
c. Activation overpotential. The oxidation and reduction equations in order to occur require *activation energy*, which is not the same for the two processes. With the passage of currents one of the two predominates leading to a difference in voltage between the electrode and electrolyte, which is called *activation overpotential*.

The total overpotential is the sum of those resulting from the above three mechanisms leading to the polarization of the electrode-electrolyte interface.

We may classify electrodes as polarizable and nonpolarizable. In polarizable electrodes no actual charge crosses the electrode-electrolyte interface when a current is applied. In this case, the current in the interface is a displacement current and thus the electrode behaves like a capacitor. Electrodes made of noble metals (e.g. platinum) approximate the behaviour of a perfect polarizable one. In nonpolarizable electrodes current passes freely across the electrode-electrolyte interface. No energy is required to make this transition. An electrode approximating a perfect nonpolarizable one is the Ag/AgCl.

11.3.3 Equivalent Circuit of an Electrode

From the electronics point of view, the characteristics of the electrode can be represented by the equivalent circuit shown in Fig. 11.4, where E_g is the half-cell potential, R_s is the total series resistance of the electrolyte and the leads, while R_d and C_d represent, respectively, the resistive and reactive components of the electrode-electrolyte interface. Clearly, at low frequencies, when $R_d \ll 1/\omega C_d$ the impedance of the model is R_d+R_s, while at high frequencies, when $R_d \gg 1/\omega C_d$ the impedance is R_s. It should be mentioned though, that both R_d and C_d vary with frequency and the current density at the electrode surface.

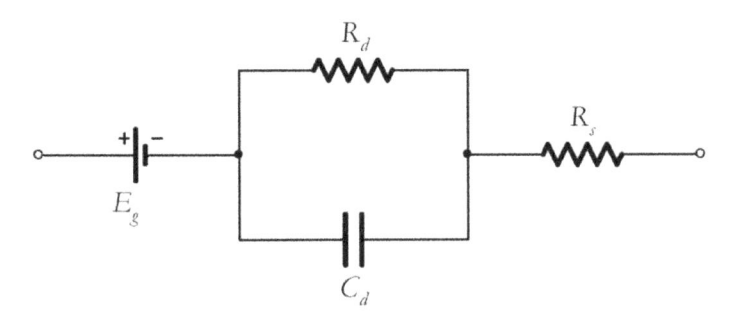

Figure 11.4. Equivalent circuit of an electrode

Silver-Silver Chloride electrode sensor, when it is in contact with a chloride gel, has been found to possess the following useful characteristics:

- Low and stable electrode potential
- Low level of intrinsic noise
- Low interface resistance
- It is relatively nonpolarizable, meaning that it has a relatively large value of exchange current density. In other words, it has a very low value of charge transfer resistance R_d.

11.3.4 Types of Commercial Electrodes

Most electrodes, which are used to detect biosignals can be classified in

- skin electrodes,
- wire electrodes and
- microelectrodes

The last ones can be realized in metal or in glass (micropipets).

Commonly used in *EEG* recordings are gold-plated electrodes with the advantages of high conductivity, being inert and reusable. They are designed so that they can be securely attached to hair-free areas of the scalp by a strong adhesive, elastic bandages, or wire mesh. Gold electrodes may

Detection and Measurement of Biosignals 295

also be used for *EMG* and *ECG* recordings. However, most commonly used for *ECG* and *EMG* measurements are silver-silver chloride electrodes. Silver-silver chloride electrodes are made of silver disks either coated electrolytically by silver chloride or by sintered together particles of silver and silver chloride to form the metallic structure of the electrode.

Disposable electrodes are made similar to silver-silver chloride ones, but because they should be of lower cost, the use of silver is minimized.

Other types of electrodes include those made of conductive polymers, those made of other metals such as stainless-steel or brass and those made of carbon. These electrodes have the disadvantages of having higher resistivity and being noisier than those of noble metals. However, they are inexpensive, reusable, and can be used when the biopotential signal is not very low. Some of them can be used also for electric stimulation and other specific research applications.

Finally, needle electrodes are used when it is required to record the signal from the organ itself. These are invasive electrodes used to record signals from muscles or muscle fibers.

Indicatively, three types of electrodes are shown in Fig. 11.5.

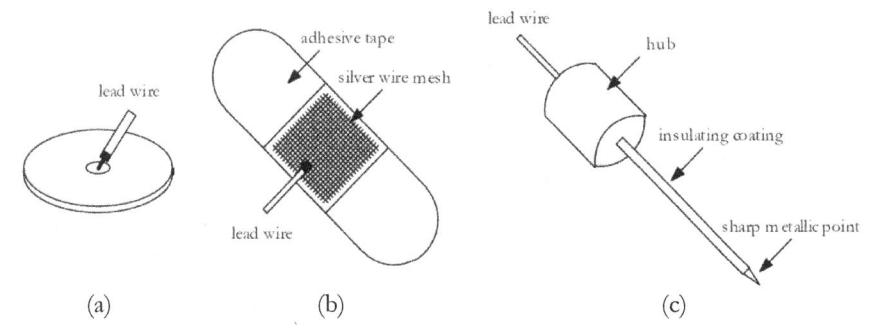

Figure 11.5. Three types of electrodes (a) metal disk electrode, (b) flexible body-surface electrode, and (c) needle electrode for use in transdermal biopotential measurements.

11.3.5 Skin Impedance and the Electrode-Tissue Model

Skin impedance, evidently, adds to the total impedance of skin-electrode compound. It can be reduced immensely by stripping or rubbing the epidermis with abrasive. Also, skin impedance can be reduced using conductive paste, which is most effective if it is allowed enough time (about seven minutes at least) to diffuse into the skin.

Before reaching the skin-electrode interface, the ionic current has to flow from the place of its generation through tissues, the subcutaneous layer, the dermis and the epidermis, as shown in Fig. 11.6(a). Electrically, this route can be represented by the equivalent shown in Fig. 11.6(b), where R_{tissue} is the tissue resistance and R_{sp} and C_{sp} are the resistive and capacitive components of the subcutaneous to epidermis interface (compound). This electrical equivalent should be connected in series with that in Fig. 11.4, if we want to follow the current flow from the place of its generation to the measuring instrument.

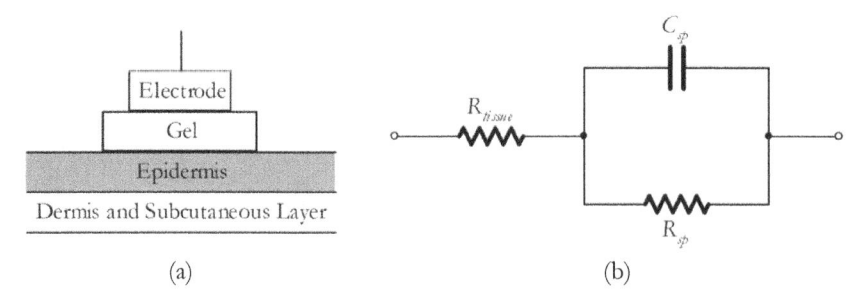

Figure 11.6. (a) Route of the ionic current before reaching the skin-electrode interface. (b) Electrical equivalent.

Of great concern for the measurement of biosignals are the artifacts that occur when the electrode moves with respect to the electrolyte. This is because the concentration of the electrolyte changes and results in a shift in the half-cell potential. The developed potential difference during the movement of the electrode is called *motion artifact* and this can cause large changes in the voltage recorded by the electrode.

11.4   Biopotential Amplifiers

It has been explained in Section 10.11 that biosignals have low amplitudes. Therefore, they should be amplified. Thus, the signals from the electrodes should be transferred to an amplifier, which must have suitable characteristics. These required characteristics arise from the fact that the biosignal on its way to the amplifier input gets corrupted by unwanted signals which are due to interference by electromagnetic fields around the patient and the measuring device and by various artifacts in the skin-electrode interface, as explained in Section 11.3. The interference induces common-mode signals that may be much higher in amplitude than the biosignal itself. This presence of common-mode signals calls for an amplifier with differential input and high *CMRR*. On the other hand, for various reasons, e.g. movement of the electrodes, the electrode-skin impedance does not remain constant and this affects the differential signal. To show this, consider the following example. Let the *CM* signal be $10mV$, the amplifier input resistance $1M\Omega$ and the skin-electrode resistance $20k\Omega$. This situation is shown in Fig. 11.7(a).

The voltage at each input terminal of the amplifier will be

Detection and Measurement of Biosignals

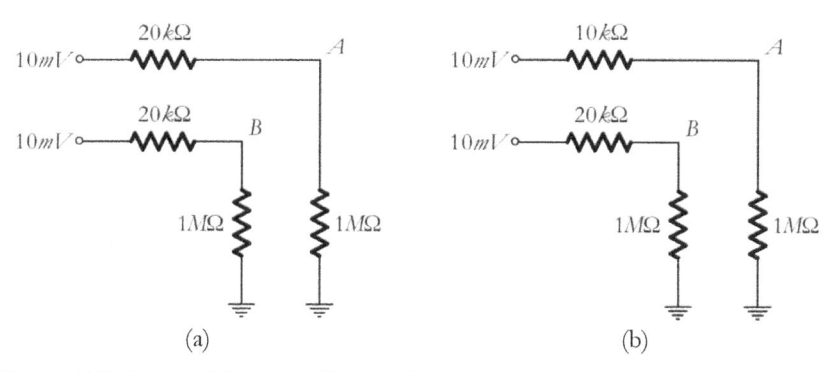

Figure 11.7. For studying the effect of the input resistance of the bioamplifier on the size of the input common-mode signal

$$V_A = V_B = \frac{1M\Omega \times 10mV}{20k\Omega + 1M\Omega} = \frac{10}{1.02}mV = 9.804mV$$

Thus, the difference V_A-V_B will be zero. However, if one of the skin-electrode resistances were changed to $10k\Omega$, say that connected to the input A, as shown in Fig. 11.7(b), the two voltages at the amplifier inputs would be

$$V_A = \frac{1M\Omega \times 10mV}{10k\Omega + 1M\Omega} = \frac{10}{1.01}mV = 9.901mV$$

While again

$$V_B = 9.804mV$$

Then

$$V_A - V_B = 9.901mV - 9.804mV = 0.097mV$$

Now consider the case when the amplifier input resistance is $100M\Omega$. Then, with the skin-electrode resistances as in Fig. 11.7(b) we will have

$$V_A = \frac{100M\Omega \times 10mV}{10k\Omega + 100M\Omega} = \frac{1000}{100.01}mV = 9.999mV$$

$$V_B = \frac{100M\Omega \times 10mV}{20k\Omega + 100M\Omega} = \frac{1000}{100.02}mV = 9.998mV$$

and

$$V_A - V_B = 9.999mV - 9.998mV = 0.001mV$$

Thus, the difference of the two CM voltages at the inputs of the amplifier will be of the order of the weakest biosignal or much smaller than this. Therefore, the bioamplifier must have as high an input resistance as possible.

Consequently, the main characteristics of a bioamplifier should be the following:

- High input impedance
- High and stable differential gain over all frequencies in the biosignal
- High CMRR

Additional desired characteristics are

- Low output impedance and
- Protection of the inputs from high currents

Clearly, the first four characteristics match those of an instrumentation amplifier, which can be properly protected by one of various schemes using, for example, resistors in series and Zener or combinations of usual diodes in parallel etc. The high currents at the inputs can be created by defibrillators (see Section 12.4.1) or by static shocks. The limit in the input voltage is set in order to avoid the possibility of the opamps to become saturated when the input voltage exceeds a certain value. With these additions, the IA of Fig. 3.9 will become as shown in Fig. 11.8. The two series resistances at the input terminals are usually about $10k\Omega$. The diode combinations connected in parallel to the inputs limit the input voltages, positive or negative, to twice the turn-on voltage of the diodes, which in the case of silicon diodes the limit is set to $\pm 1.4 V$.

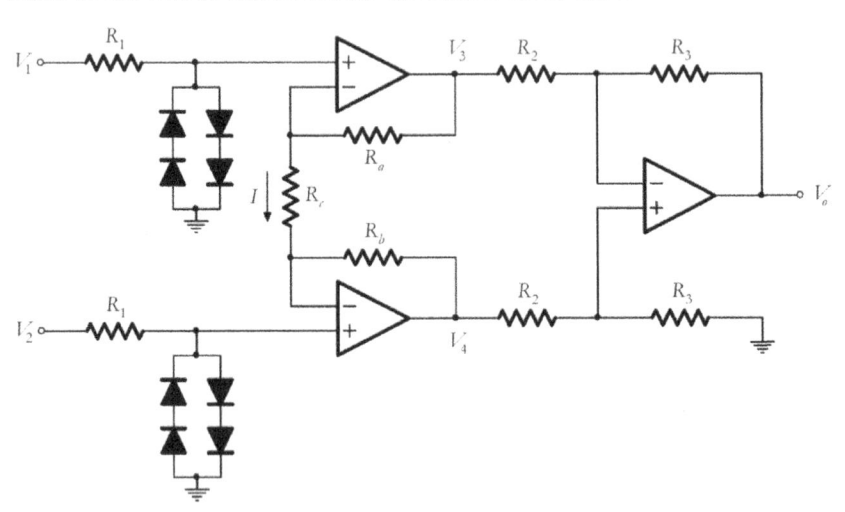

Figure 11.8. Instrumentation amplifier

As explained later below, because of the power lines ($220V$ or $120V$) running near the patient and the leads at the inputs of the amplifier, small displacement currents appear at the amplifier inputs. These currents give rise to common-mode voltages, which may be higher than the biosignals. To reduce these common-mode signals, a relatively low resistance path of the order of $10k\Omega$ is provided from a high-impedance point to the common ground.

Thus an instrumentation amplifier will be transformed to a bioamplifier as shown in Fig. 11.9.

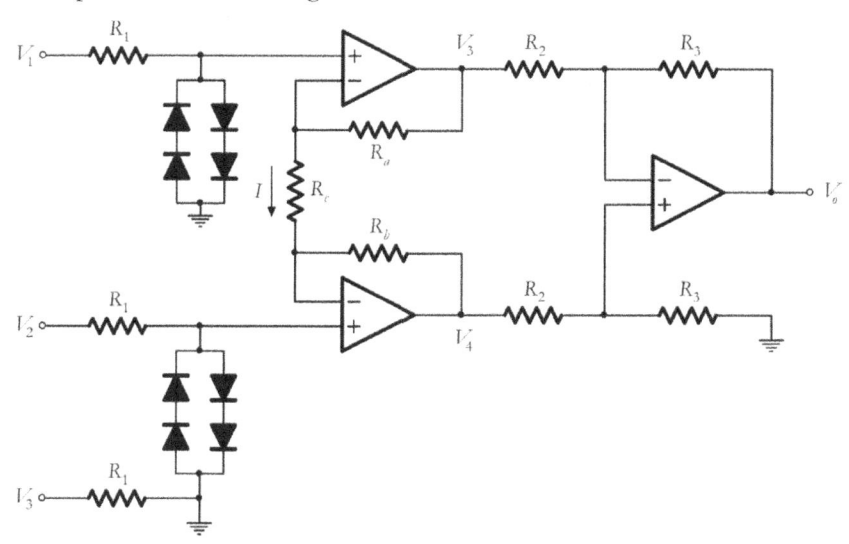

Figure 11.9. A bioamplifier circuit

A most important point should be mentioned here. When the bioamplifier is to be used to get a biosignal from a patient, in order to protect him from a microshock that could be lethal, as is explained in Chapter 12, the patient has to be isolated from the ground. In this case, batteries should be used as power supplies in the circuit. Then the common point of the amplifier is the common point of the batteries as shown in Fig. 11.10. To indicate this, the symbol of *the patient ground* and consequently, that of the bioamplifier is different from the usual ground symbol, as shown in Fig. 11.11, the latter being connected to the real ground which has the earth potential always considered to be zero volts.

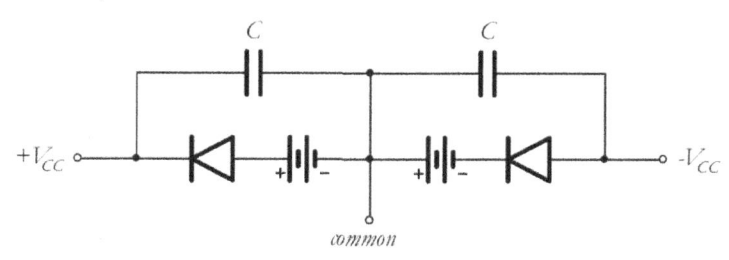

Figure 11.10. Use of batteries as power supplies

Figure 11.11. Symbols of (a) patient's ground and (b) usual ground

In the arrangement in Fig. 11.10 the diodes are used in order to prevent wrong voltage connections while the capacitors, about 1μF, to prevent high voltage spikes at the amplifiers.

11.5 ECG Detection

ECG signals are weak, ranging from $0.5mV$ to $5.0mV$, while their range of frequencies is between $0.5Hz$ to $100Hz$ (standard clinical ECG application). These weak signals may be corrupted by various kinds of noise, the main sources of which are the following:

- Interference 50 (60) Hz from power lines of the mains through capacitive coupling.
- Artifacts due to electrode movement causing variations in the skin-electrode impedance.
- Interference by EMG signals generated by muscle contraction during the measurement of ECG.
- Respiration of the patient, which affects the skin-electrode impedances.
- Any type of electromagnetic interference coming from neighboring equipment or other fields usual at high frequencies.

These noise sources have to be suppressed for accurate ECG measurements. Naturally, the signals from the electrodes are passed to the inputs of the bioamplifer as explained above. All common-mode signals, for example those arising from the interference by electromagnetic fields, are suppressed by the bioamplifier but not fully eliminated. Then, a useful method for further common-mode rejection is to apply negative feedback to the system patient-bioamplifier, namely, *right-leg-drive*, which is explained next.

11.5.1 Right-Leg-Drive for ECG Measurement

As was stated in Section 11.4, the use of a low impedance path from the output V_3 to ground, shown in Fig. 11.9, reduces the CM voltage, which is mainly due to capacitive coupling of the $220V$ ($120V$) power lines to the patient and the leads to the amplifier inputs. The situation is schematically described in Fig. 11.12. In this figure Z_1, and Z_2 represent the

Detection and Measurement of Biosignals

contact impedances of the electrode-skin tissue interfaces and they are of the order of $10k\Omega$ (see Sections 11.3.3 and 11.3.5 ignoring the presence of the parallel capacitors since the biosignals are low frequency signals). The displacement currents through the capacitors C_b, C_1, C_2 will change the useful signal from V_{ib} to, say V_{id}. Thus, the output of the amplifier with differential gain A_d will have a voltage

$$V_{od} = A_d V_{id}$$

instead of

$$V_{ob} = A_d V_{ib}$$

Clearly V_{od} will contain the useful signal V_{ob} plus a CM voltage due to the displacement currents.

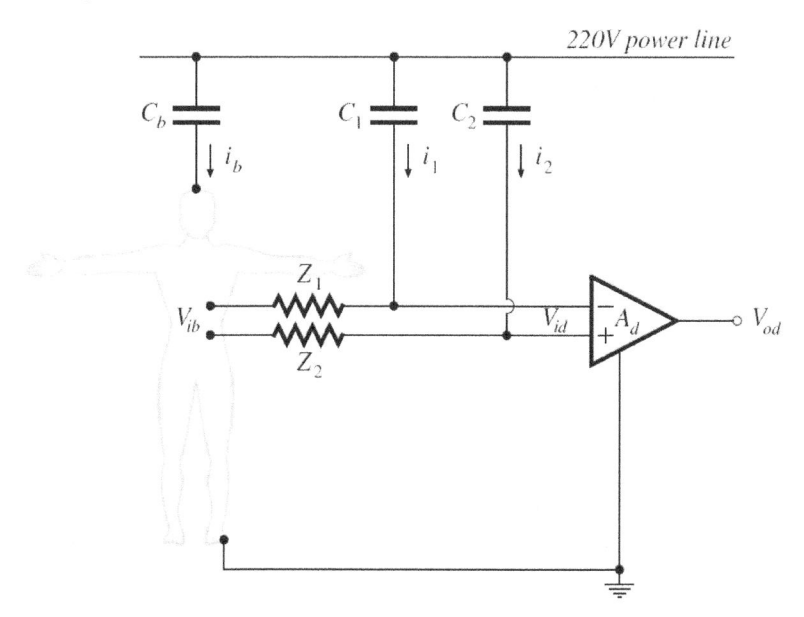

Figure 11.12. Capacitive coupling of the power line to the patient

To reduce this CM voltage, a low impedance path to ground is provided, as was explained in the previous section. In order to reduce the CM signal further, negative feedback is applied from the amplifier to the right leg of the patient provided that this leg is not grounded. Fig. 11.13 shows this right-leg-drive by means of an inverting amplifier. In this figure, the average of voltages V_3 and V_4 by means of the voltage divider, R_d, R_d is amplified by the inverting amplifier and drives the right leg of the patient, which, evidently, should not be grounded. To avoid the possibility of a large current driving the right leg, which will cause a possible microshock to the patient (see Chapter 12), a large resistance R_P is inserted between the inverting amplifier output and the patient's right leg. It can be shown with detailed study that, indeed, this scheme leads to further reduction of the *CM* voltage at the output of the bioamplifier. This is actually expected due to the well - known effect of negative feedback on distortion and internal noise in an amplifier (see Section 8.5.1). Actually, the isolation amplifier AD294A is suitable for providing right-leg drive in *ECG* measurements.

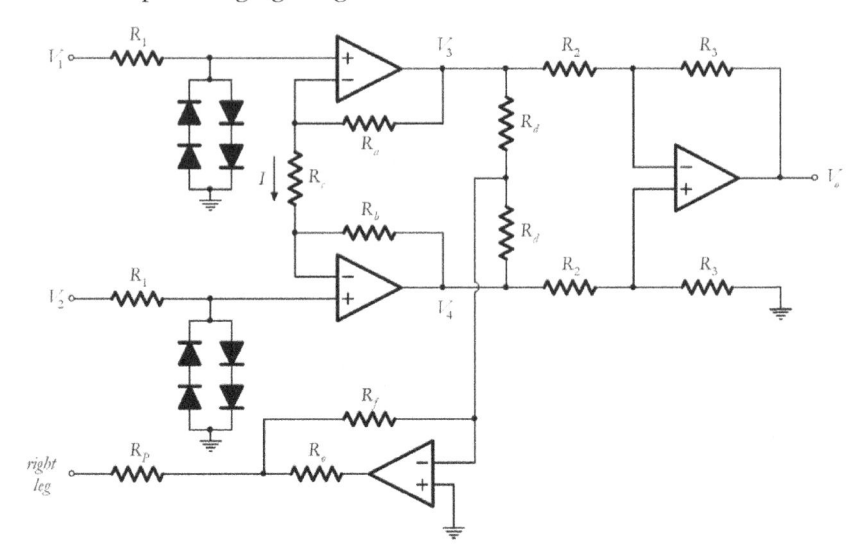

Figure 11.13. Right-leg drive by means of an inverting amplifier

11.6 Problems in Detecting *EEG* Signals

EEG signals have amplitudes between 10 and 100μV, which are much less than those of *ECG* signals. Thus, the sources of noise that corrupts the *ECG* become more severe in the case of the *EEG*. On top of these, the *ECG* signals as well as any movement of the eyes affect the *EEG* signal. Luckily the heart is much farther from the electrodes detecting the *EEG* than the brain and at the same time the *ECG* peaks are regular and can be recognized and disregarded. The same can be said for the signal produced by a heart pacemaker. This signal will be superimposed on the *EEG*, but it will also be recognized and overlooked.

On the other hand, the voltage between the front and the back of the eyeball is few *mV*. This causes potentials on the skin that change with the eye movement and this way the *EEG* signal gets affected.

To reduce the effect of all these noise sources on the *EEG* signals special precautions are required which can be summarized as follows:

It is best to construct a shielded room (free of electric fields) even though this is expensive. The door must be conductive (i.e. covered by copper) and must be electrically connected by contacts of its periphery to its frame. Windows must be avoided, because glass is not conductive. Incandescent lamps should be used instead, not fluorescent lamps, because the latter produce electric noise. A conductive grid must be incorporated in the walls, the ceiling and the floor. These grids must be interconnected and connected to the frame of the door, forming this way a complete Faraday cage, which should have its own ground (metallic triangle deep in ground).

Fortunately, in most cases, the construction of such a shielded room is unnecessary, because the correct construction and use of the electrophysiological instruments reduce the noise adequately. The following are some bits of practical advice.

- The instrument must always be grounded, i.e. the metallic cover and chassis of the device must be grounded.
- The electric supply cable to the device and the cables connecting the patient to the device must be shielded by a conductive sheath, which should be grounded. This protection almost eliminates all electrical interferences.
- Fluorescent lamps must be at distances greater than $2m$ from the patient. Use of incandescent lamps is preferable.
- Electric cables must not be near the patient or near the cable that connects the patient to the device, because, even if they are inside metallic tubes, their magnetic field causes interference.
- The patient must not be near a transformer, even the transformer of the device, therefore, he must be at a distance from the device.
- Replace injured cables and wrong installations, because they are sources of interference.
- The electric cables in or on the walls of the measurement room must run in metallic and grounded pipes, not in plastic pipes.
- The distance between the position of measurement and any source of interference should be as long as possible.

11.7 *EMG* Detection

ECG produced by the cardiac muscle is in fact an *EMG* although the term EMG is reserved for recordings of the skeletal muscles only. Essentially the same apparatus used in the *ECG* detection is used for the *EMG* detection. However, their magnitude as well as their frequency range is wider than those of the *ECG*. *EMG* amplifiers are usually capacitively coupled to the electrodes, while their low and high three *dB* frequencies are at $100Hz$ and $3kHz$ respectively.

EMG activity can be viewed in the time domain or the frequency domain. The time domain signal is sometimes integrated to obtain a smooth curve (see Section 10.6), which gives a measure of the quantity of electricity associated with the muscle action potential, and is an indication of the condition of a muscle during contraction. Similar results can be obtained by passing the *EMG* time signal through a true *RMS* voltmeter the indication of which is a smooth positive voltage proportional to the square root of the time average of the square of the time signal, and low-pass filtering (simple *RC* circuit, see Section 1.9). Other time average means can also be applied such as full-wave rectification and low-pass filtering to smooth the waveform. It has been shown that the latter processing of the signal gives an indication of the muscle activity approximately related to the force being generated at the location of the *EMG* electrode.

The *EMG* signal can be obtained by the voluntary contraction of a muscle or by external stimulation. Often the external stimulation is preferred to the voluntary contraction. The reason is that with electrical stimulation all the muscle fibers fire at nearly the same time while the voluntary contraction is spread over a longer time, because the motor units do not fire all at the same time. Also in the voluntary contraction, each motor unit may produce several action potentials depending upon the signal sent from the central nervous system.

References

[1]. Northrop R. B., *Analysis and Application of Analog Electronic Circuits to Biomedical Instrumentation*, CRC Press, 2004.

[2]. Company-Bosch E., *ECG Front-End Design is Simplified with MicroConverter*, Analog Devices, Analog Dialogue, vol. 37, November 2003.

[3]. Semmlow J., *Circuits, Signals, and Systems for Bioengineers (A MATLAB-Based Introduction)*, Elsevier, 2005.

[4]. Webster J. G., *The Measurement, Instrumentation, and Sensors Handbook*, CRC Press, 1999.

[5]. Norman R. A., *Principles of Bioinstrumentation*, John Wiley & Sons, New York, 1988.

[6]. Sansen W., *Microelectronics and Biosensors*, Notes for the PG course on Biomedical Engineering and Medical Physics, University of Patras, Medical Physics Laboratory, 1994.

[7]. Kane J. W. & Sternheim M. M., *Physics SI version*, John Wiley & Sons Inc., 1980.

[8]. Cameron J. R., Skofronic J. G. & Grant R. M., *Physics of the Body*, A Wiley Interscience Publication, 1999.

[9]. Guerrero F. P., Zoreda Bartolome J. L. & Gomez Aguilera E. J., *Fundomentos de Bioengienieria*, Volumen I, Universidad Politecnica de Madrid, TEB, Octubre 1988.

Detection and Measurement of Biosignals

Problems

11.1. The operational amplifier in the circuit in Fig. P.11.1 has low bias current and high input impedance. The circuit is to be used in the measurement of an ECG signal. V_1 consists of a $2mV$, $50Hz$ noise signal and an ECG signal of $+3mV$, while V_2 consists also of a $2mV$, $50Hz$ signal and a $-2mV$ ECG signal. Derive an expression for V_o as a function of V_1 and V_2 and calculate

   a. the size of the $50Hz$ noise in V_o and
   b. the size of ECG signal in V_o

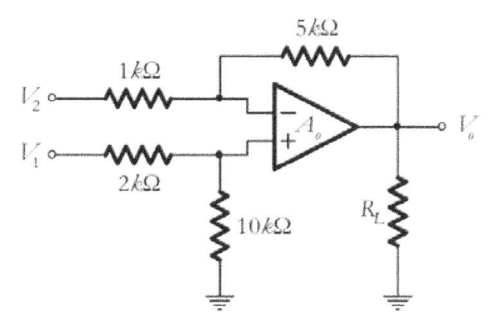

Figure P.11.1.

(Answers: a. 0, b. $25mV$)

11.2. In the circuit shown in Fig. P.11.2 the operational amplifiers are ideal, while the input voltages are dc voltages. Determine the value of the output voltage v_o. Derive the formula you will use in the calculation.

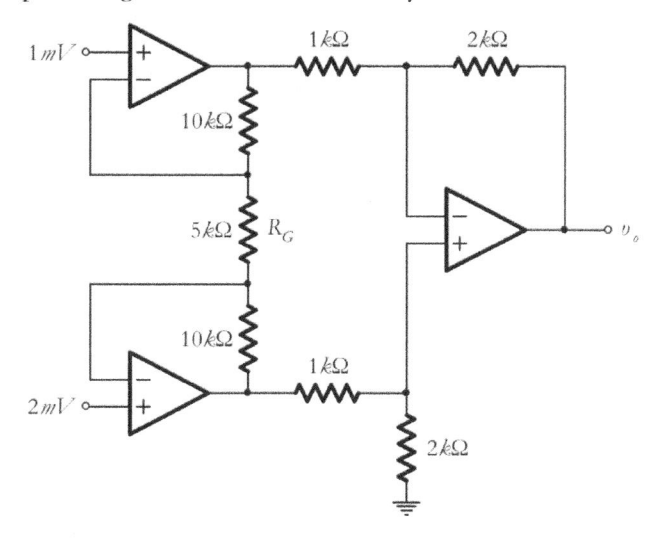

Figure P.11.2.

(Answers: $6mV$)

11.3. With V_1 and V_2 as in problem 11.1 instead of $2mV$ and $1mV$ respectively applied in the inputs of circuit in Fig. P.11.2 calculate

    a. the size of the $50Hz$ noise in v_o and
    b. the size of ECG signal in v_o

(Answers: a. 0, b. $50mV$)

Chapter 12

Bio-instruments and Safety

12.1 Introduction

Application of electricity to human body can be accidental or intentional. In the first case it can be dangerous, because the result could be even fatal. However, the intentional application of electricity to human body has been exploited by medical doctors, physiotherapists and others, among which people concerned with the beauty of the body. From among the many such intentional applications we select very few to look at in this chapter. All of them belong to the category of bio-instruments, which in general should be used safely.

In this chapter, first, we examine the physiological effects of electricity to human body. Then, we talk about the electrical shock and reasons for happening as well as ways for avoiding it. Next, we introduce some useful intentional applications, namely, the defibrillator and pacemaker for treating heart malfunctioning, diathermy for producing heating effects to various parts of the body and *TENS*, the Transcutaneous Electrical Nerve Stimulation devices used mainly by physiotherapists and at some hospitals. Finally, we mention some devices with relatively simple circuitry that the reader could design and build himself with the knowledge in electronics he acquired from the first part of this book.

12.2 Physiological Effects of Electricity

Physiological phenomena can be caused by electricity, if the human body becomes part of an electric circuit. The passage of an electric current through the human body may produce three different effects, namely,

a. Heating of tissues
b. Effects on nerves and muscles
c. Electrochemical burns

Depending on the value of the current, and, in particular, on the distribution of its density the effects can be therapeutic or dangerous for the human body, even fatal.

Consider the case when a voltage is applied between the hands of a human person. Depending on the magnitude of this voltage and the skin wetness, different values of the current will occur. The person will perceive the current, if its intensity is between 1 and $10mA$ at a frequency of 50 (60) Hz. At the same frequency, for values of current between 10 and $20mA$, 50% of men will not be able to let the wires go, because of a sustained contraction of muscles. These limits are lower for women. Note that these currents are higher at low and high frequencies. As the current increases further, the subject feels pain and in some cases faints. If the current is around $100mA$, the portion of it that passes through the heart will cause ventricular fibrillation, which can be fatal if not corrected. The result of ventricular fibrillation is rapid, irregular and ineffectual contraction of the ventricles. It is important to realize that the value of the current for fibrillation decreases as the time it is applied increases. To restore normal regular contraction of the ventricles, defibrillation is usually applied.

The defibrillator forces a high current pulse of very short duration (few *ms*) through the heart, which causes all heart muscle fibers to contract at the same time, thus restoring normal pumping of the heart (see Section 11.5.6).

Continuous currents above $6A$ can cause temporary respiratory paralysis and serious burns, depending on the individual, the dampness of the skin and the contact of the skin with the conductor.

12.3 Electrical Shock

We use the term electric shock mainly when the passage of an electric current through the human body is unintentional and is perceivable, specifically hazardous even fatal.

We can distinguish the possibility of the occurrence of an electric shock due to an electrode touching the human body externally and due to an electrode that has been placed under the skin. The first case is characterized as *macroshock,* while the latter is characterized as *microshock.* In the case of macroshock the current has to pass through the high resistance of the skin, while in the case of microshock the current is applied under the skin and thus can be led straight to vital organs of the body and in particular to the heart.

12.3.1 Macroshock

To get a more clear idea of the factors that can cause macroshock, we describe first the installation of electricity in buildings, where people live and work.

Electric power transfer from the place of its production to long distances is effected at high voltage – low current for economy reasons. This is transformed to low voltage – high current at the place of its consumption, as shown in Fig. 12.1.

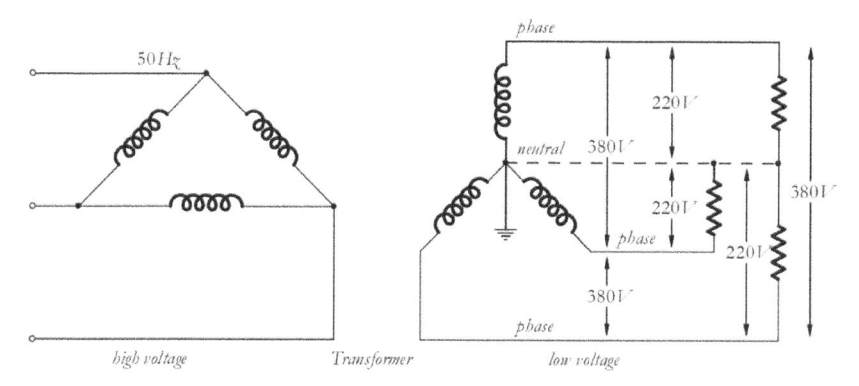

Figure 12.1. Transformation of energy from high voltage – low current into low voltage – high current for consumption

Usually, in a small house or flat electrical power is supplied by two wires, one for the phase and one for the neutral. However, following regulations, to each socket on the wall three wires should reach, one for the phase, one for the neutral and one for the ground. The house is provided with a central ground installation, that is, a thick metal wire dug deeply into the earth, to which the ground wire of each socket should be connected. For safety, a fuse must be provided in each phase wire (Fig. 12.2).

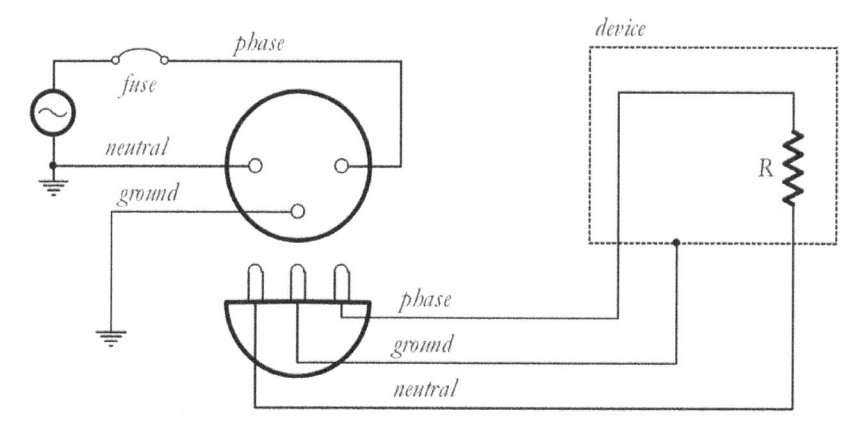

Figure 12.2. Safe connection of a device to power

If the socket is supplied by only two wires (phase and neutral) (Fig. 12.3), or, if the cable from the plug to the device has only two wires, there is a danger for a macroshock in case there is a leak from the phase to the external metallic cover of the device (Fig. 12.3). On the other hand, if the device is not earthed (grounded) and the insulation is faulty a macroshock may also happen, if a person touches the outside metallic cover of the device. The existence of the third wire, that is grounded, protects the people from macroshock, while in case of leakage the blowing of the fuse will interrupt the supply of power.

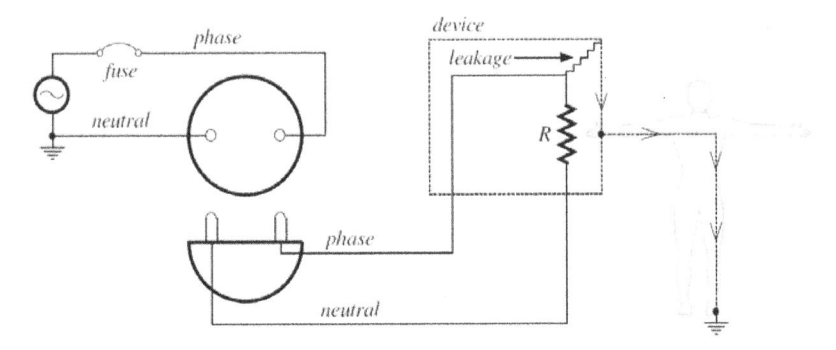

Figure 12.3. In case of leakage the two-wire installation can become dangerous

If the resistance R between the device cover and the ground is appreciable and the leakage current I is high, then the cover will be at a voltage $V = I \cdot R$, which may be dangerous. Regulations dictate that such voltages between uncovered surfaces must not exceed $500 mV$ for ordinary wards and $100 mV$ for special wards, like Intensive Care Units (*ICU*). For this reason, all conductive surfaces around the patient of intensive care must be connected to his ground point near the head of the bed. All those private ground points are connected to one ground of the *ICU*, which is connected to general ground of the building.

12.3.2 Microshock

Microshock, as was defined above, refers to passing a current inside the body under the skin. In this case, it does not have to pass through the high resistance of the skin and it is led through the arteries straight to the heart. Then, ventricular fibrillation can be caused by very low currents, in the order of $20 \mu A$. For example, it can be fatal, if it is led through the body to a wire of a catheter that is inserted through a vessel into the heart of a patient.

12.4 Intentional Application of Electricity to Human Body

The above discussion on electric shock refers to the hazardous results that the unintentional application of electricity to human body can cause. However, under the proper precautions, application of electricity for therapeutic purposes is extensively used nowadays. Electrical neurostimulation has been carried out in a variety of sites to recuperate several organ functions. Among these, pacemakers and defibrillators for cardiatic irregularities, cortical stimulators to create vision for the blind, cochlear implants for profoundly deaf persons, brain stimulators for Parkinson, etc. Also, many other devices dedicated to respiration, legs and hands movements are available. More particularly, research on bladder control for urine voiding and to prevent incontinence is being conducted. These intentional applications of electricity are briefly explained below.

12.4.1 Defibrillator

Normal heart pumping is associated with normal heart muscle contraction. A sudden change of the rhythm (arrhythmia) is due to uncoordinated ventricular pumping caused by ventricular fibrillation. The arrhythmias may occur as a result of a heart attack or a variety of serious injuries or illnesses. Eventually this will lead to stoppage of the heart pumping action resulting in the death of the patient.

The heart beating of a patient in an *Intensive-Care Unit*, *ICU*, is normally monitored continuously so that, if a sudden change in the rhythm of the heart attack patient occurs, immediate therapeutic action can be applied to save the patient's life. This is achieved by using the *defibrillator*.

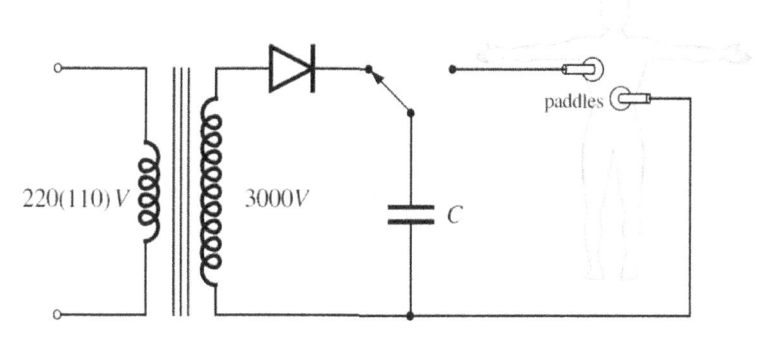

Figure 12.4. A simple defibrillator

The defibrillator, Fig. 12.4, is effectively an instrument allowing for the application of a short high-voltage pulse of the order of $3000V$ by means of two electrodes in the form of paddles, which are placed by the doctor on the patient's chest above and below the heart. Firm pressure is applied to the paddles to ensure good electrical contact with the skin. While the shock is discharged the patient often jumps, because motor nerves and skeletal muscles are simultaneously unavoidably stimulated. The electrodes are metallic of about $7.5cm$ in diameter coated with conductive paste. The handles of the paddles are plastic and electrically insulated to avoid causing any electrical shock to the operator. The application of the electric pulse to the patient's chest results in a current pulse of about $10A$ height and $5ms$ duration.

The effect of the defibrillator is to reset the electrical activity of the heart to an organized rhythm or to change a very rapid and ineffective cardiatic rhythm to a slower, more effective, one. Using a direct current in the form of an unsynchronized countershock produces a sustained simultaneous excitation of all cardiatic muscle fibers, which can terminate ventricular fibrillation and other abnormal rhythms.

Today automated external defibrillators are available for use by persons with minimal medical training in emergencies, when medical professionals are unavailable. Also miniaturization has led to the development of the implanted internal defibrillator, in which a microcomputer monitors the heartbeat by using an electrode. Upon detecting a minor arrhythmia, it activates a built-in conventional pacemaker (see below) to re-stabilize the rhythm of the heart. If that fails, it delivers a small defibrillating electrical shock and, in the extreme case, it resorts to a far stronger shock to reset the heart rate.

12.4.2 Pacemaker

As was described in Section 10.7, the rhythm of the normal heart beating is kept by electric impulses generated by the sinoatrial node, SN node, which is located in the wall of the right atrium. These impulses are transferred, through special conduction tissues on the walls of the atria, causing them to contract and pump blood. The same electric impulses reaching the atrioventricular node, AV node, are 'relayed' to the ventricles and, through special conduction tissues on their walls, cause the contraction and thus the pumping action of the ventricles.

Slow rates in heart beating can be caused by diseases and aging, which affect the SA node, the conduction tissues and the AV node. In such situations use of a pacemaker is a necessity, since there is no medicine available in oral form that can be taken regularly to increase the heart rate. The heart pacemaker is a battery operated electronic device that is used to maintain normal heart beating, when the heart is not functioning properly of bradycardia. Actually, pacemakers can be used permanently, if the slow heart rate becomes chronic, or is believed to be inevitable. However, if the abnormality in the heart rate is believed to be temporal (lasting only days) and is caused by correctable conditions a temporary pacemaker is used.

A permanent pacemaker consists of a timing electronic device, a circuitry that detects electrical signals from the heart, a battery and the lead(s). The timing device sets the pacing rate and, when the heart beats too slowly or stops beating, the pacemaker takes over generating electrical signals that force the heart to beat at the rate, which has been set by the doctor.

Permanent pacemakers are implanted inside the body in the form of a chamber, which includes the circuitry and its own battery that is durable (7-10 years before they need to be replaced). The lead(s) is (are) platinum wire(s) insulated with silicon or polyurethane. Pacemakers having only one lead are called *single-chamber* pacemakers and those having two leads are called *dual-chamber* pacemakers.

The lead of the pacemaker is inserted through a vein in the chest into the heart with its tip placed in contact with the inner wall of the right atrium or the right ventricle with the visual guidance of x-rays or under fluoroscopic control. The other end of the lead is connected to the pacemaker chamber. Then the chamber is implanted under the skin usually in the left or right upper chest near the collarbone.

Pacemakers can be equipped with rate of activity response features that measure the body's metabolic activity using suitable sensors. This way they can increase the rate of heartbeats by increasing the rate of pacing during exercise or stress, when the body requires more blood supply. Some pacemakers nowadays incorporate multiple sensors to more accurately measure the body's metabolic activity. Also pacemakers are incorporated in implantable defibrillator devices with the whole device being capable of treating both fast and slow rhythms in the same patient. Modern pacemakers are well protected from most household electrical appliances in good condition such as radio, television, stereos, microwave ovens, computers etc. However, *Magnetic Resonance Imaging* (*MRI*) having a strong magnetic field can interfere with pacemakers and thus patients with pacemakers should not undergo *MRI* scanning. Also digital cellular phones can interfere with pacemakers and should not be carried near the chest. Other precautions should also be discussed with the doctor.

12.4.3 Diathermy

Diathermy is heating parts of the human body and it is mainly achieved through the application of high frequency electricity. The frequency of the applied voltage (or current) can be 10*kHz*, long-wave diathermy or 30*MHz*, short-wave diathermy. Long-wave diathermy is not used any more, while short-wave diathermy can be in the microwave region i.e. at *2450*MHz.

In short-wave diathermy the heating effect is produced by the absorption of electromagnetic energy introduced into the body. At the frequency of 30*MHz* this introduction can be achieved by taking advantage of either the electric or the magnetic field of the electromagnetic wave. The absorption mechanism is then different as it can be realized and also explained here.

When the electric field is used, the resulting alternating electric forces cause the ions of the tissue to move back and forth, thus, acquiring kinetic energy, part of which is dissipated during the collisions of the ions with the molecules in the tissue. The dissipated energy and, consequently, the resulting heating effect, called *joule heating*, are nearly proportional to the square of the current and also depend on the properties of the tissue.

The required high-frequency electric field is produced by the capacitor of an LC tuned circuit. The tissue to be heated is placed between the plates of the capacitor, which gives the name of this method as *capacitance method of diathermy*. The tuned circuit is schematically shown in Fig. 12.5. In the ideal case a current impulse causes energy $LI^2/2$ to be stored as magnetic field in the inductance, which then produces a current that charges the capacitor C to the highest voltage V across its plates. Then, when all magnetic energy from L has been transferred to the capacitance C and stored as electric energy $CV^2/2$, the capacitor starts discharging through L turning again the stored electric energy to magnetic energy in L. The processes continue with the energy being transferred from L to C and from C to L continuously. Thus the circuit oscillates with the frequency of oscillations being $1/\sqrt{LC}$ when L and C are ideal, i.e. when no energy loss is encountered. However, in practice L and C are not ideal and the unavoidable energy loss should be compensated for externally (or by placing a negative resistance across the LC circuit). On the other hand, the produced magnetic field inside the inductor can be used to heat the tissue by placing the tissue either inside or near the inductor. In this case, which is called *inductance diathermy*, the alternating magnetic field produces Eddy currents in the tissue that heat it. Consequently, this leads to power loss, which is replaced externally.

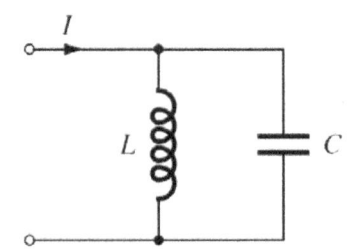

Figure 12.5. LC tuned circuit

The efficiency of short-wave diathermy is limited, when fatty layers are interleaved between the tissue to be heated and the surface of the body. A lot of energy in this case is dissipated in the fatty layers. More suitable in this case is to use microwave diathermy. Usually the frequency of the microwaves used is 2450*MHz* or about 12*cm* wavelength. The microwave radiation comes from an antenna that is placed near the part of the body to be treated. The emitted energy is partly reflected from the skin and the rest transmitted to the tissue, which absorbs it producing heat. Unlike the short-wave mechanisms of heat production, in the case of microwave diathermy, it is believed, that heat production is associated with the existence of water in the tissue. The water molecules permanently behave as electric dipoles due to a slight displacement of the centers of charge in the molecules. What the electric field of the microwave electromagnetic radiation does is to try to align these electric dipoles with it. The tissue absorbs the required energy for this alignment producing heat.

Short-wave diathermy is used in the treatment among others of arthritis, traumatic injuries and strains, while microwave diathermy is used in heating joints, tendon sheets and muscles.

High-frequency electricity is also used for cauterizing open wounds and in electrosurgery.

In all applications mentioned here, care should be taken to avoid overheating, which can cause serious hazards. To this purpose certain standards have been established concerning, for example, the duration of exposure to radiation as well as the radiation level.

12.4.4 Transcutaneous Electrical Nerve Stimulation

Transcutaneous Electrical Nerve Stimulation (TENS) devices are, typically, pulse generators, which through leads and electrodes (patches) send current pulses to the skin. They are commonly used by physiotherapists and at some hospitals to treat various forms of pain. Small *TENS* devices are also available for commercial sale to patients.

The electrical signals that a *TENS* device sends to the skin through the patches placed on the skin are painless. The patient feels just a tingling effect when the device is on. The pulses produced by the generator can be of different forms like the ones shown in Fig. 12.6. These are monophasic giving a non-zero mean value of the current or biphasic, for some of which the mean value of current is zero. Zero net current flow may prevent the build-up of ion concentrations beneath the electrodes, thus preventing adverse skin reactions due to polar concentrations.

Alternative patterns of pulse delivered by most *TENS* devices are as shown in Fig. 12.7. These signals are burst, of pulses, continuous, amplitude modulated, frequency modulated, pulse width modulated or random all of them being usually monophasic.

The operator of a *TENS* device can control the amplitude, the frequency and the width of the pulses thus determining the amount of current delivered to the patients skin via the leads and the electrodes (putches).

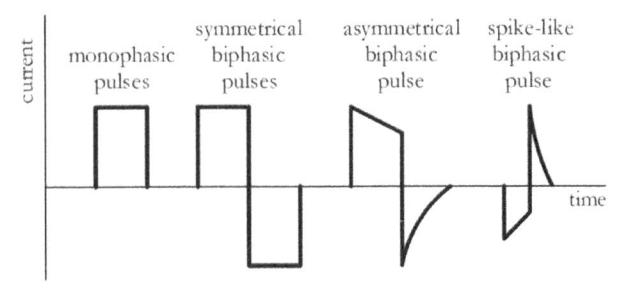

Figure 12.6. Monophasic and biphasic *TENS* current pulses

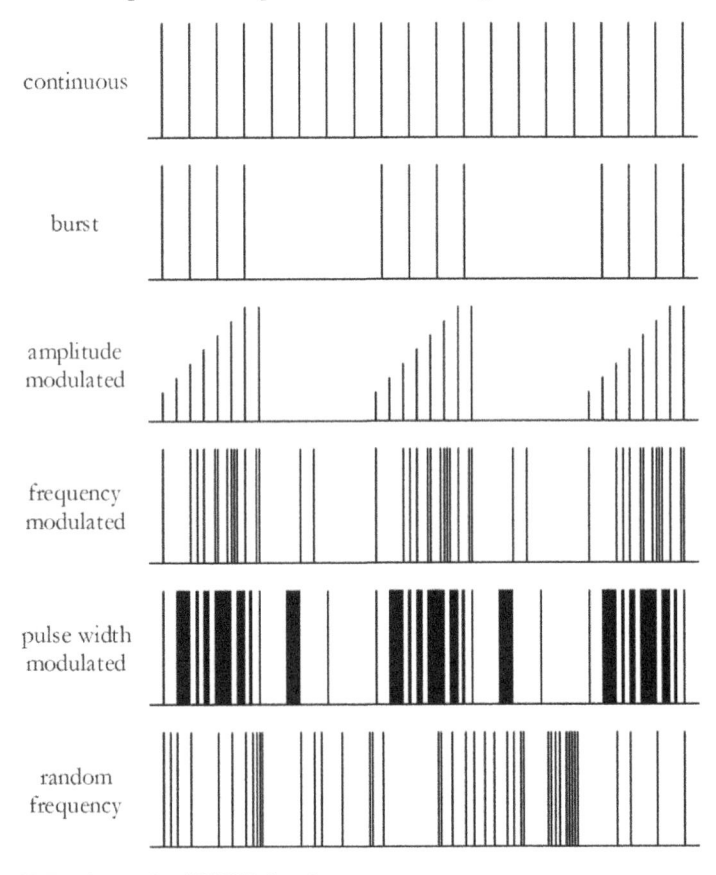

Figure 12.7. Alternative *TENS* signals

TENS devices are thought to reduce the severity of pain. In fact they are effective, whilst they are on and, often, for a while afterwards. However, they do not cure pain. There are two theories about how a *TENS* device works to reduce pain. According to one, *TENS* promotes release of endorphins that are body's natural pain - killers. According to the other theory *TENS* signals interfere with pain signals in the spinal cord. However, in some cases both theories may apply simultaneously.

Three types of *TENS* devices are available: The *Conventional TENS* type, the *Acupuncture-like TENS* type (*AL-TENS*) and the *Intense TENS* type. Each of these is most suitable for activating different populations of nerve fibers due to the different types of pulses it produces. Thus *Conventional TENS* type devices activate large diameter non-noxious coetaneous afferents. The *Acupuncture-like TENS* type (*AL-TENS*) devices activate small diameter non-noxious muscle afferents, while the *Intense TENS* type devices activate small diameter "pin-prick" coetaneous afferents. Also the positions of the electrodes on the skin is different for the different *TENS* type.

Forms of pain that can be treated by a *TENS* device are as follows:

- Chronic musculoskeletal (for example arthritis or back pain)
- Pain that occurs following an operation or surgery
- Cancer pain
- Pain during childbirth
- Reducing nausea following chemotherapy
- Phantom limb pain (pain in an already amputated arm or leg, as if it were still there)

TENS devices can also produce non-analgesic effects and are used as antianemic and for restoration of blood flow to ischemic tissues and wounds.

TENS devices should not be used while driving or operating machinery near an open wound or broken or irritated skin. Use of *TENS* is not recommended on pregnant women as it may induce contractions, on individuals with cardiac pacemakers, because the electrical pulses may interfere with the operation of the pacemaker and on patients with epilepsy or severe allodymia.

It is recommended that *TENS* electrodes should not be placed over the throat, eyes or carotid sinus (the area on the neck just below the ear and near the jaw where the carotid artery lies). If the *TENS* device is not used appropriately, there is a risk of increasing pain rather than easing it.

Technical Electrical Characteristics of *TENS* Devices

Pulse waveform, which is fixed: Monophasic
Symmetrical biphasic

Pulse amplitude (adjustable): Asymmetrical biphasic
1-50 mA into 1 $k\Omega$ load

Pulse duration (often fixed):	10-1000 μs
Pulse frequency (adjustable):	1-250 *pps*
Pulse pattern:	Continuous, burst (random frequency, modulated amplitude, modulated frequency, modulated pulse duration)
Channels:	1 or 2
Batteries:	*PP*3 (9*V*) rechargeable

Most devices deliver constant current output.

An astable multivibrator circuit, like the one designed using the integrated circuit 555, is usually the main part of a *TENS* device.

12.4.5 Implantable Smart Medical Devices

Microelectronics allows the design of implantable smart medical devices (*SMD*). Such devices are nowadays used in several medical applications and share many features and basic components.

The prime concern for implants intended for chronic use is safety. This is because a higher power consumption device increases tissue temperature as it dissipates heat, but also implies higher electromagnetic field density through tissues when powering up the device transcuntaneously. In addition to keeping device's power consumption to a minimum, some features to ensure safe stimulation are desired.

12.5 Additional Simple Bio-instruments

There is a large number of relatively simple devices that the reader can try to build himself based on the content of Part *I* of this book. Such devices consist of relatively simple circuits. Actually there is an abundance of such circuits that one can find on the web. Some of such devices are the following:

- Electronic stethoscope
- Photo-plethysmograph
- Heart rate monitor
- Skin resistance meter
- Breathing rate meter
- Pulse rate monitor

and a lot of others

Of those we give some hints on how one could proceed to design and build the first two on the above list.

12.5.1 Electronic Stethoscopes

A stethoscope is used by the doctor to listen to the heartbeat of a person. Standard stethoscopes do not provide amplification and this limits their use. By providing amplification a stethoscope can be useful to many other people apart from doctors, like home mechanics for example.

Usually an electronic stethoscope will be built out of a microphone, an amplifier with high input impedance, adequate gain and low output impedance plus a pair of earphones connected to the amplifier output terminals. Improved electronic stethoscopes will include a low-pass filter in their structure to reject high frequency background noise. As the pulse heartbeat is of low frequency, a second-order Butterworth low-pass active filter is proved adequate for this purpose (see Chapter 4).

12.5.2 Plethysmograph

Heart beating causes a pressure wave to move out along the arteries at a few meters per second, appreciably faster than the actual flow of blood. This pressure wave can be felt at the wrist, but it also causes an increase in the blood volume in the tissues and this can be detected by the plethysmograph. Apart from the existing devices on the market, one can build his own device based on the use of a photoelectric detector. The advantage of such a method will be the fact that the heart beating can be recorded without the need to make direct electrical connections to the body.

The sensor consists of a light source and a photodetector. Variation of blood volume alters the amount of light reaching the detector after its fall from the source onto the tissue. The detectable light can be either the reflected or that passing through the finger or earlobe. Use of infrared light is most proper for this application. The detected signal is amplified and filtered to reject unwanted noise.

This is a noninvasive method for measuring volume changes in parts of the body caused by blood being pumped in and out. The light source is an *IR LED* and the photodetector a photodiode or a phototransistor. A useful opamp can be the LM358 (dual opamp) which works on a single battery 5-9V. Only that the amplifier common has to be raised to over 1V from the ground using two silicon diodes in series.

12.6 Gamma Scintillation Counting

Gamma Scintillation Counting System is a composite instrument used in studies of gamma ray emission from radioactive sources. We describe it here, mainly, because its structure is made up from a number of circuits the reader has met going through Part *I* of this book. It can also be of interest to people working with nuclear physics applications in medicine in a hospital or in a corresponding specializing clinic. Two versions of this instrument are briefly presented here, namely, one using a Single-Channel Analyzer and the other using a Multi-Channel Analyzer. Both versions can be used in gamma ray spectroscopy, the former operating manually and the latter automatically.

12.6.1 Using a Single–Channel Analyzer

The structure of the Gamma Scintillation Counting System using a Single-Channel Analyzer, *SCA*, we describe here, is shown in Fig. 12.8 in block diagram form. It consists of the *detection stage*, the *signal conditioning and processing stage* and the *counting and recording stage*.

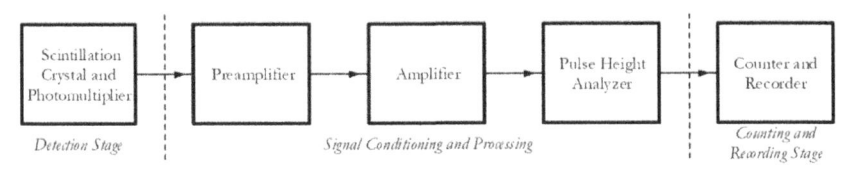

Figure 12.8. Gamma Scintillation Counting System

The detection stage consists of a crystal of NaI, which produces a light pulse for every gamma ray particle entering it. The intensity of the light pulse depends on the energy of the particle. Properly attached to this "scintillation" crystal is a photomultiplier tube. This tube has a photocathode and a series of special electrodes, called dynodes, each of which is biased at a voltage higher than its preceding one and is suitable to emit electrons by secondary emission. The final electrode in the tube is the anode that is biased at a voltage higher than that of the nearest dynode. The anode is not made to emit secondary electrons. The arrangement including the biasing of the photomultiplier tube is shown in Fig. 12.9. Then, the operation of the detection stage is as follows: Each light pulse from the scintillation crystal falls on the photocathode, which emits a number of electrons depending of the height of the light pulse. These electrons are collected by the first dynode being biased positively w.r.t. the cathode, which emits, by secondary emission, a larger number of electrons than the number of those falling on it. The secondary electrons fall on the second dynode, which multiplies them and this process of electron multiplication continues in each of the following dynodes with the anode finally collecting a much larger number of electrons for each electron initially emitted from the photocathode. If δ is the ratio of the number of the secondary electrons emitted from each dynode to the number of electrons falling on the dynode, then the initial number of electrons emitted from the cathode is multiplied by δ^n, where n is the number of dynodes. Thus an electronic pulse is obtained at the end of the detection stage the height of which is proportional to the height of the initial light pulse and, therefore, to the energy of the detected particle.

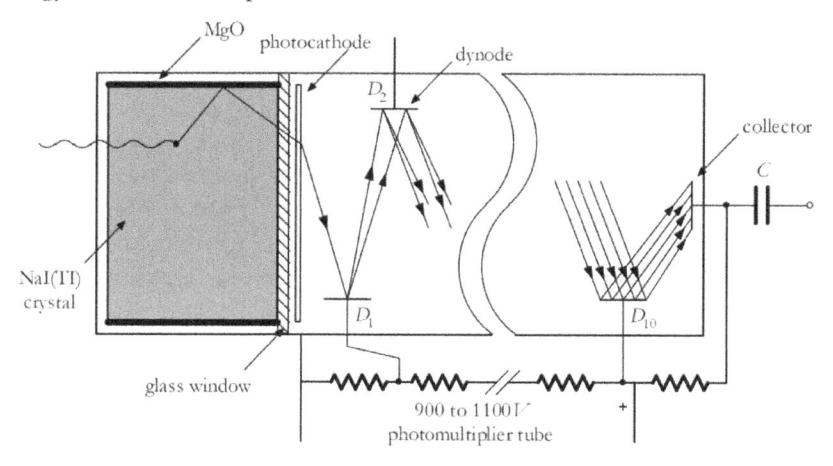

Figure 12.9. Schematic diagram of mounting the NaI scintillation crystal and the photomultiplier tube to form the detection stage

The required high *dc* voltage, which can be over $900V$ and up to $3000V$, is produced by the *dc* voltage multiplication circuit, which is shown in Fig. 12.10. This high voltage is distributed among the dynodes and the anode by the voltage divider, as shown in Fig. 12.9.

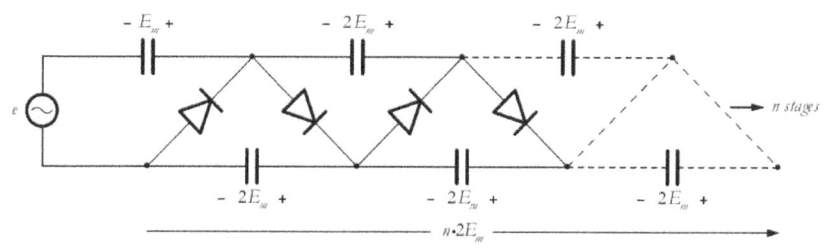

Figure 12.10. Circuit to obtain a high *dc* voltage low current (μA) from a lower *ac* voltage. The number of stages n should not be greater than 5, because for $n>5$ the output impedance becomes very high.

In the Signal Conditioning and Processing stage, the signal taken from the photomultiplier is preamplified and amplified by pulse amplifiers, which are suitable for high frequency signals. The amplifier is followed by the Pulse-Height Analyzer, a pulse discrimination circuit, which allows only pulses with heights between V and $V+\Delta V$ to pass through it (see Section 6, Fig. 6.11). The voltage window ΔV can be moved up or down, manually or automatically, to obtain the spectrum i.e. the number of pulses at each pulse height.

Finally, in the counting and recording stage the pulses coming out from the Pulse Height Analyzer are counted and their number displayed or recorded in the form of a graph giving the energy spectrum of the gamma rays entering the scintillation crystal.

12.6.2 Using the Multi-Channel Analyzer

Use of the Multi-Channel Analyzer gives the gamma ray spectrum from a radioactive source automatically. It could consist of a number of single-channel analyzers, each of which is arranged to count the number of particles having energies within a narrow window E and $E+\Delta E$, with ΔE depending on the number of channels. However, this would be a non-economical solution and it is not followed in practice. Instead the *MCA* following the same detecting stage as was presented above operates basically as follows: The pulse entering the *MCA* from the detection stage is initially converted into a digital word by an *ADC*. This digital word is the address in memory where the number of pulses having a height equal to the height of the incoming pulse is registered each time. Each address in the memory corresponds to a separate channel, i.e. there are as many channels as the available addresses in the memory for this purpose. Evidently, the number of channels in the *MCA* depends on the resolution of the *ADC*.

The registration of a pulse of a given height in the corresponding address in the memory increases its content by one and this is achieved as follows: After the conversion of the height of a pulse into a memory address, the system control unit gives the command for the transfer of the memory content in that location to the arithmetic unit (*AU*) of the system. In the *AU* the addition of 1 to the previously registered number is performed and the new number, following a new command from the control unit, is transferred to the same address in the memory. At the end of this transfer a command is sent to the *ADC* to proceed in the conversion of a new input pulse. Clearly, during the process of registering a pulse in the memory the *ADC* does not perform any conversion.

The *MCA* memory can be magnetic (non-volatile) or semiconductor (volatile). In the second case the memory can be static with each cell being a flip-flop, or dynamic with the memory cells being the *MOSFET* capacitance. The cells are connected in a way that the numbers of pulses are registered for example in *BCD* code. By the proper instruction the memory content in each location is transferred to the output of the *MCA*. Then the spectrum can be displayed on the screen of a *CRT* or a more modern display system or be plotted by a printer or the number from each channel be printed.

It is clear that the use of a *MCA* instead of a *SCA* reduces substantially the time required for obtaining the spectrum of a gamma ray emission, particularly so when the emission rate is low. The important characteristics of a *MCA* are the number of channels, the processing speed for each pulse, the capacity of each channel, its linearity, its discrimination etc.

References

[1]. *The Columbia Electronic Encyclopaedia*, Sixth Edition, Columbia University Press, 2003.

[2]. *World of the Body (site)*, The Oxford Companion to the Body, Oxford University Press, 2003.

[3]. Webster J. G., Editor, *Medical Instrumentation: Application and Design*, 3rd Edition, Wiley, 1998.

[4]. Sawan M., Yamu Hu & Coulombe J., *Wireless Smart Implants Dedicated to Multichannel Monitoring and Microstimulation*, IEEE Circuits and System Magazine, Vol. 5, No. 1, 2005, p. 21-38.

[5]. Jennings D., Flint A., Turton B. C. H. & Nokes L. D. M., *Introduction to Medical Electronics Applications*, Edward Arnold, 1995.

[6]. Northrop R. B., *Analysis and Application of Analog Electronic Circuits to Biomedical Instrumentation*, CRC Press, 2004.

[7]. Company-Bosch E., *ECG Front-End Design is Simplified with MicroConverter*, Analog Devices, Analog Dialogue, vol. 37, November 2003.

[8]. Semmlow J., *Circuits, Signals, and Systems for Bioengineers (A MATLAB-Based Introduction)*, Elsevier, 2005.

[9]. Webster J. G., *The Measurement, Instrumentation, and Sensors Handbook*, CRC Press, 1999.

[10]. Norman R. A., *Principles of Bioinstrumentation*, John Wiley & Sons, New York, 1988.

[11]. Sansen W., *Microelectronics and Biosensors*, Notes for the PG course on Biomedical Engineering and Medical Physics, University of Patras, Medical Physics Laboratory, 1994.

[12]. Kane J. W. & Sternheim M. M., *Physics SI version*, John Wiley & Sons Inc., 1980.

[13]. Cameron J. R., Skofronic J.G. & Grant R.M., *Physics of the Body*, A Wiley Interscience Publication, 1999.

[14]. Giokaris P., *Clinical Electrotherapy*, Litsas Medical Editions, 1998 (in Greek).

Problems

12.1. In the three-phase transformer circuit in Fig. 12.1, explain why the voltage between any two phases is $380V$ when the voltage between each phase and the neutral is $220V$. All voltages are given in *rms* values.

12.2. Determine the peak value of the voltage in the capacitor in Fig.12.4, if the transformer is supplied by $220Vrms$ and the turns-ratio is 1:10. The diode is considered to be ideal.

12.3. Determine the frequency of the sinusoidal oscillations produced by the tuned circuit in Fig.12.5, when $L=2mH$ and $C=5nF$.

12.4. Explain the operation of the *dc* multiplying circuit in Fig.12.10.

Appendix A

More on Circuit Analysis

A.1 Nodal Analysis of a Circuit

We will explain the method of *Nodal Analysis* by means of an example. Consider the circuit in Fig. A.1(a), which for simplicity contains ideal sources and resistances but the method is general when capacitances and inductances are also included. It is simpler to use conductances instead of resistances to avoid the presence of denominators in the equations. The nodes have been numbered clearly. The objective is to determine the node voltages. For simplicity, one node is selected as reference and, arbitrarily, its voltage is taken to be zero. Then the voltages of the other nodes will be in reference to this node, the reference node. Here we select node 4 as the reference node. Our task then is to determine the values of v_1, v_2 and v_3, the voltages at nodes 1, 2 and 3 respectively. Applying *KCL* and taking the currents entering a node as negative and those leaving a node as positive we get for node 1:

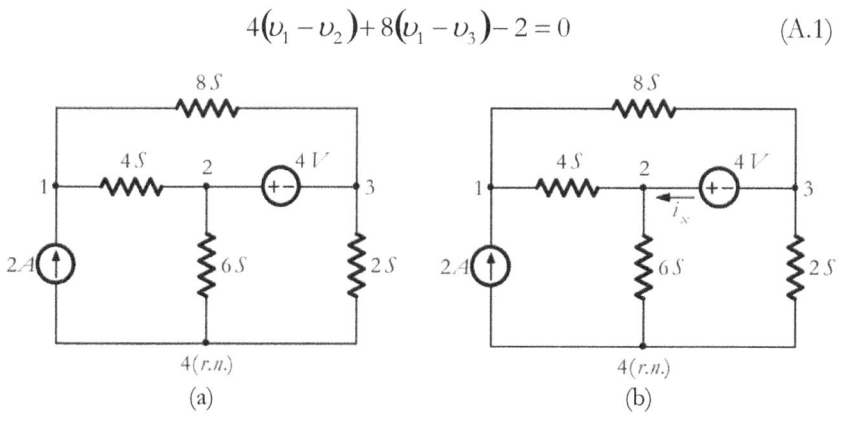

$$4(v_1 - v_2) + 8(v_1 - v_3) - 2 = 0 \tag{A.1}$$

Figure A.1.

At nodes 2 and 3 an ideal voltage source is connected. Its current depends on its load and let it be i_x, as shown in Fig. A.1(b). Then the equations for nodes 2 and 3 will be respectively as follows:

node 2 $\qquad 4(v_2 - v_1) + 6v_2 - i_x = 0 \qquad$ (A.2)

node 3 $\qquad 8(v_3 - v_1) + 2v_3 + i_x = 0 \qquad$ (A.3)

A fourth equation can be, due to the IVS of $4V$

$$v_2 - v_3 = 4 \qquad (A.4)$$

We have four equations with four unknowns, but these can be reduced to 3 by adding Eq. (A.2) and (A.3) when i_x is eliminated. Thus we can get

$$4(v_2 - v_1) + 6v_2 + 8(v_3 - v_1) + 2v_3 = 0$$

or

$$-12v_1 + 10v_2 + 10v_3 = 0 \qquad (A.4)$$

Then the equations are 3 with 3 unknowns as follows:

$$12v_1 - 4v_2 - 8v_3 = 2$$
$$-12v_1 + 10v_2 + 10v_3 = 0$$
$$v_2 - v_3 = 4$$

Solving these equations is found that

$$v_1 = -1.25V$$
$$v_2 = 1.25V$$
$$v_3 = -2.75V$$

The current in any branch can then be easily found knowing the voltage across it. For example the current in the conductance between nodes 1 and 2 will be

$$4(v_1 - v_2) = 4(-1.25 - 1.25) = -10A$$

In conclusion, the basic steps in applying Nodal Analysis in a circuit are as follows:

1. Arrange the circuit in a simple, neat diagram in which the element values are shown. Conductance values are preferable to resistance values.

2. For a circuit with n nodes select one as the reference node, usually one to which most branches are connected, while the rest of the

nodes take a number from 1 to $n-1$. Voltages $v_1, v_2, \ldots, v_{n-1}$ are assigned to corresponding nodes. These voltages are measured relative to the reference node.

3. Apply Kirchhoff's current law at each node, except for the reference node, to obtain $n-1$ equations the solution of which will give the values of voltages $v_1, v_2, \ldots, v_{n-1}$ w.r.t. the reference node. The current in each branch is determined by the voltages of the nodes to which it is connected and the conductance of the branch.

4. Step 3 is straight forward if the only sources in the circuit are current sources. However, if there are any voltage sources in the circuit an unknown current i_x, i_y etc is assigned for each source while the voltage of each voltage source is related to the voltages $v_1, v_2, \ldots, v_{n-1}$.

A.2 Mesh Analysis

The method of *Mesh Analysis* is a systematic application of KVL in a circuit in order to determine the branch currents. To avoid having a large number of equations equal to the number of branches, the concept of mesh current is introduced. We explain this by means of an example. Consider the circuit in Fig. A.2(a) containing a voltage and a current source as well as resistances. Resistance values are preferable in this case.

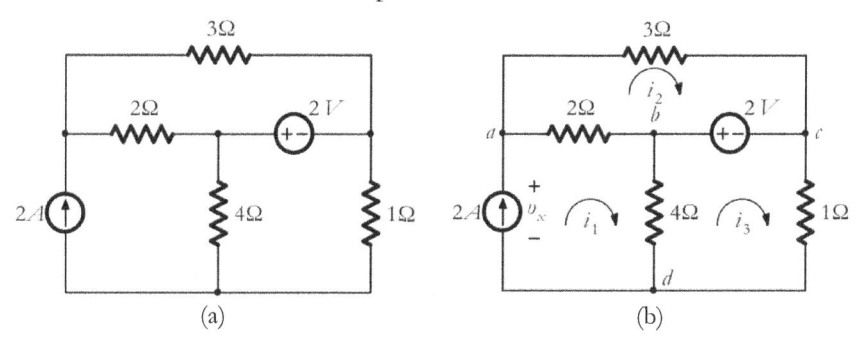

Figure A.2.

The mesh current is defined as the current that flows along the perimeter of the mesh only. By convention we take all mesh currents flowing clockwise, as shown in Fig. A.2(b). Clearly, if we determine all mesh currents, the current in each branch can be easily found. For example the current in the 2Ω resistance in Fig. A.2(b) will be $i_1 - i_2$. Applying KVL in mesh No. 1 (*abda*) we assume that the voltage of the current source is v_x with the polarity shown. Then we will have

$$2(i_1 - i_2) + 4(i_1 - i_3) - v_x = 0 \quad (A.5)$$

For mesh No. 2 (*acba*) we have

$$3i_2 + 2(i_2 - i_1) - 2 = 0 \tag{A.6}$$

For mesh No. 3 (*bcdb*)

$$4(i_3 - i_1) + 1i_3 + 2 = 0 \tag{A.7}$$

There are 3 equations but 4 are required since v_x in Eq. (A.5) is not known. However, in mesh No. 1 i_1 is equal to the current of the current source i.e.

$$i_1 = 2A$$

Then Eqs. (A.6) and (A.7) simplify to

$$5i_2 = 6$$

$$5i_3 = 6$$

giving

$$i_2 = i_3 = 1.2A$$

Therefore the branch currents are as follows:

In the 2Ω resistance $i_1 - i_2 = 0.8A$

In the 3Ω resistance $i_2 = 1.2A$

In the 4Ω resistance $i_1 - i_3 = 0.8A$

In the 1Ω resistance $i_3 = 1.2A$

The node voltages can be determined next and this would be simpler by assuming arbitrarily the voltage at a node, e.g. that of node *d*, to be zero. For example, for $V_d = 0$ then V_b will be

$$V_b = (i_1 - i_3)4 = 0.8 \times 4 = 3.2V$$

In applying mesh analysis the number of meshes chosen should be minimal to avoid excessive numbers of equations.

The circuit analyzed here is planar and the number of meshes required was easy to choose. However, if the circuit is non-planar the number of meshes is not easy to obtain. In such a case it is easier to apply Nodal analysis instead of Mesh Analysis.

Mesh analysis was demonstrated for simplicity for a circuit containing sources and resistors. However, as in the case of nodal analysis,

More on Circuit Analysis 329

the method is applicable also when impedances instead of pure resistances are present in the circuit.

A.3 Linearity, Superposition, Source Transformation

A two terminal circuit element is *linear*, if the voltage across its terminals and the current in it are linearly related. Thus the elements, resistance, capacitance and inductance are linear. A circuit is *linear* if all its elements are linear.

For a linear circuit the principle of *superposition* holds. This states the following:

In any linear circuit, which contains ideal independent sources, the voltage or the current in one of its elements can be determined by the algebraic addition of all the individual voltages or currents in the element, that are due to each independent source, when only this is acting with all the other independent voltage sources having been short-circuited and all the other independent current sources open-circuited.

When the circuit contains non-ideal voltage and/or current sources, in applying the principle of superposition each source is substituted by its internal resistance.

As an example of applying the superposition principle consider the circuit in Fig. A.3 and determine the voltage at node A.

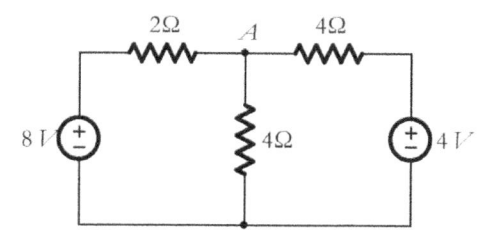

Figure A.3.

a. Short-circuiting the $4V$ voltage source with only the $8V$ source acting gives:

$$V'_A = \frac{2}{2+2} \cdot 8 = 4V$$

b. Short-circuiting the $8V$ voltage source with only the $4V$ source acting gives:

$$V''_A = \frac{\frac{2\times4}{2+4}}{\frac{2\times4}{2+4}+4} \cdot 4 = \frac{\frac{8}{6}}{\frac{8}{6}+4} \cdot 4 = \frac{8\times4}{8+24} = \frac{32}{32} = 1V$$

Therefore the total voltage V_A when both sources are acting is as follows:

$$V_A = V'_A + V''_A = 4 + 1 = 5V$$

The reader should check this result by using any other method of analysis, for example mesh analysis.

Source Transformation

Practical voltage and current sources are not ideal. A voltage source behaves as ideal for very small currents while for higher currents the voltage across its terminals is lower than its voltage V_{oc} on open-circuit. This behaviour is explained, if we consider that the practical voltage source consists of an ideal voltage source V_{oc} in series with a resistance R_{sv}, as shown in Fig. A.4(a), which we call *internal resistance* of the source. Connecting a resistance R_L across its terminals we will have for the current and the voltage in R_L

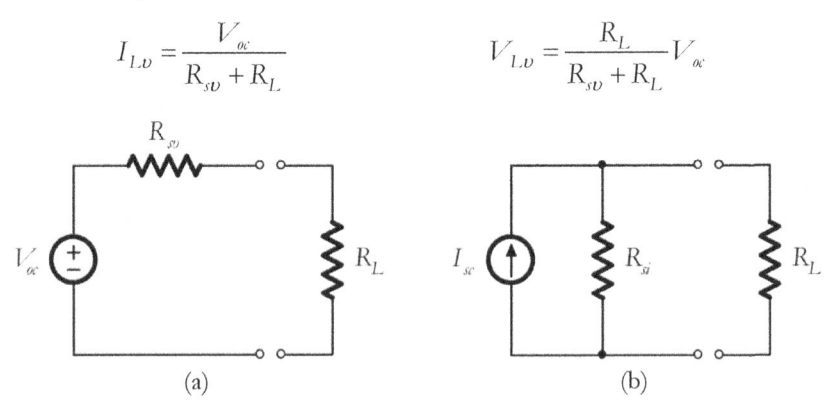

$$I_{Lv} = \frac{V_{oc}}{R_{sv} + R_L} \qquad V_{Lv} = \frac{R_L}{R_{sv} + R_L} V_{oc}$$

Figure A.4. (a) Model of a practical voltage source and (b) model of a practical current source

Clearly, for $R_L >> R_{sv}$ we get

$$I_{Lv} \approx \frac{V_{oc}}{R_L} \qquad \text{and} \qquad V_{Lv} \approx V_{oc}$$

So for $R_L >> R_{sv}$ the practical voltage source behaves as ideal. However, if $R_L << R_{sv}$, I_{Lv} becomes

$$I_{Lv} \approx \frac{V_{oc}}{R_{sv}}$$

Since V_{oc} and R_{sv} are characteristics of the voltage source and independent of R_L as long as $R_L << R_{sv}$ the practical voltage source behaves as an ideal current source.

Similarly a practical current source can be modeled by an ideal current source I_{sc} in parallel with an internal resistance R_{si}. I_{sc} is the short-circuit current in the source. Connecting a resistance R_L across the terminals of the current source the current and the voltage in R_L will be as follows:

$$I_{Li} = \frac{R_{si}}{R_{si} + R_L} I_{sc} \qquad V_{Li} = \frac{R_{si} R_L}{R_{si} + R_L} I_{sc}$$

Then for $R_L << R_{si}$ we get

$$I_{Li} \approx I_{sc} \qquad \text{and} \qquad V_{Li} \approx R_L I_{sc}$$

Thus, for $R_L << R_{si}$ the practical current source behaves as ideal. However, if $R_L >> R_{si}$ V_{Li} becomes

$$V_{Li} \approx R_{si} I_{sc}$$

But R_{si} and I_{sc} are the characteristics of the current source meaning that, when $R_L >> R_{si}$ voltage V_{Li} remains constant independent of R_L as long as $R_L >> R_{si}$. Thus, the practical current source, under the condition $R_L >> R_{si}$, behaves as an ideal voltage source.

We can conclude from the above discussion that a practical voltage or current source under certain conditions can behave either as ideal voltage source or as ideal current source.

Two sources are called equivalent if they supply any load with the same voltage and the same current i.e. when

$$I_L = \frac{V_{oc}}{R_{sv} + R_L} = \frac{R_{si}}{R_{si} + R_L} I_{sc}$$

and

$$V_L = \frac{R_L}{R_{sv} + R_L} V_{oc} = \frac{R_{si} R_L}{R_{si} + R_L} I_{sc}$$

which is the same as the previous one.

For the two sources to be equivalent the following should be valid:

$$R_{sv} = R_{si} = R_s$$

$$V_{oc} = R_s I_{sc}$$

where R_s is the internal resistance of anyone of the two sources.

Thus, the voltage source in Fig. A.5(a) is equivalent to the current source in Fig. A.5(b).

(a) (b)

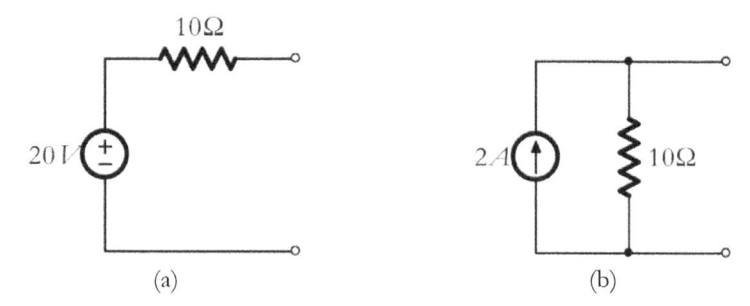

Figure A.5.

Clearly for equivalent practical voltage and current sources the current of the current source is the current of the voltage source on short-circuit and the voltage of the voltage source is the voltage of the current source on open-circuit.

A.4 Thévenin's and Norton's Theorems

Thévenin's Theorem states the following:

"Any linear two-terminal circuit that consists of resistances and practical sources can be substituted by an ideal voltage source V_T in series with a resistance R_T. The voltage V_T of the source is the open-circuit voltage of the circuit, while the resistance R_T is the resistance of the two terminal circuit when all sources in it have been substituted by their internal resistances."

As an example, consider the two terminal circuit inside the broken lines in Fig. A.6(a) and applying Thévenin's Theorem determine the current i_L.

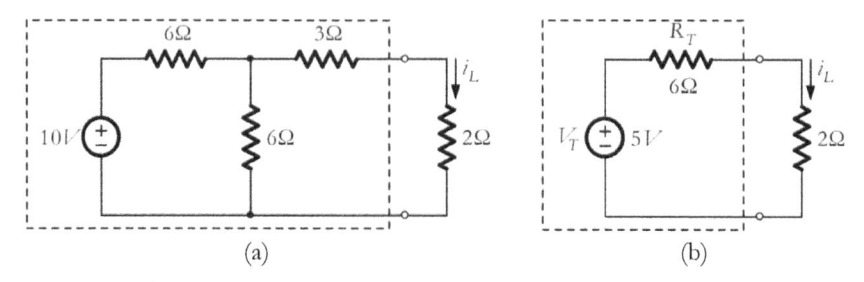

(a) (b)

Figure A.6.

Clearly, the Thévenin voltage V_T with resistance of 2Ω removed is

$$V_T = \frac{6}{6+6} \cdot 10 = 5V$$

and the Thévenin resistance R_T

$$R_T = 3 + 3 = 6\Omega$$

Then

$$i_L = \frac{5}{6+2} = \frac{5}{8} A$$

Although the above statement of Thévenin's Theorem refers to resistive circuits, it is also valid for any linear circuit containing other kinds of elements i.e. inductances and capacitances.

Norton's Theorem

Norton's Theorem states the following:

"Any linear two-terminal circuit that consists of practical sources and resistances can be substituted by an ideal current source I_N in parallel with a resistance R_N. The current I_N of the source is the current in short-circuiting the two terminals and the resistance R_N is the resistance between the two terminals when all sources in the circuit have been substituted by their internal resistances."

According to Norton's Theorem R_N is equal to R_T and using source transformation

$$I_N = \frac{V_T}{R_T} = \frac{V_T}{R_N}$$

Following this the Norton equivalent of the circuit in Fig. A.6(a) is as shown in Fig. A.7. Then the current i_L will be

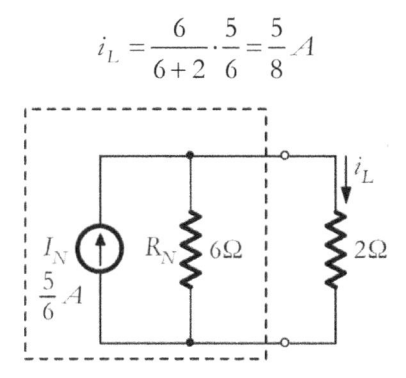

$$i_L = \frac{6}{6+2} \cdot \frac{5}{6} = \frac{5}{8} A$$

Figure A.7.

Again Norton's Theorem also holds for any kind of linear impedances instead of resistances only.

Appendix B

Fourier Transform

B.1 Sinusoidal Excitation

Consider the sinusoidal voltage

$$\upsilon(t) = V_m \cos(\omega t + \varphi) \qquad (B.1)$$

where V_m is the amplitude, $\omega = 2\pi f$ the frequency and $\omega t + \varphi$ the argument of $\upsilon(t)$ with φ being the phase angle. This is a periodic function with period

$$T = \frac{1}{f} = \frac{2\pi}{\omega} \qquad (B.2)$$

given in seconds (s), f in Hertz (Hz) and ω in rad/s.

The response of a circuit consisting of resistors, capacitors and/or inductors can be found by applying KVL and/or KCL, as was done in the simple cases of the differentiator and integrator. The result will be a differential equation, which, however, depending on the numbers of inductors and capacitors present in the circuit can be rather involved.

The study of the response of such a circuit can be greatly simplified by using the concept of *phasor*, which is introduced below.

B.2 Fourier Series

Consider a periodic signal with period T

$$x(t) = x(t \pm nT) \qquad n=0, 1, 2, \dots$$

If $x(t)$ additionally has only a finite number of discontinuities in any finite period and the integral

$$\int_{\alpha}^{\alpha+T} |x(t)| dt$$

is finite, where α is an arbitrary real number, then, $x(t)$ can be expanded into the infinite trigonometric series

$$x(t) = \frac{a_o}{2} + \sum_{n=1}^{\infty} (a_n \cos n\omega t + b_n \sin n\omega t) \tag{B.3}$$

where $\omega = 2\pi/T$. This series is generally referred to as the *Fourier Series*.

The coefficients $a_o, a_n, b_n, n=1,2,\ldots,\infty$ can be determined as follows:

$$a_o = \frac{2}{T} \int_0^T x(t) dt \tag{B.4}$$

$$a_n = \frac{2}{T} \int_0^T x(t) \cos n\omega t\, dt \tag{B.5}$$

$$b_n = \frac{2}{T} \int_0^T x(t) \sin n\omega t\, dt \tag{B.6}$$

B.3 Fourier Transform

Consider the function $f(t)$ for which the integral

$$\int_{-\infty}^{\infty} |f(t)| dt \quad \text{or} \quad \int_{-\infty}^{\infty} |f(t)|^2 dt$$

is finite. Then the *Fourier Transform* of $f(t)$ denoted as $\mathcal{F}[f(t)]$ is defined as follows:

$$\mathcal{F}[f(t)] = F(j\omega) = \int_{-\infty}^{\infty} f(t) e^{-j\omega t} dt \tag{B.7}$$

The *Inverse Fourier Transform* of $F(j\omega)$ denoted as $\mathcal{F}^{-1}[F(j\omega)]$ is defined as:

$$\mathcal{F}^{-1}[F(j\omega)] = f(t) = \frac{1}{2\pi} \int_{-\infty}^{\infty} F(j\omega) e^{j\omega t} d\omega \tag{B.8}$$

In general $F(j\omega)$ is a complex quantity which can be written as

$$F(j\omega) = \operatorname{Re} F(j\omega) + j \operatorname{Im} F(j\omega) \tag{B.9}$$

where Re and Im mean real part and imaginary part respectively. The magnitude $|F(j\omega)|$ gives the *amplitude spectrum* of $F(j\omega)$ as

$$A(\omega) = |F(j\omega)| = \{[\operatorname{Re} F(j\omega)]^2 + [\operatorname{Im} F(j\omega)]^2\}^{1/2} \qquad (B.10)$$

The quantity

$$\varphi(\omega) = \arg F(j\omega) = \tan^{-1} \frac{\operatorname{Im} F(j\omega)}{\operatorname{Re} F(j\omega)} \qquad (B.11)$$

gives the *phase spectrum* of $F(j\omega)$.

These spectra are continuous in contrast to the spectrum of a Fourier Series which is a line spectrum. Thus $F(j\omega)$ can be written in the form

$$F(j\omega) = A(\omega)e^{j\varphi(\omega)} \qquad (B.12)$$

This expression of $F(j\omega)$ is known as the *Phasor*. It should be stressed at this point that not all functions of $j\omega$ are phasors. A phasor is the Fourier Transform of a time domain function.

It is usual in literature to denote a phasor in bold capital letters like **V** or **I** for the phasors of a voltage or a current respectively.

Example B.M.1

Using MATLAB find the Fourier transform of the function

$$f(x) = e^{-x^2}$$

Code

```
>> syms x
>> f=exp(-x^2);
>> F=fourier(f)

F =

pi^(1/2)*exp(-1/4*w^2)
```

that means

$$F(j\omega) = \sqrt{\pi}\, e^{-\frac{\omega^2}{4}}$$

The solution can be confirmed by taking the Inverse Fourier Transform of the result.

```
>> f1=ifourier(F)

f1 =

exp(-x^2)
```

B.4 Fourier Transform and the R, L, C Elements

Consider a resistance R in which voltage $v(t)$ and current $i(t)$ both have Fourier Transforms $V(j\omega)$ and $I(j\omega)$ respectively. Taking the Inverse Fourier Transform gives:

$$v(t) = \frac{1}{2\pi} \int_{-\infty}^{\infty} V(j\omega) e^{j\omega t} d\omega \tag{B.13}$$

$$i(t) = \frac{1}{2\pi} \int_{-\infty}^{\infty} I(j\omega) e^{j\omega t} d\omega \tag{B.14}$$

Substituting in the equation

$$v(t) = Ri(t) \tag{B.15}$$

gives

$$\frac{1}{2\pi} \int_{-\infty}^{\infty} V(j\omega) e^{j\omega t} d\omega = R \frac{1}{2\pi} \int_{-\infty}^{\infty} I(j\omega) e^{j\omega t} d\omega \tag{B.16}$$

Dropping $1/2\pi$, $e^{j\omega t}$ and the integral being common to both sides, gives

$$V(j\omega) = RI(j\omega) \tag{B.17}$$

Thus Ohm's law is valid for the Fourier Transforms of voltage and current in the resistance.

In the case of an inductance we have

$$v_L(t) = L \frac{di(t)}{dt} \tag{B.18}$$

Then

$$\frac{di}{dt} = \frac{1}{2\pi} \int_{-\infty}^{\infty} \frac{d}{dt} I(j\omega) e^{j\omega t} d\omega = \frac{1}{2\pi} \int_{-\infty}^{\infty} j\omega I(j\omega) e^{j\omega t} d\omega \tag{B.19}$$

Substituting v_L from Eq. (B.18) and di/dt from Eq. (B.19) gives

$$\frac{1}{2\pi} \int_{-\infty}^{\infty} V_L(j\omega) e^{j\omega t} d\omega = \frac{1}{2\pi} \int_{-\infty}^{\infty} j\omega L I(j\omega) e^{j\omega t} d\omega \tag{B.20}$$

Therefore

$$V_L(j\omega) = j\omega L I(j\omega) \tag{B.21}$$

Fourier Transform

Since $j\omega L$ is the ratio of voltage over current it is measured in Ω and is called the impedance of the inductance.

Finally consider the case of a capacitance in which

$$i(t) = C\frac{dv_C}{dt} \qquad (B.22)$$

Following the previous procedure we easily get that

$$I(j\omega) = j\omega C V(j\omega) \qquad (B.23)$$

Here $j\omega C$ is measured in Siemens and is called the admittance of the capacitance while $1/j\omega C$ is the impedance of the capacitance. It is clear that the time domain relationships between voltage and current for the elements R, L and C are transformed to simple relationships in the real frequency domain by means of the Fourier Transform. They can then be used to analyze any R, L, C circuit when the excitation is sinusoidal as this is shown below.

B.5 Use of Phasors in the Analysis of a Linear Circuit

Another way of developing the previous results, once the concept of the phasor has been fully understood, is the following: Consider the sinusoidal voltage

$$v(t) = V_m \cos(\omega t + \varphi) \qquad (B.24)$$

which can be written in the following form

$$\begin{aligned}v(t) &= \mathrm{Re}\{V_m[\cos(\omega t + \varphi) + j\sin(\omega t + \varphi)]\} = \\ &= \mathrm{Re}[V_m e^{j(\omega t + \varphi)}] = \mathrm{Re}[V_m e^{j\varphi}(e^{j\omega t})]\end{aligned} \qquad (B.25)$$

Clearly, the term $e^{j\omega t}$ will be common to all sinusoidal quantities, voltages and currents, in the same frequency, if the circuit is linear. On the other hand, the term $V_m e^{j\varphi}$ is characteristic of $v(t)$ since it gives the amplitude V_m and the phase φ. This term

$$\mathbf{V} = V_m e^{j\varphi} = V_m \angle\varphi = V_m(\cos\varphi + j\sin\varphi) \qquad (B.26)$$

is the *phasor* **V** of $v(t)$. It is a complex quantity independent of time. In each sinusoidal quantity of frequency ω there corresponds a phasor. For example the phasor **I** of the current

$$i(t) = I_m \cos(\omega t - \theta) \qquad (B.27)$$

will be

$$\mathbf{I} = I_m e^{-j\theta} = I_m \angle -\theta \tag{B.28}$$

In obtaining the phasor of any sinusoidal quantity of ω one should remember that

$$\sin \omega t = \cos(\omega t - 90°) \tag{B.29}$$

Also that this discussion refers to linear circuits only.

Applying the phasor transformation to equation

$$\upsilon(t) = Ri(t) \tag{B.30}$$

since R is linear, we get

$$V_m \angle \varphi = RI_m \angle \theta \tag{B.31}$$

or

$$\mathbf{V} = R\mathbf{I} \tag{B.32}$$

with $\varphi = \theta$.

The equation

$$\upsilon(t) = L\frac{di}{dt} \tag{B.33}$$

is written as

$$V_m e^{j(\omega t + \varphi)} = jL\omega I_m e^{j(\omega t + \theta)} \tag{B.34}$$

or

$$\mathbf{V} = j\omega L\mathbf{I} \tag{B.35}$$

The presence of j in Eq. (B.35) means that $\mathbf{I}$ lags $\mathbf{V}$ by 90°, since

$$j = e^{j\frac{\pi}{2}} = \cos\frac{\pi}{2} + j\sin\frac{\pi}{2} \tag{B.36}$$

Similarly it is found that the relationship

$$i(t) = C\frac{d\upsilon}{dt} \tag{B.37}$$

is transformed to

$$\mathbf{I} = j\omega C\mathbf{V} \tag{B.38}$$

with $\mathbf{V}$ now lagging $\mathbf{I}$ by 90°.

From Eq. (B.35) we get

Fourier Transform

$$j\omega L = \frac{V}{I} \tag{B.39}$$

which is the impedance of L. On the other hand from Eq. (B.38) we get

$$j\omega C = \frac{I}{V}$$

which is the admittance of C with $1/j\omega C$ being its impedance.

Kirchhoff's laws are equally valid when using phasors. As an example consider the circuit in Fig. B.1(a)

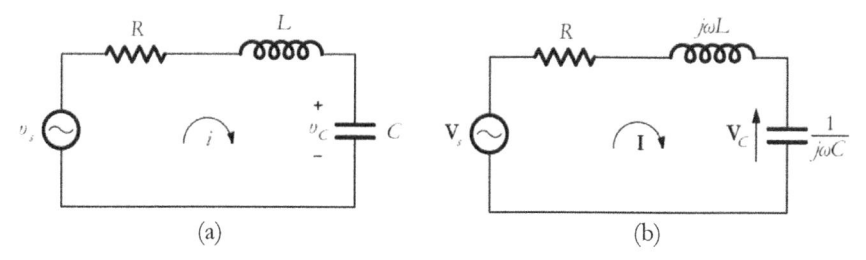

(a) (b)

Figure B.1.

where $v_s = V_m \cos\omega t$. Using phasors this circuit is transformed to that in Fig. B.1(b). Applying KVL we get

$$IR + j\omega LI + \frac{1}{j\omega C}I = V_s \tag{B.40}$$

Solving for **I**

$$I = \frac{V_s}{R + j\omega L + \dfrac{1}{j\omega C}} = \frac{j\omega C V_s}{1 - \omega^2 LC + j\omega CR} \tag{B.41}$$

We can also find that

$$V_C = \frac{I}{j\omega C} = \frac{V_s}{1 - \omega^2 LC + j\omega CR} \tag{B.42}$$

Let

$$I = I_m \angle \varphi \qquad V_C = V_{Cm} \angle \theta$$

Then from Eq. (B.40)

$$I_m = \frac{\omega C V_m}{\sqrt{(1 - \omega^2 LC)^2 + \omega^2 C^2 R^2}} \tag{B.43}$$

and

$$\varphi = -\tan^{-1}\frac{\omega CR}{1-\omega^2 LC} + 90° \quad (B.44)$$

On the other hand from Eq. (B.41) we have

$$V_{Cm} = \frac{V_m}{\sqrt{(1-\omega^2 LC)^2 + \omega^2 C^2 R^2}} \quad (B.45)$$

and

$$\theta = -\tan^{-1}\frac{\omega CR}{1-\omega^2 LC} \quad (B.46)$$

Eq. (B.40) can be written as follows:

$$\frac{\mathbf{V}_s}{\mathbf{I}} = R + j\omega L + \frac{1}{j\omega C} = Z(j\omega) \quad (B.47)$$

Obviously, the quantity $Z(j\omega)$ is the total series impedance in the circuit. Although, it is a function of $j\omega$ it is not a phasor for the reason that **V** and **I** are, as was explained above in Section B.3.

References

[1]. Hayt W. H. Jr & Kemmerly J. E., *Engineering Circuit Analysis*, McGraw-Hill, New York, 1971.
[2]. Kuo F. F., *Network Analysis and Synthesis*, Wiley, New York, 1962.

Appendix C

Laplace Transform

C.1 Complex Excitation

Consider an excitation function of the form

$$f(t) = Ke^{st} \tag{C.1}$$

where K and s are complex constants, independent of time. We say that such a function is characterized by the complex frequency $s = \sigma + j\omega$.

It can be easily shown that any other type of excitation can be represented by Eq. (C.1). Thus for

$$\begin{aligned}
\upsilon(t) &= V_0 & K &= V_0 & \sigma &= 0, & \omega &= 0 \\
\upsilon(t) &= V_0 e^{\sigma t} & K &= V_0 & \sigma &\neq 0, & \omega &= 0 \\
\upsilon(t) &= V_0 \cos(\omega t + \theta) & K &= V_0 & \sigma &= 0, & \omega &\neq 0 \\
\upsilon(t) &= V_0 e^{st} & K &= V_0 & \sigma &\neq 0, & \omega &\neq 0
\end{aligned}$$

Following the procedure presented in Appendix B for the sinusoidal excitation, it can be shown that for linear components, when complex excitation is applied, Eqs. (B.11), (B.14) and (B.16) become

$$\mathbf{V} = R\mathbf{I} \tag{C.2}$$

$$\mathbf{V} = sL\mathbf{I} \tag{C.3}$$

$$\mathbf{V} = \frac{1}{sC}\mathbf{I} \tag{C.4}$$

The quantities $\mathbf{V}$ and $\mathbf{I}$ are again called phasors. Accordingly, sL and $1/sC$ are the impedance of L and of C respectively. The inverse of these, i.e. $1/sL$ and sC are the admittance of L and the admittance of C respectively.

Thus the circuit in Fig. B.1(a), for the complex excitation V_s is transformed to that in Fig. C.1.

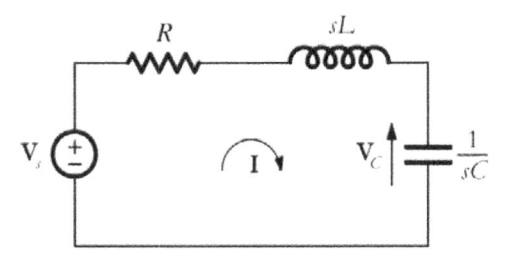

Figure C.1.

With Kirchhoff's laws being valid we can write

$$\mathbf{V}_s = R\mathbf{I} + sL\mathbf{I} + \frac{1}{sC}\mathbf{I} \tag{C.5}$$

Then from Eq. (C.5) we get

$$\mathbf{I} = \frac{\mathbf{V}_s}{R + sL + \dfrac{1}{sC}} = \frac{sC\mathbf{V}_s}{1 + sCR + s^2 LC} \tag{C.6}$$

and

$$\mathbf{V}_C = \frac{\dfrac{1}{sC}}{R + sL + \dfrac{1}{sC}} \mathbf{V}_s \tag{C.7}$$

or

$$\mathbf{V}_C = \frac{\mathbf{V}_s}{1 + sCR + s^2 LC} \tag{C.8}$$

Comparing Eqs. (C.6), (C.8) to (B.19) and (B.20) it follows that the latter are obtained from the former by setting $s = j\omega$.

Therefore, using the concept of complex frequency, Kirchhoff's equations can be written as algebraic equations.

Coming back to Fig. B.1(a), we can write in the time domain Kirchhoff's voltage law as follows:

$$\upsilon_s = Ri + L\frac{di}{dt} + \frac{1}{C}\int_0^t i\, dt + \upsilon_C(0) \tag{C.9}$$

where $\upsilon_C(0)$ is the voltage of the capacitor at $t = 0$. Letting $\upsilon_C(0) = 0$, Eq. (C.9) is written as

$$v_s = Ri + L\frac{di}{dt} + \frac{1}{C}\int_0^t i\,dt \qquad (C.10)$$

It can be seen that if we set

$$s = \frac{d}{dt} \qquad (C.11)$$

$$\frac{1}{s} = \int dt \qquad (C.12)$$

we can obtain Eq. (C.5) straight from Eq. (C.10) provided that the quantities v_s and i have been written as phasors $\mathbf{V}_s$ and $\mathbf{I}$. On the other hand, Eq. (C.10) can be obtained from Eq. (C.5) using Eqs. (C.11) and (C.12) and writing $\mathbf{V}_s$ and $\mathbf{I}$ in their time domain symbols. This way a circuit can be analyzed in the complex frequency domain, to avoid writing down differential equations, just using algebraic equations, which is much simpler particularly so, if the numbers of inductors and capacitors are relatively high.

C.2   Complex Frequency and Laplace Transform

Assume that v_s is an arbitrary function of time. Eq. (C.10) can be easily solved using the *Laplace Transform*, which is defined as follows:

$$\mathcal{L}[v_s] = V_s(s) = \int_0^\infty v_s e^{-st}\,dt \qquad (C.13)$$

under the condition that

$$\int_0^\infty |v_s(t)| e^{-\sigma t}\,dt < \infty \qquad (C.13a)$$

With zero initial conditions $i_L(0)$ and $v_C(0)$, taking the Laplace Transform of both sides of Eq. (C.10) we get

$$V_s(s) = RI(s) + LsI(s) + \frac{I(s)}{sC} \qquad (C.14)$$

Therefore the unknown current $I(s)$ will be

$$I(s) = \frac{V_s(s)}{R + sL + \dfrac{1}{sC}} \qquad (C.15)$$

or

$$I(s) = \frac{sCV_s(s)}{s^2 LC + sCR + 1} \qquad (C.16)$$

which is the same as Eq. (C.6).

Table C.1. Laplace Transform pairs

Laplace Transform $F(s)$	Function of time $f(t)$ ($t>0$)	
1	$\delta(t)$	unit impulse function
$1/s$	$u(t)$	unit step function
$1/s^2$	t	unit ramp function
$1/s^{n+1}$	$t^n/n!$	n: positive integer
$1/(s+a)$	e^{-at}	
$1/(s+a)^2$	te^{-at}	
$\dfrac{1}{(s+a)(s+b)}$	$\dfrac{e^{-at}-e^{-bt}}{b-a}$	
$\dfrac{1}{(s+a)^n}$	$\dfrac{1}{(n-1)!}t^{n-1}e^{-at}$	n: positive integer
$\dfrac{1}{s(s+a)}$	$\dfrac{1}{a}\left(1-e^{-at}\right)$	
$\dfrac{1}{s(s+a)(s+b)}$	$\dfrac{1}{ab}\left(1-\dfrac{b}{b-a}e^{-at}+\dfrac{a}{b-a}e^{-bt}\right)$	$a \neq b$
$\dfrac{1}{s(s+a)^2}$	$\dfrac{1}{a^2}\left[1-(1+at)e^{-at}\right]$	
$\dfrac{\omega}{s^2+\omega^2}$	$\sin \omega t$	
$\dfrac{s}{s^2+\omega^2}$	$\cos \omega t$	
$\dfrac{\omega_n^2}{s^2+2\zeta\omega_n s+\omega_n^2}$	$\dfrac{\omega_n^2}{\sqrt{1-\zeta^2}}e^{-\zeta\omega_n t}\sin \omega_n\sqrt{1-\zeta^2}\,t$	
$\dfrac{\omega_n^2}{s\left(s^2+2\zeta\omega_n s+\omega_n^2\right)}$	$1+\dfrac{1}{\sqrt{1-\zeta^2}}e^{-\zeta\omega_n t}\sin\left(\omega_n\sqrt{1-\zeta^2}\,t-\phi\right)$ where $\phi=\tan^{-1}\dfrac{\sqrt{1-\zeta^2}}{-\zeta}$	
$\dfrac{\omega_n^2}{s\left(s^2+\omega_n^2\right)}$	$1-\cos \omega_n t$	
$\dfrac{s}{\left(s^2+\omega_n^2\right)^2}$	$\dfrac{1}{2\omega_n}t\sin \omega_n t$	
$\dfrac{1}{s^2(s+a)}$	$\dfrac{1}{a^2}\left(at-1+e^{-at}\right)$	

Laplace Transform

To get $i(t)$ from Eq. (C.16) and, provided that $V_s(s)$ can be found, we take the *Inverse Laplace Transform*, which is defined as follows:

$$\mathcal{L}^{-1}[I(s)] = i(t) = \frac{1}{2\pi j} \int_{C-j\infty}^{C+j\infty} I(s) e^{st} ds \qquad (C.17)$$

where C is a real positive quantity, which is greater than the σ convergence factor in Eq. (C.13a).

In practice it is not usually necessary to try to find the above integral, because the second part of Eq. (C.16) can be matched using the table of Laplace Transform pairs, given as Table C.1. In case of higher order functions the roots of the denominator polynomial are determined these being called the *poles* of the function. Then the function is expanded in partial fractions, which are of low orders and can be matched with one of the Laplace Transform pairs of table C.1.

Table C.2. Element Volt-Ampere relationship with (a) time-varying excitation, (b) exponential excitation and (c) sinusoidal excitation

Element	Time-varying excitation $\upsilon = \upsilon(t)$ $i = i(t)$	Exponential excitation $\upsilon = Ve^{st}$ $i = Ie^{st}$	Sinusoidal excitation $\upsilon = V_m \cos(\omega t + \theta)$ $i = I_m \cos(\omega t + \alpha)$
Resistance (conductance)	$\upsilon = Ri$, $R = \frac{1}{G}$, $i = G\upsilon$	$V = RI$, $R = \frac{1}{G}$, $I = GV$	$V = RI$, $R = \frac{1}{G}$, $I = GV$
Inductance	$\upsilon = L\frac{di}{dt}$, $i = \frac{1}{L}\int \upsilon dt$	$V = sLI$, $I = \frac{1}{sL}V$	$V = j\omega LI = jX_L I$, $I = \frac{1}{j\omega L}V = jB_L V$
Capacitance	$\upsilon = \frac{1}{C}\int i\, dt$, $i = C\frac{d\upsilon}{dt}$	$V = \frac{1}{sC}I$, $I = sCV$	$V = \frac{1}{j\omega C}I = jX_C I$, $I = j\omega CV = jB_C V$
Passive network	differential equation	$Y(s) = \frac{1}{Z(s)}$, $V = ZI$, $I = YV$	$Y(s) = \frac{1}{Z(s)}$, $V = ZI = (R + jX)I$, $I = YV = (G + jB)V$

As an example consider that $\upsilon_s(t)$ is the unit step function, whose Laplace Transform taken from Table C.1 is $1/s$. Additionally let

$$I(s) = \frac{2V_s(s)}{(s+1)(s+2)}$$

Substituting for $V_s(s)$ gives

$$I(s) = \frac{2}{s(s+1)(s+2)}$$

This function can be split into partial functions as follows:

$$I(s) = \frac{1}{s} - \frac{2}{s+1} + \frac{1}{s+2}$$

Then using Table C.1 it is found that

$$i(t) = 1 - 2e^{-t} + e^{-2t}$$

Table C.2 summarizes the behaviour of the passive elements R, L and C for the three types of excitation namely, a. time varying, b. exponential and c. sinusoidal excitation.

Example C.M.1

Using MATLAB find the time domain function $f(t)$ corresponding to

$$F(s) = \frac{s^2 + 8s + 12}{s^3 + 4s^2 + 3s}$$

Code

$F(s)$ can be split into partial functions as follows:

```
>> num=[1 8 12];          % coefficients of nominator polynomial
>> den=[1 4 3 0];         % coefficients of denominator polynomial
>> [r,p,k]=residue(num,den)   % see Appendix D

r =                       p =

   -0.5000                   -3              k =
   -2.5000                   -1
    4.0000                    0              []
```

which means

$$F(s) = \frac{-0.5}{s+3} + \frac{-2.5}{s+1} + \frac{4}{s}$$

Then using Table C.1 it is found that

$$f(t) = 4 - 2.5e^{-t} - 0.5e^{-3t}$$

Another approach could be the following:

```
>> syms s
>> F=(s^2+8*s+12)/(s^3+4*s^2+3*s);
>> f=ilaplace(F)                        % inverse Laplace Transform of F(s)

f =

4+exp(-2*t)*(-3*cosh(t)-2*sinh(t))
```

which means

$$f(t) = 4 - e^{-2t}\left[3\cosh(t) + 2\sinh(t)\right]$$

and can be written as

$$f(t) = 4 - e^{-2t}\left[3 \cdot \frac{1}{2}\left(e^{t} + e^{-t}\right) + 2 \cdot \frac{1}{2}\left(e^{t} - e^{-t}\right)\right] = 4 - 2.5e^{-t} - 0.5e^{-3t}$$

All the above can be confirmed by taking again the Laplace Transform of the result:

```
>> syms t
>> f=4-2.5*exp(-t)-0.5*exp(-3*t);
>> F=laplace(f)                         % Laplace Transform of f(t)

F =

4/s-5/2/(1+s)-1/2/(s+3)
```

that is

$$F(s) = \frac{-0.5}{s+3} + \frac{-2.5}{s+1} + \frac{4}{s}$$

Appendix D

MATLAB and Simulink Tutorial

D.1 Computer Aided Analysis

Knowledge of the behavior of electronic circuits requires the simultaneous solution of a number of equations. The easiest problem is that of finding the *dc* operating point of a linear circuit, which requires one to solve a set of equations derived from Kirchhoff's law and the equations from Ohm's law for the branches. The analysis of circuits that contain nonlinear elements adds another level of complexity. Only small circuits can be solved by hand calculations, which yield only approximate results. Another level of complexity is added when one has to predict the behavior in time or frequency of an electrical circuit. The nonlinear equations become differential equations, which can be solved by hand only under such approximations as small-signal approximation or other limiting restrictions.

Computer-Aided Design (CAD) is a process, which uses a computer system to assist in the creation, modification, verification, and display of a design. Computer aided circuit analysis programs can provide information about circuit performance that is almost impossible to obtain with laboratory prototype measurements. Such programs help the user to simulate and analyze the circuit design graphically on the computer without having to build a physical circuit. *CAD* tools provide a simple, cost-effective means of confirming the intended operation prior to circuit construction and of verifying new ideas that could lead to improved circuit performance. Designers can quickly visualize the discrepancies between the simulation results and the measurements from their prototypes and can identify problem areas or tune performance during the development process.

In the early 1960's military requirements led to the development of mathematical simulation of components (capacitors, semiconductors, etc.)

to determine their response to pulsed x-ray and gamma radiation. These simulation studies were subsequently extended to small circuits to study their response to the same radiation conditions. This work resulted in the early circuit analysis programs (*ECAP* – Electronic Circuit Analysis Program, *SCEPTRE* – System for Circuit Evaluation and Prediction of Transient Radiation, etc). Later program capabilities included *ac*, *dc* and transient performance simulation with and without radiation effects. RF and microwave circuit simulation capabilities, sensitivity analysis, noise analysis, Monte Carlo, worst-case analysis, Fourier analysis and optimization analysis capabilities were also eventually added. They became more user friendly and included high resolution graphics. These circuit analysis and simulation tools eventually became available for the ubiquitous *PC*s.

Early simulations were run overnight on large mainframe computers while it was cumbersome to input the data and the graphics outputs were poor. These simulation programs quickly moved to engineering workstations and their capabilities were significantly enhanced by such features as simulation integration and schematic capture.

Schematic capture and *circuit simulation* are two essential tools in computer aided circuit analysis and design. Schematic capture is an electronic means of graphically constructing a circuit in a format that can be subsequently read by some software for circuit simulation and circuit board layout. Modern packages allow the user to go directly from the schematic to the circuit simulation. Device model libraries are usually supplied with the simulation program. More extensive integrated circuit libraries are also frequently available when needed from the manufacturers themselves. Semi-automated processes for creating model libraries may also be included.

The most widely known and used circuit simulation program is *SPICE* (**S**imulation **P**rogram with **I**ntegrated **C**ircuit **E**mphasis). *SPICE* was originally an analogue circuit simulator software package that was developed by the Electronics Research Laboratory at the University of California at Berkeley in the early 1970's. Several commercial versions of *SPICE* have since been produced that include a wide variety of useful extensions of the original program. Commercially supported versions of *SPICE* are divided into two types: mainframe versions and *PC*-based versions. Generally, mainframe versions of *SPICE* are intended to be used by sophisticated integrated-circuit designers, who require large amounts of computer power to simulate complex circuits. *SPICE* is de facto university and industry standard for analog circuit simulation, and it is indispensable for simulating medium-scale and very large-scale integrated circuits.

MATLAB (**MAT**rix **LAB**oratory) is numeric computation software for engineering and scientific calculations. MATLAB is being used to teach circuit theory, filter design, random processes, control systems and communication theory. MATLAB matrix functions are shown to be versatile in doing analysis of data obtained from electronics experiments. The graphical features of MATLAB are especially useful for display of frequency response of amplifiers and illustrating the principles and concepts of semiconductor physics. The interactive programming and versatile graphics of MATLAB is especially effective in exploring some of the characteristics of devices and electronic circuits. MATLAB has a large collection of toolboxes for a variety of applications. A toolbox consists of functions that can be used to perform some computations in the toolbox domain. Such a toolbox is Simulink.

Simulink is a graphical extension of MATLAB for the modelling and simulation of systems. In Simulink, systems are drawn on screen as block diagrams. Many elements of block diagrams are available (such as transfer functions, summing junctions, etc.), as well as virtual input devices (such as function generators) and output devices (such as oscilloscopes). Simulink is integrated with MATLAB and data can be easily transferred between the programs.

D.2 Basics of MATLAB

In this section some basics of MATLAB, concerning the solution of the examples and problems of this book, are briefly introduced. There are many books in the literature that the interested reader may consult for more details such as [8].

The MATLAB statements are entered at the prompt on the command window (Fig. D.1). Typing a statement and pressing Enter MATLAB automatically displays the results on screen. Ending the line with a semicolon, MATLAB performs the computation but does not display any output. This is very useful in the case of large matrices.

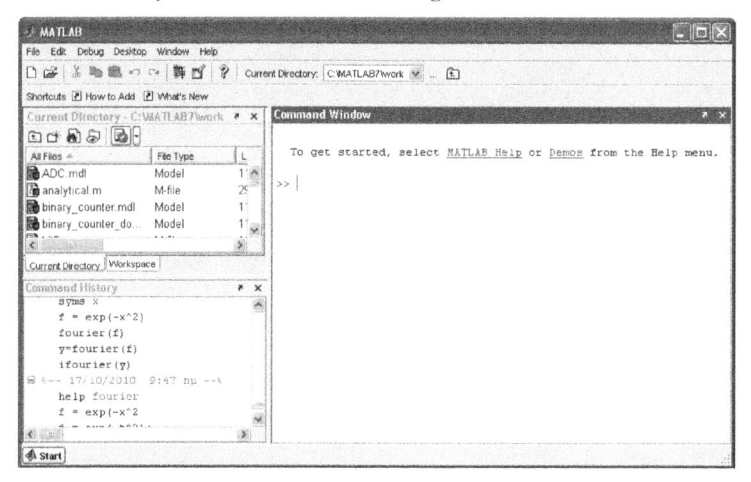

Figure D.1.

Comments can be put within a line by typing % followed by the comment text. MATLAB treats all the information after the % on that line as a comment and ignores it. A help text for a function appears in the command window when typing *help* followed by the function name. Some useful commands are *clc* for clearing the command window, *clear* for clearing variables and functions from memory and *whos* for listing the current variables.

D.2.1 Matrices and Arrays

Matrices can be entered into MATLAB entering an explicit list of elements in square brackets. The elements of a row must be separated with blanks while a semicolon indicates the end of each row. For example

>> A=[12 -4 -8; -12 10 9; 0 1 -1]

and MATLAB displays

A =

```
    12    -4    -8
   -12    10     9
     0     1    -1
```

The sum of each column of the above matrix can be found by using *sum* and then MATLAB returns a row vector containing the sums of the columns:

>> sum(A)

ans =

```
     0     7     0
```

The inverse of a matrix can be found using *inv* as in the example:

>> inv(A)

ans =

```
   0.2262    0.1429   -0.5238
   0.1429    0.1429    0.1429
   0.1429    0.1429   -0.8571
```

Very useful is the function *zeros* for generating matrix of zeros for initialization reasons. For example

>> zeros(3,2)

ans =

```
     0     0
     0     0
     0     0
```

The element in row i and column j of a matrix is denoted as $A(i,j)$. For example, the element of the second row and third column of the above matrix A is

```
>> A(2,3)
ans =

    9
```

Using the colon operator one can obtain portions of a matrix. For example, the following expression gives all the elements of the first row of the above matrix A.

```
>> A(1,:)
ans =

    12    -4    -8
```

Another way for creating arrays without typing all the elements is to use the colon operator. The following expression creates a row vector containing the integers from 10 to 20

```
>> B=10:20

B =

    10   11   12   13   14   15   16   17   18   19   20
```

while the following expression creates a row vector containing the numbers from 10 to 20 with increment 4.

```
>> C=10:4:20

C =

    10   14   18
```

The last expression is very useful in defining the time or frequency range in the examples and problems of this book.

Attention must be paid to the variable names, because MATLAB is case sensitive; it distinguishes between uppercase and lowercase letters. So the variable R_1 is not the same with the variable r_1.

MATLAB uses the familiar arithmetic operators: + for addition, - for subtraction, * for multiplication, / for division and ^ for power.

Attention must be paid to arithmetic operations on arrays, which are executed element by element. In this case a dot must be used before the multiplication, division or power operator. In the following example the two arrays are multiplied element by element.

```
>> a=[1 2 3];
```

```
>> b=[2 2 4];
>> c=a.*b
ans =

     2     4    12
```

MATLAB provides a large number of mathematical functions such as trigonometric functions (*sin* for sine, *asin* for inverse sine, *cos* for cosine, *acos* for inverse cosine, *tan* for tangent, *atan* for inverse tangent, etc), exponential functions (*exp* for exponential, *log* for natural logarithm, *log10* for base 10 logarithm, *sqrt* for square root, etc) and complex functions (*abs* for absolute value, *angle* for phase angle, etc). Special functions provide values of useful constants such as *pi* for 3.14. Some examples are the following:

```
>> sin(pi/6)
ans =

    0.5000
>> log10(100)
ans =

     2
>> sqrt(2)
ans =

    1.4142
```

D.2.2 Graphics

The function *plot(x,y)* produces a graph of vector *y* versus vector *x* while *xlabel* and *ylabel* label the axes and *title* adds a title on the top of the graph. Multiple x-y pair arguments may be plotted on the same graph with $plot(x_1,y_1,x_2,y_2,...)$. In this case MATLAB uses different colors for each graph. Color, line style and markers may be specified using *plot(x,y,'...')* adding a suitable string. Typing *help plot* one can find more details. Text can be inserted anywhere in the figure by typing *text* and defining its location in axes units.

The command *hold on* enables adding plots to an existing graph. Graphing functions automatically open a new figure window if there are no other windows already on the screen. If a figure window already exists, MATLAB uses it for graphics output. By typing *figure* a new figure window opens for the graph to be plotted. By typing *clf* (clear figure) the figure window clears from an existing graph.

The command *subplot(m,n,p)* partitions the figure window in an $m \times n$ matrix of small subplots and selects the *p*th subplot for the current plot. In this way multiple plots can be displayed in the same window.

MATLAB sets the axis limits to a graph automatically. One can specify his own limits by typing *axis([xmin xmax ymin ymax])*. The command *grid on* sets grid lines to the graph.

All the above could be showed with the following example:

```
>> x=0:0.001:2*pi;
>> y1=sin(x);
>> y2=cos(x);
>> subplot(2,1,1)        % partitions the figure window in 2×1 subplots
                         % and selects the first one
>> plot(x,y1)
>> axis([0 2*pi -1.5 1.5])
>> grid on
>> xlabel('x')
>> ylabel('y')
>> title('y=sinx')
>> text(5.5,0.5,'sinx')
>> subplot(2,1,2)        % selects the second subplot of the partitioned
                         % figure window
>> plot(x,y2)
>> axis([0 2*pi -1.5 1.5]);grid on;xlabel('x');ylabel('y');title('y=cosx');
text(5.5,0.5,'cosx');
```

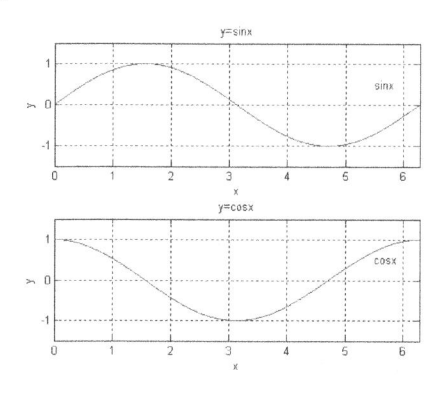

Figure D.2.

Very useful are the functions *semilogx* and *semilogy* for semilogarithmic plots. *Semilogx* creates a plot using a base 10 logarithmic scale for the *x* axis and a linear scale for the *y* axis.

D.2.3 Programming

A logical expression can be evaluated with the *if* statement which executes a group of statements when the expression is true. The optional *elseif* and *else* execute alternate groups of statements. The group of statements is terminated with an *end* keyword, which matches the *if*. For example, the following expression

$$y = \begin{cases} x^2 - 2 & \text{if} \quad x < 2 \\ 3x & \text{if} \quad 2 \leq x \leq 8 \\ 5 & \text{if} \quad x > 8 \end{cases}$$

is given in MATLAB as

\>> if x<2

y=x^2-2;

elseif x>=2 & x<=8

y=3*x;

else

y=5;

end

The *for* loop repeats a group of statements for a predetermined number of times. The group of statements is terminated with an *end* keyword. For example, the following example calculates the square roots of all the integers between 1 and 10:

\>> for i=1:10

y(i)=sqrt(i)

end

y =

   1.0000   1.4142   1.7321   2.0000   2.2361   2.4495   2.6458   2.8284   3.0000   3.1623

The user can create his own functions by creating *m-files* using a text editor. They can accept input arguments and return output arguments. The name of the m-file should be the same with that of the function. The first line of the m-file starts with the keyword *function* followed by the function name and order of arguments. The rest of the file is the executable

MATLAB code defining the function and is terminated with an *end* keyword. For example, the following m-file under the name add.m adds two numbers:

function w=add(x,y)

w=x+y;

end

and is called in command window as

\>\> add(2,3)
```
ans =

    5
```

D.2.4 Useful Functions

Below in Table D.1 we present briefly all the functions that have been used in the examples and problems of this book.

Table D.1. Useful functions

Function	Description
base2dec('strn',base)	Conversion of the string number *strn* of the specified *base* into its decimal (base10) equivalent
bin2dec('strn')	Conversion of binary to decimal number
[mag,phase]=bode(num,den,w)	Bode frequency response, magnitude and phase (in degrees) at frequencies *w* (in rad/sec)
[z,p,k]=buttap(n)	Zeros, poles and gain of an order *n* Butterworth analog low-pass filter prototype
[num,den]=butter(n,Wn,'s')	Order *n* low-pass analog Butterworth filter with angular cutoff frequency *Wn rad/sec*
[num,den]=butter(n,Wn,'high','s')	Order *n* high-pass analog Butterworth filter with angular cutoff frequency *Wn rad/sec*
[num,den]=butter(n,[w1 w2],'s')	Band-pass analog Butterworth filter with passband $w1<\omega<w2$
[num,den]=butter(n,[w1 w2],'stop','s')	Band-stop analog Butterworth filter with stopband $w1<\omega<w2$
[z,p,k]=cheb1ap(n,Rp)	Zeros, poles and gain of an order *n* Chebyshev analog low-pass filter prototype with *Rp dB* of ripple in the passband
[z,p,k]=cheb2ap(n,Rs)	Zeros, poles and gain of an order *n* Inverse Chebyshev analog low pass filter prototype with stopband ripple *Rs dB* down from the passband peak value

$[num,den]=cheby1(n,Rp,Wn,'s')$	Order n low pass analog Chebyshev filter. The magnitude response of the filter is $-Rp$ dB at frequency Wn rad/sec
$[num,den]=cheby1(n,Rp,Wn,'high','s')$	Order n high pass analog Chebyshev filter. The magnitude response of the filter is $-Rp$ dB at frequency Wn rad/sec
$[num,den]=cheby1(n,Rp,[w1\ w2],'s')$	Band pass analog Chebyshev filter with passband $w1<\omega<w2$
$[num,den]=cheby1(n,Rp,[w1\ w2],'stop','s')$	Band stop analog Chebyshev filter with stopband $w1<\omega<w2$
$[num,den]=cheby2(n,Rs,Wn,'s')$	Order n low pass analog Inverse Chebyshev filter. The magnitude response of the filter is $-Rs$ dB at frequency Wn rad/sec
$[num,den]=cheby2(n,Rs,Wn,'high','s')$	Order n high pass analog Inverse Chebyshev filter. The magnitude response of the filter is $-Rs$ dB at frequency Wn rad/sec
$[num,den]=cheby2(n,Rs,[w1\ w2],'s')$	Band pass analog Inverse Chebyshev filter with passband $w1<\omega<w2$
$[num,den]=cheby2(n,Rs,[w1\ w2],'stop','s')$	Band stop analog Inverse Chebyshev filter with stopband $w1<\omega<w2$
$dec2base(d,base)$	Conversion of the nonnegative integer d to the specified *base*
$dec2bin(d)$	Binary representation of d
$dec2hex(d)$	Conversion of decimal to hexadecimal number
$[num,den]=ellip(n,Rp,Rs,Wn,'s')$	Order n low pass analog Elliptic filter. The magnitude response of the filter is $-Rp$ dB at frequency Wn rad/sec
$[num,den]=ellip(n,Rp,Rs,Wn,'high','s')$	Order n high pass analog Elliptic filter. The magnitude response of the filter is $-Rp$ dB at frequency Wn rad/sec
$[num,den]=ellip(n,Rp,Rs,[w1\ w2],'s')$	Band pass analog Elliptic filter with passband $w1<\omega<w2$
$[num,den]=ellip(n,Rp,Rs,[w1\ w2],'stop','s')$	Band stop analog Elliptic filter with stopband $w1<\omega<w2$
$[z,p,k]=ellipap(n,Rp,Rs)$	Zeros, poles and gain of an order n Elliptic analog low pass filter prototype with Rp dB of ripple in the passband and a stopband Rs dB down from the passband peak value
$fourier(f)$	Fourier transform of the symbolic scalar f
$freqs(num,den,w)$	Frequency response of analog filter specified by coefficient vectors *num* and *den* at the angular frequencies in *rad/sec* specified in real vector w

hex2dec('hex_value')	Conversion of hexadecimal to decimal number
ifourier(F)	Inverse Fourier transform of the scalar symbolic object F
ilaplace(F)	Inverse Laplace transform of the scalar symbolic object F
laplace(f)	Laplace transform of the scalar symbol f
length(X)	Length of vector X
logspace(a,b,n)	Logarithmically spaced vector, n points between decades 10^a and 10^b
[numt,dent]=lp2hp(num,den,Wo)	Transformation of analog low pass filter prototype into high pass filter with cutoff angular frequency Wo
[numt,dent]=lp2lp(num,den,Wo)	Transformation of analog low pass filter prototype into low pass filter with cutoff angular frequency Wo
max(A)	Largest element in vector A
[t,Y]=ode23('odefun',[t0 tf],y0)	Integration of a system of ordinary differential equations (ODEs) described by *odefun* from time $t0$ to tf with initial conditions $y0$ (see the example below)
quad('fun',a,b)	Numerical method for computing the definite integral of function *fun* from a to b
[r,p,k]=residue(num,den)	Residues, poles and direct term of partial fraction expansion of the ratio of two polynomials *num*, *den*
syms x	Creation of symbolic variable with the name x
tf(num,den)	Creation of a continuous-time transfer function with numerator *num* and denominator *den*
[num,den]=zp2tf(z,p,k)	Conversion of a factored transfer function (zero-pole-gain) to rational transfer function

Very useful is the function *ode23* for solving a set of ordinary differential equations. For example consider the following set of equations that describes the famous Lorenz oscillator, which is a chaotic oscillator.

$$\frac{dx}{dt} = 10(y - x)$$
$$\frac{dy}{dt} = -xz + 28x - y$$
$$\frac{dz}{dt} = xy - \frac{8}{3}z$$

The set of equations is described in lorenz.m m-file below and is called in MATLAB via *ode23*. The integration will take place from time 0 to 200 with initial conditions −5 for x, 0 for y and 10 for z. As a result the famous Lorenz chaotic attractor (known as Lorenz butterfly) is derived in Fig. D.3.

```
function dx=lorenz(t,x)
dx=zeros(3,1);
dx(1)=10*(x(2)-x(1));
dx(2)=-x(1)*x(3)+28*x(1)-x(2);
dx(3)=x(1)*x(2)-8*x(3)/3;
end
>> [t,Y]=ode23('lorenz',[0 200],[-5 0 10]);
>> plot(Y(:,1),Y(:,3));xlabel('x');ylabel('z');
```

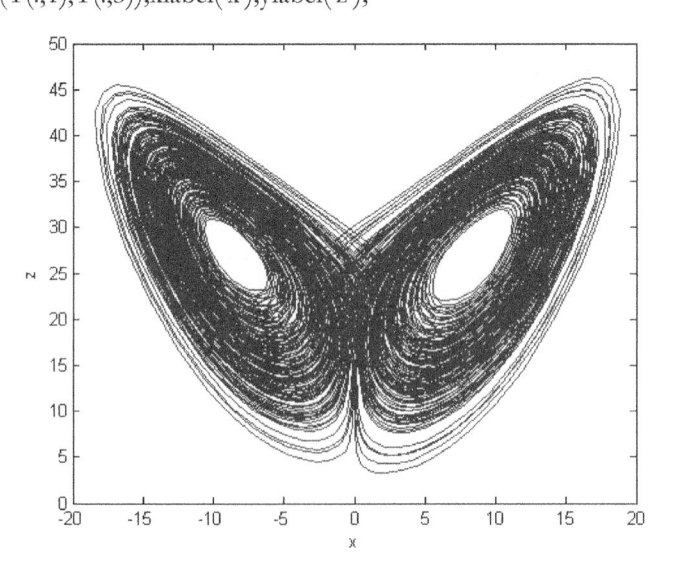

Figure D.3. The famous Lorenz butterfly

D.3 Basics of Simulink

Simulink provides a graphical user interface for building models as block diagrams using drag and drop mouse operations. It includes a comprehensive block library of sinks, sources, linear and nonlinear components, connectors, etc as well as the user can create his own blocks. After defining a model, it can be simulated either from the Simulink menus or by entering commands in MATLAB command window.

D.3.1 Building a Model

Simulink opens either by typing *simulink* in MATLAB or by clicking the icon. Then, the Simulink Library Browser of Fig. D.4(a) appears. A new model (Fig. D.4(b)) can be created by selecting New Model from the menu.

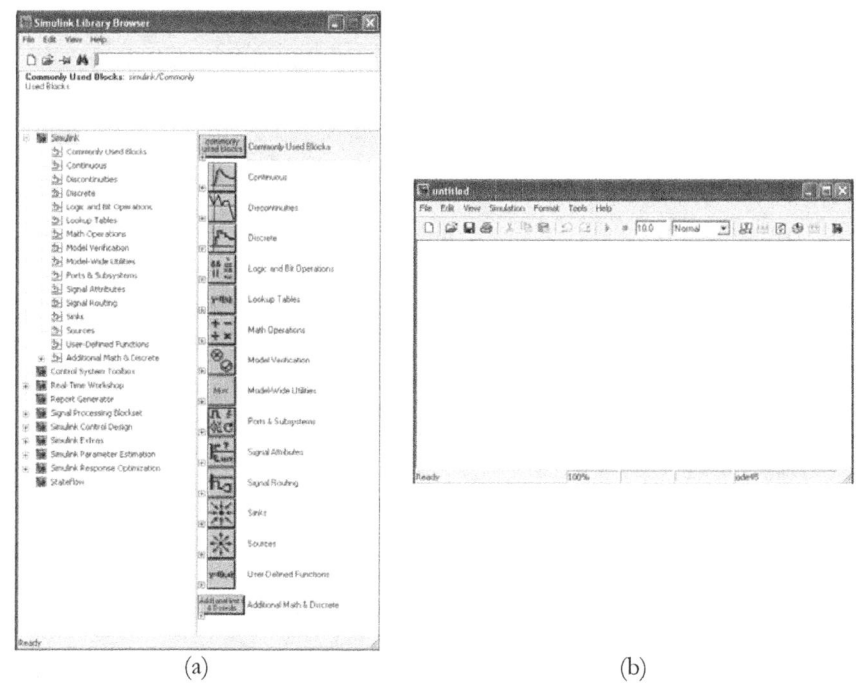

(a)   (b)

Figure D.4. (a) The Simulink Library Browser and (b) a new empty model

Let's build for example a model for confirming the truth table of a NAND gate. We drag and drop a *logical operator* block from the *logic and bit operations* library, two *constant* blocks from the *sources* library and one *display* block from the *sinks* library into the model window. Now it looks something like Fig. D.5(a). Port symbols pointing out or to a block denote output ports or input ports respectively. The blocks can be connected by positioning the pointer over an output port, then holding down the mouse button, moving the cursor to an input port and finally releasing the mouse button. When the blocks are connected, the port symbols disappear and the model looks like Fig. D.5(b). The logical operator block has to operate as NAND gate. By double clicking it a block parameters window appears (Fig. D.6(a)) where the NAND operation can be selected. By double clicking the constant blocks (Fig. D.6(b)) the constant values **0** or **1** can be selected. In this case the input values are defined inside the blocks. Another way is to set as **A** and **B** the constant values inside the blocks and define them as **0** or **1** in MATLAB command prompt.

Figure D.5. (a) Building the model and (b) the final model

Figure D.6. Parameters selection for (a) logical operator block and (b) constant block

The model can be saved with any desired name and simulated by clicking ▶ or by choosing Start from the Simulation menu. The saved model can run simply by typing its name in MATLAB command prompt without opening Simulink.

D.3.2 Useful Blocks

Below in Table D.2 we present briefly all the blocks that have been used in the examples and problems of this book.

Table D.2. Useful blocks

Block	Name	Library	Description
Memory	*Memory*	Discrete	Outputs its input from the previous time step applying a one integration sample-and-hold to its input signal
Zero-Order Hold	*Zero-order hold*		Samples and holds its input for the specified sample period
Combinatorial Logic	*Combinatorial logic*	Logic and bit operations	Implements a truth table by specifying a matrix that defines block outputs for every combination of inputs
AND Logical Operator	*Logical operator*		Performs the specified logical operation (AND, OR, NAND, NOR, XOR, NOT) on its inputs. The number of inputs can be specified inside the block.

MATLAB and Simulink Tutorial

Add	*Add*	Math operations	Performs addition or subtraction on its inputs
Subsystem	*Subsystem*	Ports & subsystems	Represents a system within another system. You can add blocks to the subsystem by opening it and drag and drop blocks into its window.
Data Type Conversion	*Data type conversion*	Signal attributes	Converts an input signal to a specified data type. Very useful for conversion to Boolean type for use as input to logic gates.
	Mux		Combines several input signals into a single output. The number of inputs is specified inside the block.
	Demux	Signal routing	Extracts the components of an input signal and outputs them as separate signals. The number of outputs is specified inside the block.
Manual Switch	*Manual switch*		Selects one of its two inputs to pass through to the output by double clicking it
Display	*Display*		Shows the value of its input
Scope	*Scope*	Sinks	Displays signals generated during simulation with respect to simulation time by double clicking it. It can have multiple *y*-axes (one per port) with common time range. The range of input values displayed can be adjusted by right clicking the axes.
Clock	*Clock*	Sources	Outputs the current simulation time at each simulation step
Constant	*Constant*		Generates a constant value. It must be defined as Boolean type when it is going to be used with flip-flop and combinatorial logic blocks.

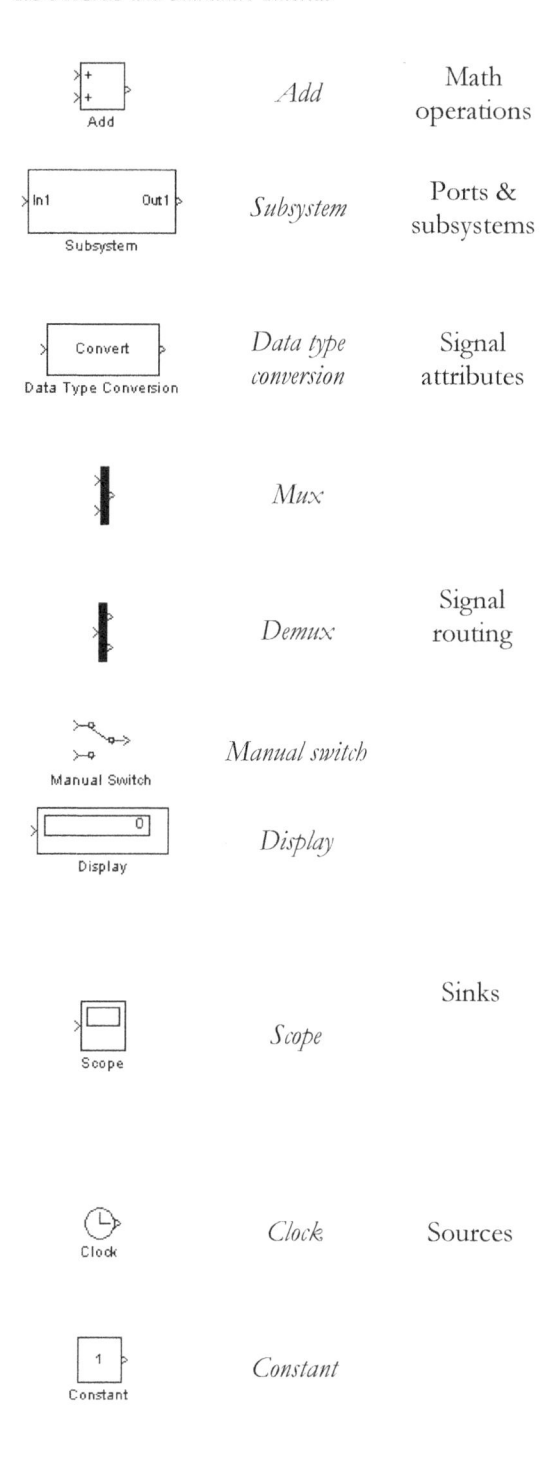

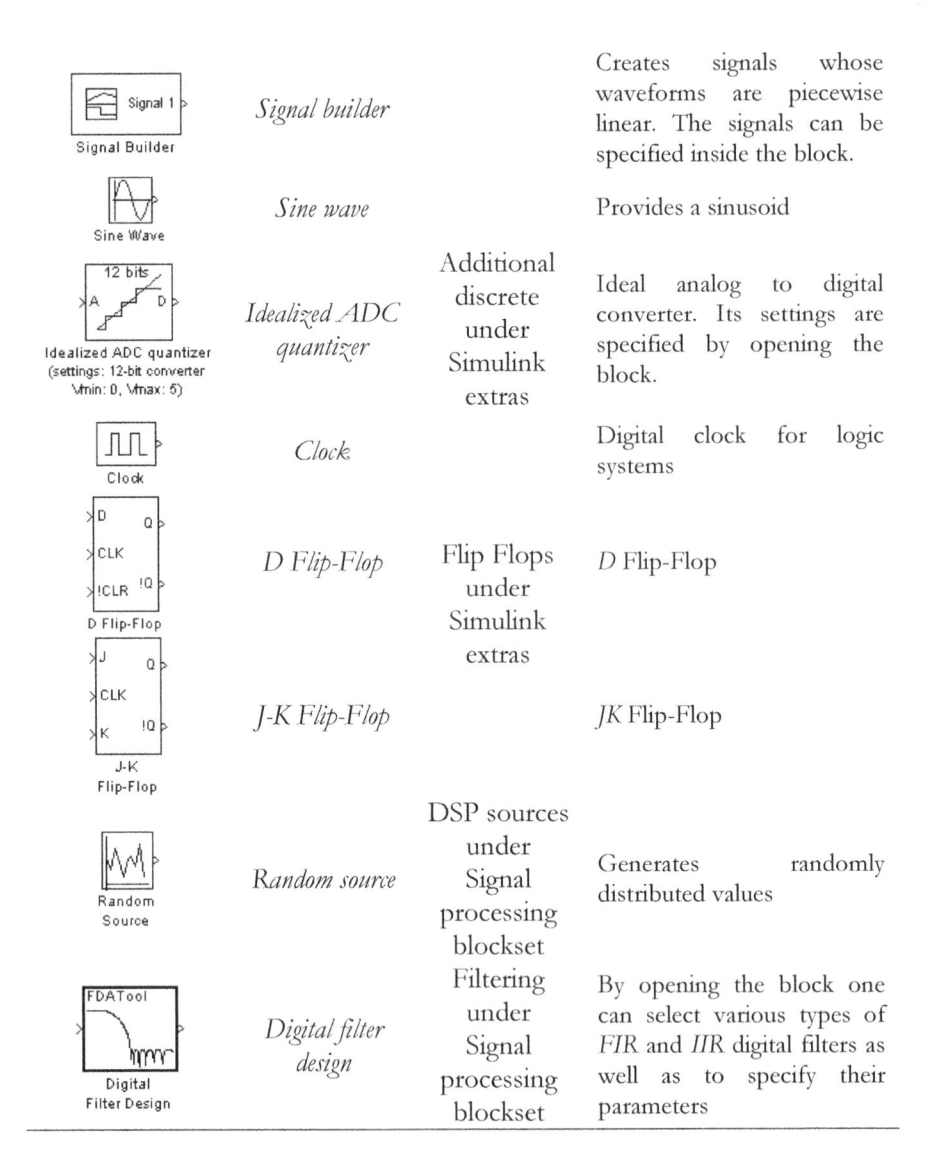

Block	Name	Library	Description
Signal Builder	*Signal builder*		Creates signals whose waveforms are piecewise linear. The signals can be specified inside the block.
Sine Wave	*Sine wave*		Provides a sinusoid
Idealized ADC quantizer (settings: 12-bit converter Vmin: 0, Vmax: 5)	*Idealized ADC quantizer*	Additional discrete under Simulink extras	Ideal analog to digital converter. Its settings are specified by opening the block.
Clock	*Clock*		Digital clock for logic systems
D Flip-Flop	*D Flip-Flop*	Flip Flops under Simulink extras	D Flip-Flop
J-K Flip-Flop	*J-K Flip-Flop*		JK Flip-Flop
Random Source	*Random source*	DSP sources under Signal processing blockset	Generates randomly distributed values
Digital Filter Design	*Digital filter design*	Filtering under Signal processing blockset	By opening the block one can select various types of FIR and IIR digital filters as well as to specify their parameters

References

[1]. Burrus C. S., McClellan J. H., Oppenheim A. V., Parks T. W., Schafer R. W. & Schuessler H. W., *Computer-Based Exercises for Signal Processing Using MATLAB*, Prentice-Hall Intl, Englewood Cliffs, New Jersey, USA, 1994.

[2]. Attia J. O., *Electronics and Circuit Analysis Using MATLAB*, CRC Press LLC, Boca Raton, Florida, 1999.

[3]. Karris S. T., *Circuit Analysis I with MATLAB Applications*, Orchard Publications, Fremont, California, 2004.

[4]. Karris S. T., *Circuit Analysis II with MATLAB Applications*, Orchard Publications, Fremont, California, 2003.
[5]. Karris S. T., *Signal and Systems with MATLAB Computing and Simulink Modeling*, 3rd Edition, Orchard Publications, 2007.
[6]. Karris S. T., *Digital Circuit Analysis and Design with Simulink Modeling and Introduction to CPLDs and FPGAs*, 2nd Edition, Orchard Publications, 2005.
[7]. Karris S. T., *Introduction to Simulink with Engineering Applications*, Orchard Publications, 2006.
[8]. *Getting Started with MATLAB 7*, The MathWorks, Inc., 2007.
[9]. *Using Simulink*, The MathWorks, Inc., 1999.

Index

Accuracy of Measurements 244
Action Potential 266, 267, 274, 275
Activation
    Energy 293
    Overpotential 293
Active Filters 111
    First-order 111
    High-order 132
    Second-order 115
Adder
    Full Adder 158
    Half-Adder 158
All-pass *SAB* 127
Alternating Current, *ac* 5
Amplifiers 65
    Biopotential 296
    Differential, Difference 84
    Finite Gain 74
    General 66
    Ideal 70
    Instrumentation 87, 298
    Isolation 236
    Negative-input Capacitance 98
    Operational, opamps 72
    Transconductance, *OTA* 90
Amplitude Modulation 219
Amplitude Spectrum 337
Analog Comparator 217
Analog-to-Digital Converter 141, 207, 215
Applications of opamps 73
    Characteristics 79
    Constant Current Source 78
    Integration of a Voltage 77
    Inverting 74
    Non Inverting 74
    Summation of Voltages 77
Arithmetic-Logic Unit, *ALU* 202
Arithmetic Systems
    Binary 141
    Hexadecimal 145
    Octal 145
ASCII Code 161
Atrioventricular Node 278
Atrium 278
Autonomic Nervous System 265
Axon 266
    Electrical Properties of 269

Myelinated 266
Unmyelinated 266
Axoplasm 269

Band-pass *SAB* 122
Band-reject *SAB* 125
Bi*CMOS* 199
Binary Adder 158
Binary Coded Decimal, *BCD* 160
Binary Codes
    1's Complement 143
    2's Complement 143
    ASCII 161
    BCD 160
    Physical 141
Binary System of Numbers 141
Bioamplifier 299
Bio-instruments 307
Biopotential Amplifiers 265
Biosignals 265
    Characteristics 284
    Detection 289
    Measurement 289
Bipolar Junction Transistor, *BJT* 51
Biquad 115
    Three-Opamp 131
Bit 142
Bode Plots 24
Boolean Algebra 148
    Theorems 149
Brainstem Auditory Evoked Potential 283
Bridges 245
    Capacitance 247, 255
    Resistance 245
    Weatstone 245
Byte 142

Capacitance, Capacitor 19
Capacitance Bridge 247, 255
Capacitance Sensor 254
Central Nervous System, *CNS* 265, 266
Central Processing Unit, *CPU* 196
Characteristics of Biosignals 284
Circuit Analysis 29, 335
    Simulation 352
Clipping, Waveform 43

Clocks   179
Codes, Binary   160
Common-Mode, *CM*   86
Common-Mode Rejection Ratio   83, 87
Comparator, Digital   157
Complementary Metal-Oxide
Semiconductor, *CMOS*   197
Complex Excitation   344
Complex Frequency   345
Computer Aided Analysis   351
  ECAP   352
  MATLAB   352, 353
  Schematic Capture   352
  SEPTRE   352
  Simulink   353, 362
  SPICE   352
Conductivity   9
Converters
  ADC   141, 207, 215
  DAC   141, 208, 213
Counters   183
  Asynchronous   184
  Popular *IC*   187
  Ripple   184
  Synchronous   187
CPLD   202
Current Source   10, 15, 78
Custom Made *IC*   199

Data Conversion   207
Decibel   68
Decoders   163
Defibrillator   311
Delay, Propagation   201
Delta-Sigma, ΔΣ, Modulator   219
Demultiplexers   164
Dendrites   266
Design with *HDL*   202
Detection of signals
  ECG   300
  EEG   302
  EMG   303
Diathermy   313
Differential Amplifier   84
Differentiation of Signals   20
Differentiator   21
Digital Comparator   157
Digital Computers   194
  Arithmetic Logic Unit, *ALU*   196
  Boot   195
  Control Unit   195
  Input/Output Unit   196
  Memory   195

Digital Electronics
  Combinational   139
  Integrated Circuits   197
  Sequential   171
Digital Filters   222
  Finite Impulse Response, *FIR*   223
  Infinite Impulse Response, *IIR*   223
  Non Recursive   222
  Recursive   223
Digital *IC*   197
  CMOS   199
  Emitter-Coupled Logic, *ECL*   198
  Fanout, Fanin   201
  Transistor-Transistor Logic, *TTL*   197
Digital Processing   208
Digital-to-Analog Converter   141, 208, 213
Diode
  Ideal   37
  Light-Emitting   59
  Practical   48
  Semiconductor   48
  Zener   50
Displacement Measurement Transducer   254
Dual-In-line Package   57

Einthoven Triangle   280
Electrical
  Noise   225
  Safe Installation   309
  Shock   308
Electrically Alterable *ROM*, *EAROM*   194
Electricity
  Intentional Application of   310
  Physiological Effects   307
Electric Signal Representation   5
Electrocardiogram, *ECG*   278
Electrodes
  Electrode-Electrolyte Interface   291
  Electrode-Tissue Model   295
  Equivalent Circuit   293
  Polarization   292
  Skin Impedance   295
  Types of Commercial   294
Electroencephalogram, *EEG*   282
Electromyogram, *EMG*   277
Electronic Circuit Analysis Program   352
Electronic Measuring System   243, 290
Electronic Stethoscope   319
Electrooculogram, *EOG*   283
Electroretinogram, *ERG*   283
Emitter-Coupled Logic   198
Encoders   162
Epidermis   296
Erasable *PROM*   194

Index

Error Detectors 167
Evoked Potential, *EP* 283
   Brainstem Auditory 283
   Somatosensory, *SEP* 283
   Visual, *VEP* 283
Exclusive – OR Gate 154
Exclusive – NOR Gate 154
External Noise 226

Feedback in Amplifiers 92
   Effects of 94
Fibrillation 308
Field-Effect Transistor, *FET* 53
Filters 24
   Active 111
   All-pass *SAB* 127
   Analog 107
   Band-pass 109, 122
   Band-reject 109, 125
   Characteristics 107
   Digital 222
   First-order 111
   High-order 132
   High-pass 108, 119
   Low-pass 24, 108, 115
   Passive 111
   Second-order 115
Finite Impulse Response, *FIR* 223
Flash *ADC* 217
Flicker Noise 228
Flip-Flops 173
   Characteristics 178
   D 175
   JK 176
   Master-Slave 178
   Race Problem in 177
   R-S 173
   T 174
Flow Measurements
   Gas Flow 259
   Liquid Flow 259
Flow Meters
   Electromagnetic 260
   Ultrasonic 260
Fourier Series 335
Fourier Transform 335, 336
Frequency Modulation 219
Frequency Transformation 111
Full-Adder 158
Full-Wave Rectifier 40
Functional Cells 199

Gamma-Ray Counting 319

Gamma Scintillation Counting 319
Gas-Flow Measurements 259
Gate-Arrays 199, 202
   PGA 202
Gate Characteristics 200
General Amplifier 66
Graphics, *MATLAB* 356

Half-cell Potential 291
Half-wave Rectifier 37
Hall-effect Sensors 260
Hardware Description Language, *HDL* 202
Hexadecimal Number System 145
High-pass Biquad 108, 119
High-pass Filter 108, 119
Human Heart 279

Ideal Amplifier 70
Ideal Current Source 11
Ideal Diode 36
Ideal Voltage Source 10
Inductance, Inductor 28
Inductive Sensor 256
Infinite Impulse Response, *IIR* 223
Inherent Noise 226
Input Offset Current 80
Input Offset Voltage 80
Instrument
   Linearity 245
   Resolution 245
   Sensitivity 244
Instrumentation Amplifier 87, 90
Integrated Circuits 56
   Analog 56
   Digital 56, 197
Integration of Signals 20, 77
Integrator 22
Intensive Care Unit, *ICU* 310, 311
Interference Noise 226
Interstitial Fluid 266
Inverse Transform
   Fourier 336
   Laplace 345
Isolation Amplifiers 236

J*K* Flip-Flop 174

Kirchhoff's Current Law 12, 341
Kirchhoff's Voltage Law 11, 341

Laplace Transform 344

Laplace Transform Pairs 346
Large-Scale Integration 200
Latches 171
Least Significant Bit 143
Light Dependent Resistor 57
Light-Emitting Diode, LED 59
Linear Variable Differential Transformer, LVDT 256
Linearity 245
Liquid-flow Meters 259
Logic Function Implementation 153
Logic Gates 151
Low-pass Filter 24, 108, 115

Macroshock 309
Magnetic Resonance Imaging, MRI 313
Magnetocardiogram, MCG 284
Magnetoencephalogram, MEG 284
MATLAB Basics 353
   Graphics 356
   Matrices and Arrays 354
   Programming 358
   Useful Functions 359
Measurements of
   Gas-flow 259
   Liquid-flow 259
   Optical 262
   Pressure 257
Medium-Scale Integration 200
Memories 190
   Dynamic 192
   Random-Access, RAM 191
   Read-Only 193
   Static 192
MESFET 200
Mesh Analysis 30, 327
Metal-Oxide Semiconductor FET 53
Microcomputer 197
Microprocessing Unit, MPU 197
Microshock 310
Modulation
   Amplitude 219
   Delta-Sigma 219
   Frequency 219
   Phase 219
   Pulse-code 219
MOSFET 53
Motion Artifacts 296
Motor
   Neurons 266
   Units 277
Multi-Channel Analyzer, MCA 322
Multiplexers 164
Multivibrators

Astable 180
Monostable 180
Myelin 266

Negative Logic 147
Nernst Equation 268
Nerves
   Afferent 266
   Efferent 266
Nervous System 265
   Autonomic 265, 266
   Central 265, 266
   Peripheral 265, 266
Netlist Language 202
Neuron 266
   Structure 266, 267
Nodal Analysis 30, 335
Noise
   $1/f$ 226, 228
   Characteristics 230
   Effect of Feedback on 235
   External 226
   Figure 231
   Flicker 228
   In Analog Signals 225
   In Digital Signals 226
   Inherent 226
   Interference 226
   Margin 201
   Measures 230
   Model of an Opamp 233
   Pink 228
   Reduction of 233
   Shot 226, 228
   Thermal or Johnson 226, 227
   White 227, 230
Non-Recursive Digital Filter 222
Norton's Theorem 333
Number Representation
   Parallel 148
   Serial 148
Number Systems 141
Nyquist Frequency 209

Octal Number System 145
One-Shot 180
Operational Amplifier, opamp 72
Operational Transconductance Amplifier 90
Optical Measurements 262
Optocouplers 60
Optoelectronics 57
Optoisolators 60
Oscillator 100

Index

Overpotential 293
Oversampling Converters 218
Overshoot 84

Pacemaker 312
Parity Checkers 167
Passive Filters 111
Patient Ground 299
Peripheral Nervous System 265, 266
Phase Modulation 219
Phase Spectrum 337
Phasor 337
Photoconductors 57, 262
Photodiodes 58
Photoemission 59
Photomultiplier 321
Phototransistor 58
Photovoltaic 59, 262
Physiological Effects of Electricity 307
Piezoresistivity 257
Pink Noise 228, 230
Plethesmograph 319
Polarization of Electrodes 292
Poles of a Function 347
Positive Logic 147
Potential
    Action 266, 267, 274, 275
    Evoked 283
    Resting 267
Potentiometers 17
Power Supplies, dc 50
    for an opamp 92
    Sensitivity 82
Precision of Measurement 244
Pressure Measurement 257
Priority Encoder 217
Product Of Sums, POS 154
Programmable Logic Array, PLA 194
Programmable Read Only Memory 193
Propagation Delay 201
Pulse-Code Modulation 219
Pulse-Height Analyzer 321

Quantization 209
    Error 209, 226, 237
    Noise 209, 237
    Step 209

Race Problem 177
Random Access Memory, RAM 191
Ranvier Nodes 266
RC-Oscillator, Wien-Bridge 100

Read-Only Memory, ROM 193
Receptor 266
Rectification of Signals 37
    Full-wave 40
    Half-wave 37
Reduction of Noise by
    Averaging 233
    Filtering 233
Registers 188
    Shift- 188
Repeatability 244
Reproducibility 245
Resistance, Resistor 8
Resistivity 9, 248
Resolution 209, 245
Response
    Amplitude 24
    Frequency 24
    Phase 24
Resting Potential 267
Right-Leg Drive 300
Ring Counters 190
Rise Time 83
R-S Flip-Flop 173

Saltatory Conduction 276
Sample-and-Hold Circuit 219
Sampler 209
Sampling 208
Scales of Integration 200
Schmitt Trigger 181
Schuann Cells 266
Second-order Filters 115
Semiconductor Diodes 48
Sensors
    Capacitive 254
    Inductive 256
    for Pressure Measurement 257
    Thermistor 249
    Thermocouples 248
    Thermoresistive 248
Shift Registers 188
    Applications of 189, 190
    Integrated Circuit 190
Shock, Electrical
    Macroshock 309
    Microshock 310
Shot Noise 226, 228
Signal
    Differentiation 20
    Integration 20
    Representation 5
Signal to Noise Ratio, SNR 230
Simulink Basics 362

Model Building 363
  Useful Blocks 364
Single-Amplifier Biquad, *SAB* 115
Single-Channel Analyzer, *SCA* 320
Sinoatrial Node 278
Sinusoidal Excitation 335
Skin Impedance 295
Slew Rate 82
Smart Medical Devices, *SMD* 318
Somatosensory Evoked Potential 283
Sources of Noise in 225
  Analog Signals 225
  Digital Signals 226
Source Transformation 330
Space Parameter 269
Spectrum
  Amplitude 337
  Continuous 6
  Line 6
  Phase 337
SQUID 284
Stability 99
Subcutaneous Layer 296
Successive Approximation *ADC* 216
Successive Approximation Register 216
Sum Of Products, *SOP* 153
Summation of Voltages 77
Superposition 329
Synchronous Counters 187
Synapse 266

Temperature Measurements 247
Temperature Sensors
  Thermistors 249
  Thermocouples 248
  Thermoresistive 248
TENS 316
T Flip-Flop 174
Theorems of Boolean Algebra 149
Thermal Noise 226, 227
Thevenin Theorem 332
Three-opamp Biquad 130

Transcutaneous Electrical Nerve Stimulation 316
Transducers 243
Transducers for
  Displacement Measurement 254
  Temperature Measurement 247
Transfer Function, Voltage 24
Transform
  Fourier 335, 336
  Laplace 344
Transformation of Sources 330
Transistor 51
  Bipolar 51
  Junction Field-Effect 51
  MOSFET 53
  Phototransistor 58
Transistor Outline Package 57
Transistor-Transistor Logic, *TTL* 197
Twin-Tee 126
Twisted-Ring Counters 190

Ultra Large-Scale Integration 200
Ultrasonic Flow Meter 260

Variable Differential Transformer 256
Ventricles 278
Very High Speed Integrated Circuits 199
Very Large-Scale Integration 200
Visual Evoked Potential 283
Voltage-Controlled Oscillator 180
Voltage-Controlled Voltage Source 70
Voltage Integration 20, 77
Voltage Margin 147
Voltage Representation of Numbers 147
Voltage Source 10, 15

Wafer-Scale Integration 200
Waveform Clipping 43
White Noise 227, 230

Made in the USA
Coppell, TX
04 November 2022